URBAN GEOMORPHOLOGY IN DRYLANDS

R. U. COOKE
D. BRUNSDEN, J. C. DOORNKAMP,
and D. K. C. JONES

WITH CONTRIBUTIONS BY
J. GRIFFITHS, P. KNOTT,
R. POTTER, and R. RUSSELL

Published on behalf of
The United Nations University
by
Oxford University Press

1982

Oxford University Press, Walton Street, Oxford OX2 6DP

London Glasgow New York Toronto
Delhi Bombay Calcutta Madras Karachi
Kuala Lumpur Singapore Hong Kong Tokyo
Nairobi Dar es Salaam Cape Town
Melbourne Auckland
and associates in
Beirut Berlin Ibadan Mexico City Nicosia

Published in the United States by
Oxford University Press, New York

British Library Cataloguing in Publication Data
Cooke, R. U.
    Urban geomorphology in drylands.
    1. Deserts
    I. Title
    551.4   G13611
    ISBN 0–19–823239–X

Library of Congress Cataloguing in Publication Data
Main entry under title:
Urban Geomorphology in drylands.
    Bibliography: p.
    Includes index.
1. Arid regions. 2. Geomorphology. 2. Cities and towns. I. Cooke, Ronald U.
    GV612.U73   333.77′09154   82–3561
    ISBN 0–19–823239–X   AACR2

Typeset by CCC, printed and bound in Great Britain by
Butler and Tanner Ltd, Frome

Dedicated To

Dr PETER G. FOOKES
Consultant Engineering Geologist

AUTHORS

Professor R. U. COOKE, Professor of Geography at University College London (in the University of London), has worked on aspects of desert geomorphology and applied geomorphology for a number of years. He has written *Geomorphology in Deserts* (Batsford, London, 1973) with A. Warren, *Geomorphology in Environmental Management* (OUP, Oxford, 1974) with J. C. Doornkamp, and *Arroyos and Environmental Change in the American Southwest* (OUP, Oxford, 1976) with R. W. Reeves, and he has published geomorphological studies on drylands in South America, North America, and the Middle East. He has been chairman of the British Geomorphological Research Group.

Dr D. BRUNSDEN, Reader in Geography at King's College in the University of London, is co-author with J. B. Thornes of *Geomorphology and Time* (Methuen, London, 1977), and co-editor with J. C. Doornkamp of *The Unquiet Landscape* (David and Charles, Newton Abbot, 1972), with J. C. Doornkamp and D. K. C. Jones of *Geology, Geomorphology and Pedology of Bahrain* (GeoBooks, Norwich, 1980), and with C. Embleton and D. K. C. Jones of *Geomorphology – present problems and future prospects* (OUP, Oxford, 1978). He has been chairman of publications for the British Geomorphological Research Group, and has served on the Executive Committee of the Engineering Group of the Geological Society of London. His research interests are focused on slope stability problems and geomorphological mapping, and he has carried out fieldwork in the UK, New Zealand, the USA, Nepal, Sri Lanka, and the Middle East.

Dr J. C. DOORNKAMP, Senior Lecturer at the University of Nottingham, has been concerned for many years with geomorphological mapping surveys, chiefly in the UK, Uganda, South Africa, Nepal, Sri Lanka, and the Middle East. His publications include joint authorship with R. U. Cooke of *Geomorphology in Environmental Management* (OUP, Oxford, 1974), with D. H. Krinsley of *Atlas of Quartz Sand Surface Textures* (CUP, Cambridge, 1973), and co-editorship of *The Unquiet Landscape* (David and Charles, Newton Abbot, 1972) with D. Brunsden, of *Evaluating the Human Environment* (Arnold, London, 1973) with J. A. Dawson, and of *Geology, Geomorphology and Pedology of Bahrain* (GeoBooks, Norwich, 1980) with D. Brunsden and D. K. C. Jones. He has been Honorary Secretary of the Institute of British Geographers.

Mr D. K. C. JONES, Lecturer in Geography at the London School of Economics in the University of London, has worked extensively on geomorphological problems in the UK, Nepal, and the Middle East. In addition to co-editing *Geology, Geomorphology and Pedology of Bahrain* (GeoBooks, Norwich, 1980) with J. C. Doornkamp and D. Brunsden, and

Geomorphology – present problems and future prospects (OUP, Oxford, 1978) with C. Embleton and D. Brunsden, he is author of *South-East England* (Methuen, London, 1980), a volume in a series on the geomorphology of Britain, and has published a number of papers on applied geomorphological subjects.

ACKNOWLEDGEMENTS

Our interest and experience in applied geomorphology in drylands owes much to the advice, support, and encouragement we have received from numerous colleagues and sponsors. We owe a particular debt to Dr Peter Fookes, consultant engineering geologist, who has never failed in his enthusiastic and enlightened assistance; to Dr Peter Bush, of Imperial College (University of London) who has provided good-natured co-operation in the field over several years; and to Mr Andrew Sharman, of Sir William Halcrow and Partners, who has made several of our field projects possible. All four authors have worked as a geomorphological survey team in several drylands, including Bahrain, Dubai, Ras Al Khaimah, and Egypt, under the sponsorship of various government agencies, and engineering consultants that have included Sir William Halcrow and Partners, Halcrow Middle East, Messrs Sandbergs, Wimpey Laboratories Ltd, and Engineering Geology Ltd.

In writing this assessment we have received substantial help from Mr J. Griffiths (King's College London, Chapter VI), Mr P. Knott (University College London, Chapter VII), Dr R. Potter (Chapter I), and Dr R. Russell (Geomorphological Services Ltd, Chapter IV). Many of the diagrams were prepared by Mr R. Halfhide, Miss Claire Wastie, and Miss G. Collins (Bedford College London) and Miss R. Beaumont and Mr G. Reynall (King's College London). The manuscript was typed by Mrs B. A. Cooke and Mrs R. Dawe, and Mr D. Garvin and Miss L. Hartwell undertook the photographic work. Several engineers and scientists have kindly provided pertinent information including Mr T. Allison and Dr M. F. Al-Kazily (Kuwait), Dr C. Bardinet (Montrouge, France), Professor J. H. Bater (Waterloo, Canada), Professor I. Douglas (Manchester, UK), Dr M. Inbar (Israel), Professor J. A. Mabbutt (Sydney, Australia), Mr G. Webster (Qatar), and Professor D. Weide (Las Vegas, USA). We would like to record our gratitude to them all.

We also thank Dr P. G. Fookes, Professor J. A. Mabbutt, and Dr A. Warren for their conscientious and constructive comments on early drafts of this book. Professor Walther Manshard, formerly Vice-Rector of the United Nations University, provided support and encouragement throughout the project.

We would like to thank the following for their kind permission to reproduce material from previously published work: Dr J. M. Bleck (Fig. VI.11), Professor D. J. Dwyer (Fig. I.7), Dr P. Fookes (Figs. V.1, V.5, V.6, V.8, V.9, V.10, V.11, VI.4, VI.7), Dr J. Griffiths (Figs. VI.13, VI.23), Sir William Halcrow and Partners (Figs. III.13, III.14, III.15, III.16, IV.5, IV.6, V.16, V.17, V.18, V.19, VI.21, VI.22), C. J. Lawrance (Fig. III.1), Professor J. A. Mabbutt (Figs. VI.18, VII.1), Dr R. Perrin and Dr C. Mitchell (Fig. III.8), Dr A. Schick (Figs. VI.8, VI.9, VI.19, VI.22), Dr S. Singh (Fig. II.5), Dr J. Thornes (Fig. VI.14), and Dr A. Yair (Table VI.12). We have been unsuccessful in contacting a small number of authors for permission to reproduce figures from their work; we apologize to them, and hope that our acknowledgements and use of their material in the text will meet with their approval.

CONTENTS

INTRODUCTION

In the winter of 1969 a series of storms in the 'drylands' of southern California caused serious soil erosion, numerous landslides, and widespread flooding. Over a hundred people were killed, thousands of homes and other buildings were ruined, the total estimated damage exceeded $200 million, the lives of several million people in the sprawling Los Angeles metropolis were disrupted, and the resources of the community had to be mobilized to cope with the emergency and rehabilitate the region at a cost of well over $8 million (US Corps of Engineers, 1969). The dominant environmental processes responsible for this crisis – erosion, slope failure, and flooding – are geomorphological problems, problems relating to the nature of the land surface and the forces that act upon it.

Such problems are not new to southern California. They have occurred many times before (in 1914, 1938, and 1963, for instance), and they have recurred since (as in 1978 and 1980). Why do these problems arise, and what can be done to prevent them, or to reduce their impact? In this region, the geomorphological problems have been studied for many years, and there has been an enormous investment in research and in a varied range of solutions. This investment has clearly begun to pay off. For example, one perhaps rather optimistic estimate suggests that damage *prevented* in 1969 as a result of existing engineeering works alone amounted to over $1.5 billion (US Corps of Engineers, 1969). But to be successful, or more successful, management must, among other things, be well informed about the nature of the geomorphological problems.

On the other side of the earth, in the drylands of the Middle East, rapidly growing urban areas, particularly in the oil-rich states, are encountering related but somewhat different geomorphological conditions and hazards. In one city, a large modern hotel has had to be partly abandoned because of unanticipated settling of foundation material; in another, a hospital building has been condemned and evacuated because its foundations have been destroyed by weathering processes. Advancing sand dunes continuously threaten and occasionally overwhelm highways, suburban developments, and oasis settlements; and in places there are dramatic shortages of suitable aggregates for the voracious construction industry. All these, and many other problems arise in part from mismanagement or misunderstanding of geomorphological conditions. How can they be avoided, managed, or controlled? What information about the ground-surface conditions do urban environmental managers in drylands require? How is such information to be collected, analysed, and presented?

This study attempts to answer such questions – its purpose is to provide a statement of value to geomorphologists, engineers, and planners on the ways in which geomorphological research can assist urban development in drylands. In addressing three different audiences, the book is written from the perspective of a team of geomorphologists who do not pretend to be engineers or planners but whose experience leads them to believe that not all planners and engineers yet recognize the potential of geomorphology for saving time and money, especially if it is used in conjunction with related environmental information provided by engineering geology and soil mechanics.

The book began as a report, prepared under the United Nations University Natural Resources Program, entitled *An Assessment of Geomorphological Problems in Urban Areas of Drylands*. The original report included substantial discussions of techniques which have been greatly reduced here to focus attention more clearly on philosophy, principles, and the systematic nature of the problems that geomorphological work can help to solve; the details of techniques will be presented elsewhere, probably in a manual.

A wealth of recent dryland studies, including numerous contributions to the debate on desertification, have placed overwhelming emphasis on rural problems – such problems as overgrazing, water supply, soil erosion, sedentarization of nomads, fuel supplies, and drought hazards. It is worth noting that over a third of all dryland inhabitants live in urban areas, and in some dryland countries the majority of the population lives in cities. Furthermore, increasing numbers of desert dwellers are migrating into urban areas – some are forced out of the wilderness, perhaps as a result of 'desertification' or by policies of 'sedentarization', whereas others are attracted by the urban prospect of security, employment, and food.

Thus urban areas are assuming increasing importance in drylands, and in recent years many of them have expanded rapidly into the desert. In the course of this expansion, urban planners and developers have often encountered serious and unexpected environmental problems. Some of these have been examined in Gideon Golany's recent book, *Urban Planning in Arid Zones* (Wiley, New York, 1978). Golany and his co-authors – writing essentially from the perspectives of architects and planners – recognize only briefly the importance of ground-surface conditions and the need to examine landforms in planning urban development in drylands. While there have been recent systematic studies of desert geomorphology (e.g. Cooke and Warren, 1973) there has been no study prior to this that has attempted to assess in depth the range of geomorphological phenomena in drylands as they relate to urban planning and development. This book therefore provides a detailed, complementary perspective on planning and development in drylands to that outlined by Golany and his colleagues, and a complementary applied perspective on desert geomorphology to that given by academic texts. The volume focuses on urban problems, especially in the drier areas of drylands, because this is the context of most of the field studies and experience of the authors. Much of the discussion is also relevant to rural development in drylands and to the general application of geomorphology to civil engineering and planning.

The first chapter of this report sets the urban scene: it provides, probably for the first time, a global perspective on the distribution and growth of large urban areas in drylands. This contribution reveals something of the magnitude of dryland urban development – it shows, for instance that there are over 350 cities in the drylands with a population of 100 000 or more, and that many of them are rapidly expanding. The chapter concludes with a general view of the planning framework within which this rapid urban growth can be controlled.

The second chapter begins by setting the geomorphological scene, through a brief introduction to the subject and the way it can be applied to drylands and urban areas. The remainder of the chapter seeks to establish, generally and through specific examples, the ways in which geomorphological work can be of value in the planning and management of urban areas in drylands. After an initial review, two fundamental themes are explored through specific examples: the relationships between geomorphological problems and management structure and responsibilities and the actual use of geomorphological information at various phases and scales in the planning process. This is followed by an assessment of the availability of information relevant to geomorphological studies in urban development in drylands. Finally, it is recognized that geomorphological evidence often forms only a part of the whole assemblage of environmental data required by planners, and ways of integrating the geomorphological data into the whole are briefly reviewed.

In Chapter III, attention is focused on the methods that have proved of value in recording, analysing, and presenting geomorphological information. These methods – which are based mainly on geomorphological mapping, morphological mapping, and land systems surveys – have now been used with success in many resource surveys at a variety of mapping scales in many different environments, and an attempt is made, based on this experience, to codify the ways in which mapping programmes suitable for surveys prior to urban development should be structured. Geomorphological maps are shown to provide a basic information source from which systematic maps – demonstrating, for instance, the spatial context of particular hazards – can be derived.

Subsequent chapters examine, in turn, the major geomorphological problems related to urban development in drylands. Chapter IV examines geomorphological aspects of the provision of aggregate and superficial materials for construction; Chapter V reviews the nature of the problems posed by the presence of surface salt and saline waters in many areas of dryland urban growth; Chapter VI explores the group of problems associated with water movement, and sediment movement by water; and in Chapter VII, problems of sand and dust movement are considered. In each of these chapters, attention is given to the geomorphological circumstances that underlie the urban problems, to the nature of the problems themselves, to ways of avoiding or managing them, and to a selection of case studies from drylands. Throughout, emphasis is placed on practical ways of approaching and solving geomorphological problems, and of presenting the information in ways that planners and engineers can understand and use.

To assess comprehensively all environmental problems in urban areas of drylands would be a monumental task, and one that would be impeded at every turn by a shortage of detailed information. In adopting a geomorphological perspective in this study, the potential field of enquiry has been deliberately restricted to those problems in which the authors have considerable first-hand experience, and where the value of a geomorphological contribution is both clear and can be accepted by many environmental managers and other decision-makers. It is argued that geomorphology can provide a valuable basis for environmental assessment of land related to urban development in drylands, and that almost all urban development in drylands can benefit from a prior knowledge of geomorphological conditions. Of course, the use to which such information is put will depend greatly on the political and cultural contexts of development. In general, however, the geomorphological information is relevant to most desert communities. For instance, encroaching sand dunes represent a problem to both a modern metropolis and an ancient oasis village: the important differences lie in the ways the different communities perceive and respond to the problem. This investigation does not presume to dictate how all dryland urban areas should respond to ground-surface conditions. But where appropriate it does make suggestions and, through numerous case studies, it does show how geomorphological information can be beneficially used.

   This book outlines only one set of approaches to evaluating environmental data relevant to urban development in drylands. The geomorphological perspective advocated here in no way precludes equally valid approaches to the study of environmental problems in urban areas of drylands adopted by other scientists, such as geologists, ecologists, pedologists, and hydrologists. Indeed, different approaches can be both stimulating and compatible. What is essential in all approaches, however, is that their procedures are clear, and their products intelligible and useful to those who have the responsibility for urban development in drylands: the major problem is one of communication.

I

URBAN DEVELOPMENT IN DRYLANDS

(a) Some Definitions

1. *Drylands*

There have been many attempts to classify and delimit the world's drylands in terms of precise climatic indices. For the purposes of this study, drylands are defined as the extremely arid, arid, and semi-arid areas mapped by Meigs (1953) and shown in Fig. I.1. This definition has many merits – it is conservative, it is widely accepted and used, and it is soundly based on the fundamental concepts of water balance and the availability of water for plant growth. Other definitions produce similar patterns (UNESCO, 1979), and differences between precise scientifically based classifications are not sufficient to justify detailed discussion here. Meigs's classification includes no less than a third of the world's land surface – a rich diversity of terrestrial environments that have in common a persistent shortage of surface water. The drylands include landscapes as different as the sand seas of the Sahara, the foggy mountainous coasts of Chile, and the limestone plateaux of South Australia; they include the high-altitude deserts of Inner Asia, and Death Valley in California, which is well below sea level; some areas, such as Iran, are tectonically active, whereas others, such as central Australia, are relatively stable. Equally important, the drylands include a great variety of cultures, patterns of human activity, and many substantial, rapidly growing urban areas.

2. *Urban development: Data sources and methods*

For the purposes of the statistical review in this chapter, attention will be focused on urban settlements in drylands with 100 000 or more inhabitants – a limit determined by data availability. But the limit is arbitrary, and it must be remembered that there are very many dryland towns and cities of smaller size, and many such settlements are growing rapidly. The UN *Demographic Yearbooks* are the principal source of data on the size of urban centres with populations of 100 000 or more. The *Demographic Yearbook 1976* (1977) is the first of two major data sources for this survey; the second is the extensive study of world urbanization 1950–70 by Davis (1969, 1972).

All cities with 100 000 or more inhabitants (at the most recent date of enumeration) that lie on or within the limit of dryland areas defined by Meigs (Fig. I.1) are included in this review. Previous workers, such as Wilson (1960, 1974) and Langford-Smith (1978) have discussed major urban centres in arid lands. For example, Wilson (1974) identified 89 major cities in the arid lands,

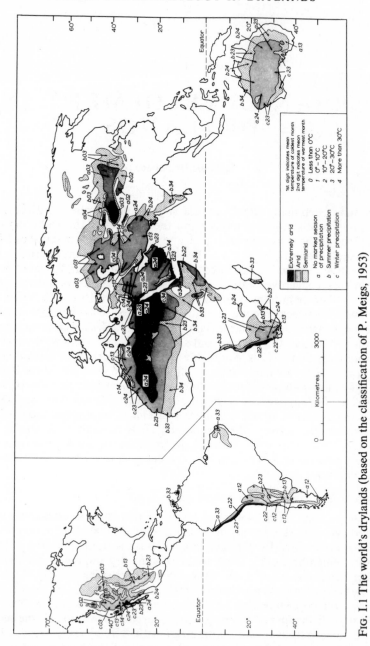

FIG. I.1 The world's drylands (based on the classification of P. Meigs, 1953)

and in an earlier study (1973) he examined coastal desert cities. But this is the first attempt at a comprehensive enumeration of dryland cities that also includes the large number of urban centres in semi-arid lands. Where cities appeared to be located on the boundary, they have been included.

(b) Location and Size of Major Dryland Cities

1. *The Global Pattern*

At the global scale, population data are available from the UN *Demographic Yearbook* for two different statistical definitions of urban settlements (United Nations, 1977, p. 19):

(i) *The city proper*, defined as a locality with legally fixed boundaries and an administratively recognized urban status which is usually characterized by some form of local government;

(ii) *Urban agglomeration*, comprising the city proper together with the suburban fringe or thickly settled territory lying outside of, but adjacent to, the city boundaries.

Wherever available, urban agglomeration data are preferred, as they most closely reflect the magnitude and importance of the entire urban settlement. It is important to note that the two definitions may vary from one country to another. In the tables of this chapter, both sets of data are listed when they are available. Initially, dryland urban areas are ranked by population size, whether this refers to the city proper or urban agglomeration data, although when both are available, the larger figure is used. The ranking is based on data from the most recent enumeration date, a date that varies from country to country. Nearly all of the statistics relate, however, to the period 1971–76, with the notable exception of China where information is from the 1950s except for cities of over a million inhabitants. While most of the data are derived from official censuses, it is important to note that some are estimates that may be of questionable reliability (United Nations, 1977).

The numbers of major dryland cities are shown cross-tabulated by continent and population size in Table I.1. Three major continental divisions are used in this table and throughout this discussion: Africa (northern and southern drylands), America (northern and southern drylands), and Asia (western and eastern divisions). No cities of 100 000 population are located within the Australian drylands as defined by Meigs.

A total of 355 major urban agglomerations were identified within the world's drylands, of which 208 were in Asia, 94 in the Americas, and 53 in Africa (Table I.1). The location of these cities is shown in Figs. I.2–I.5. On a national basis, the largest number of dryland cities occurs in India (62), followed by the USSR (55), the USA (42), Mexico (22), China (19), Pakistan (19), Egypt (18), Iran (12), Morocco (10), and Turkey (10). The size-frequency distribution of world dryland cities shows a relatively smooth progression from a preponderance of small cities to a few large ones (Table I.1). Sixty-nine dryland cities have populations exceeding half a million, and of these, over 30 contain more than a million inhabitants, and 15 contain over two million inhabitants.

2. *African dryland cities*

The 53 African dryland cities with 100 000 or more inhabitants are shown ranked by population size in Table I.2. The only two 'million' cities are in Egypt – Cairo, with 5.7 millions, and Alexandria, with 2.3 millions. Egypt has long been one of the most highly urbanized nations in Africa (Abu-Lughod, 1965) and in 1975, 43.9 per cent of the population lived in urban settlements. Other major dryland urban centres are found in Tunisia (43.4% urban in 1970), Morocco (37.9% urban in 1974), and the Sudan (20.4% urban in 1976) (Fig. I.2 and Table I.2).

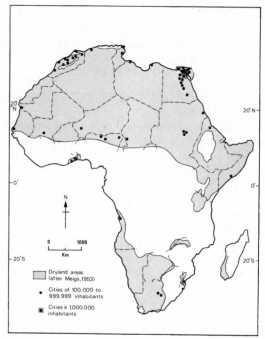

FIG. I.2 Drylands in Africa: cities with 100 000 or more inhabitants

3. *Asian dryland cities*

Within the vast Asian dryland zone, extending from Israel to China, there are over 200 cities with 100 000 or more inhabitants (Table I.3 and Fig. I.3).

TABLE I.1

Size frequency distribution of dryland cities of 100 000 or more inhabitants by continent

City size (total population)	Continent			All drylands
	Africa	Asia	America	
4 000 000+	1	3	1	5
2–4 000 000	1	3	4	8
1–2 000 000	0	12	5	17
500 000–1 000 000	4	23	12	39
250 000–500 000	17	55	25	97
100 000–250 000	30	112	47	189
Totals	53	208	94	355

TABLE I.2

Dryland cities of 100 000 or more inhabitants ranked by size of population:
Africa

Rank	City/Country	Year and source E = Estimate C = Census	City proper	Urban agglomeration
1	Cairo, Egypt	1974(E)	5 715 000	—
2	Alexandria, Egypt	1974(E)	2 259 000	—
3	Giza, Egypt	1974(E)	853 700	—
4	Accra, Ghana	1970(C)	564 194	738 498
5	Tunis, Tunisia	1966(C)	468 997	647 640
6	Dakar, Senegal	1969(E)	—	581 000
7	Luanda, Angola	1970(c)	—	475 328
8	Marrakech, Morocco	1971(C)/1973(E)	330 400	436 300
9	Fez, Morocco	1971(C)/1973(E)	321 460	426 000
10	Meknès, Morocco	1971(C)/1973(E)	244 520	403 000
11	Kano, Nigeria	1975(E)	399 000	—
12	Maputo, Mozambique	1970(C)	—	383 775
13	Suez, Egypt	1974(E)	368 000	—
14	Oujda, Morocco	1971(C)/1973(E)	155 800	349 400
15	Subra-El Khema, Egypt	1974(E)	346 000	—
16	Port Said, Egypt	1974(E)	342 000	—
17	Asmara, Ethiopia	1976(E)	340 206	—
18	Oran, Algeria	1966(C)	327 493	328 257
19	Tétouan, Morocco	1971(C)/1973(E)	137 080	308 700
20	El Mahalla El Kûbra, Egypt	1974(E)	287 800	—
21	Tanta, Egypt	1974(E)	278 300	—
22	Khartoum, Sudan	1971(E)	261 840	—
23	Omdurman, Sudan	1971(E)	258 532	—
24	Aswan, Egypt	1974(E)	246 000	—
25	Mansûra, Egypt	1974(E)	232 400	—
26	Mogadiscio, Somalia	1972(E)	230 000	—
27	Sfax, Tunisia	1966(C)	79 595	215 836
28	Safi, Morocco	1971(C)/1973(E)	129 100	214 600
29	Tripoli, Libya	1964(C)	213 506	—
30	Asyût, Egypt	1974(E)	197 200	—
31	Bamako, Mali	1972(E)	—	196 800
32	Zagazig, Egypt	1974(E)	195 100	—
33	Ismailia, Egypt	1974(E)	189 700	—
34	Agadir, Morocco	1973(E)	—	189 000
35	Maiduguri, Nigeria	1975(E)	189 000	—
36	Bloemfontein, South Africa	1970(C)	148 282	180 179
37	Ndjamena, Chad	1972(E)	—	179 000
38	Damanhûr, Egypt	1974(E)	175 900	—
39	Faiyûm, Egypt	1974(E)	167 700	—
40	Khouribga, Morocco	1973(E)	—	159 000
41	Lomé, Togo	1970(C)	148 443	—
42	Bengazi, Libya	1964(C)	137 295	—
43	Nouakchott, Mauritania	1976(C)	—	134 986
44	Minya, Egypt	1974(E)	131 200	—
45	Niamey, Niger	1975(E)	130 299	—
46	Khartoum North, Sudan	1971(E)	127 672	—
47	Settat, Morocco	1973(E)	—	125 000
48	Port Sudan, Sudan	1971(E)	110 091	—

Rank	City/Country	Year and source E = Estimate C = Census	City proper	Urban agglomeration
49	Damietta, Egypt	1974(E)	110 000	—
50	Katsina, Nigeria	1971(E)	109 424	—
51	Beni-Suef, Egypt	1974(E)	107 100	—
52	Kimberley, South Africa	1970(C)	103 789	—
53	El Jadida, Morocco	1973(E)	—	102 000

Source: United Nations, Statistical Office (1977).

TABLE I.3

Dryland cities of 100 000 or more inhabitants ranked by size of population:
Asia

Rank	City/Country	Year and source E = Estimate C = Census	City proper	Urban agglomeration
1	Peking, China	1970(E)	7 570 000	—
2	Tientsin, China	1970(E)	4 280 000	—
3	Teheran, Iran	1973(E)	4 002 000	—
4	Delhi, India	1971(C)	3 287 883	3 647 023
5	Karachi, Pakistan	1972(C)	3 498 634	—
6	Lahore, Pakistan	1972(C)	2 165 372	—
7	Hyderabad, India	1971(C)	1 607 396	1 796 339
8	Ahmedabad, India	1971(C)	1 585 544	1 741 522
9	Baghdad, Iraq	1965(C)	1 490 759	1 657 424
10	Tashkent, USSR	1976(E)	1 643 000	—
11	Ankara, Turkey	1973(E)	1 461 345	1 553 897
12	Baku, USSR	1970(C)/1976(E)	851 547	1 406 000
13	Kanpur, India	1971(C)	1 154 388	1 275 242
14	Kuibyshev, USSR	1976(E)	1 186 000	—
15	Damascus, Syria	1975(E)/1970(E)	1 042 245	923 253
16	Odessa, USSR	1976(E)	1 023 000	—
17	Taiyuan, China	1957(E)	1 020 000	—
18	Omsk, USSR	1976(E)	1 002 000	—
19	Volgograd, USSR	1976(E)	918 000	—
20	Tsinan, China	1957(E)	862 000	—
21	Saratov, USSR	1976(E)	848 000	—
22	Lyallpur, Pakistan	1972(C)	822 263	—
23	Tzepo, China	1957(E)	806 000	—
24	Tangshan, China	1957(E)	800 000	—
25	Aleppo, Syria	1975(E)	778 523	—
26	Kabul, Afghanistan	1971(E)/1975(E)	318 094	749 000
27	Lanchow, China	1957(E)	699 000	—
28	Tsitsihar, China	1957(E)	668 000	—
29	Riyadh, Saudi Arabia	1974(C)	666 840	—
30	Jaipur, India	1971(C)	615 258	636 768
31	Agra, India	1971(C)	591 917	634 622
32	Amman, Jordan	1975(E)	634 000	—
33	Hyderabad, Pakistan	1972(C)	628 310	—

Table I.3.—*continued*

Rank	City/Country	Year and source E = Estimate C = Census	City proper	Urban agglomeration
34	Rawalpindi, Pakistan	1972(C)	615 392	—
35	Esfahan, Iran	1973(E)	605 000	—
36	Shihkiachwang, China	1957(E)	598 000	—
37	Mashhad, Iran	1973(E)	592 000	—
38	Karaganda, USSR	1976(E)	570 000	—
39	Jeddah, Saudi Arabia	1974(C)	561 104	—
40	Multan, Pakistan	1972(C)	542 195	—
41	Irkutsk, USSR	1976(E)	519 000	—
42	Frunze, USSR	1976(E)	498 000	—
43	Baroda, India	1971(C)	467 422	—
44	Zhdanov, USSR	1976(E)	467 000	—
45	Trichurapalli, India	1971(C)	307 400	464 624
46	Tolyatti, USSR	1976(E)	463 000	—
47	Amritsar, India	1971(C)	407 628	458 029
48	Astrakhan, USSR	1976(E)	458 000	—
49	Dushanbe, USSR	1976(E)	448 000	—
50	Nikolaev, USSR	1976(E)	436 000	—
51	Ulyanovsk, USSR	1976(E)	436 000	—
52	Salem, India	1971(C)	308 716	416 440
53	Gwalior, India	1971(C)	384 772	406 140
54	Srinagar, Jammu & Kashmir	1971(C)	403 413	—
55	Ludhiana, India	1971(C)	397 850	401 176
56	Sholapur, India	1971(C)	398 361	—
57	Shiraz, Iran	1973(E)	373 000	—
58	Meerut, India	1971(C)	270 993	367 754
59	Mecca, Saudi Arabia	1974(C)	366 801	—
60	Gujranwala, Pakistan	1972(C)	360 419	—
61	Gaziantep, Turkey	1973(E)	275 000	352 638
62	Vijayawada, India	1971(C)	317 258	344 607
63	Jullundur, India	1971(C)/1970(E)	296 106	342 207
64	Konya, Turkey	1973(E)	227 887	324 362
65	Jodhpur, India	1971(C)	317 612	—
66	Kherson, USSR	1976(E)	315 000	—
67	Huhehot, China	1957(E)	314 000	—
68	Abadan, Iran	1973(E)	312 000	—
69	Basra, Iraq	1965(C)	310 950	310 950
70	Samarkand, USSR	1976(E)	304 000	—
71	Eskisehir, Turkey	1973(E)	243 328	303 257
72	Ahvez, Iran	1973(E)	302 000	—
73	Ulan-Ude, USSR	1976(E)	302 000	—
74	New Delhi, India	1971(C)	301 801	—
75	Rajkot, India	1971(C)	300 612	—
76	Sining, China	1957(E)	300 000	—
77	Kayseri, Turkey	1973(E)	183 128	297 328
78	Ashkhabad, USSR	1976(E)	297 000	—
79	Kurgan, USSR	1976(E)	297 000	—
80	Chimkent, USSR	1976(E)	296 000	—
81	Sevastopol, USSR	1976(E)	290 000	—
82	Ulan Bator, Mongolia	1970(E)	287 000	—
83	Simferopol, USSR	1976(E)	286 000	—
84	Aden, Yemen DR	1973(E)	264 326	285 373

Table I.3.—*continued*

Rank	City/Country	Year and source E = Estimate C = Census	City proper	Urban agglomeration
85	Taganrog, USSR	1976(*E*)	282000	—
86	Semipalatinsk, USSR	1976(*E*)	277000	—
87	Urumchi, China	1957(*E*)	275000	—
88	Guntur, India	1971(*C*)	269991	—
89	Peshawar, Pakistan	1972(*C*)	268366	—
90	Homs, Syria	1975(*E*)	267132	—
91	Paoting, China	1957(*E*)	265000	—
92	Ajmer, India	1971(*C*)	262851	264291
93	Mosul, Iraq	1965(*C*)	264146	264146
94	Ust-Kamenogorsk, USSR	1976(*E*)	262000	—
95	Aligarh, India	1971(*C*)	252314	—
96	Diyarbakir, Turkey	1973(*E*)	180237	251034
97	Pavlodar, USSR	1976(*E*)	247000	—
98	Zarka, Jordan	1975(*E*)	238000	—
99	Malatya, Turkey	1973(*E*)	144247	233977
100	Chandigarh, India	1971(*C*)	218743	232940
101	Angarsk, USSR	1976(*E*)	231000	—
102	Makhachkala, USSR	1976(*E*)	231000	—
103	Kalgan, China	1953(*C*)	229300	—
104	Tatung, China	1953(*C*)	228500	—
105	Jamnagar, India	1971(*C*)	199709	227640
106	Bhavnagar, India	1971(*C*)	225358	225974
107	Saharanpur, India	1971(*C*)	225396	—
108	Shakhty, USSR	1976(*E*)	222000	—
109	Andizhan, USSR	1976(*E*)	220000	—
110	Kuwait City, Kuwait	1970(*C*)	80405	217749
111	Namangan, USSR	1976(*E*)	217000	—
112	Tselinograd, USSR	1976(*E*)	217000	—
113	Sivas, Turkey	1973(*E*)	150267	213233
114	Kirovabad, USSR	1976(*E*)	211000	—
115	Siirt, Turkey	1973(*E*)	110498	210508
116	Kandahar, Afghanistan	1975(*E*)	209000	—
117	Bikaner, India	1971(*C*)	188518	208894
118	Ta'if, Saudi Arabia	1974(*C*)	204857	—
119	Sialkot, Pakistan	1972(*C*)	203779	—
120	Sargodha, Pakistan	1972(*C*)	201407	—
121	Ambala, India	1971(*C*)/1970(*E*)	102493	200576
122	Temirtau, USSR	1976(*E*)	200000	—
123	Medina, Saudi Arabia	1974(*C*)	198186	—
124	Jhansi, India	1971(*C*)	173292	198135
125	Volzhsky, USSR	1976(*E*)	195000	—
126	Jerusalem, Israel	1960*	167000	192000
127	Malegaon, India	1971(*C*)	191847	—
128	Rajahmundry, India	1971(*C*)	165912	188805
129	Syzran, USSR	1976(*E*)	185000	—
130	Kirkuk, Iraq	1965(*C*)	175303	183981
131	Novocherkassk, USSR	1976(*E*)	183000	—
132	Aktyubinsk, USSR	1976(*E*)	179000	—
133	Vellore, India	1971(*C*)	139082	178554
134	Rubtsovsk, USSR	1976(*E*)	171000	—
135	Ghom, Iran	1973(*E*)	170000	—
136	Sumgait, USSR	1976(*E*)	168000	—

Table I.3.—*continued*

Rank	City/Country	Year and source E = Estimate C = Census	City proper	Urban agglomeration
137	Aurangabad, India	1971(*C*)	150 483	165 253
138	Kakinada, India	1971(*C*)	164 200	—
139	Hama, Syria	1975(*E*)	162 010	—
140	Udaipur, India	1971(*C*)	161 278	—
141	Engels, USSR	1976(*E*)	159 000	—
142	Sukkur, Pakistan	1972(*C*)	158 876	—
143	Tajrish, Iran	1966(*C*)	157 486	—
144	Herat, Afghanistan	1975(*E*)	157 000	—
145	Uralsk, USSR	1976(*E*)	157 000	—
146	Quetta, Pakistan	1972(*C*)	156 000	—
147	Jammu, Jammu & Kashmir	1971(*C*)	155 338	—
148	Melitopol, USSR	1976(*E*)	155 000	—
149	Osh, USSR	1976(*E*)	155 000	—
150	Kerch, USSR	1976(*E*)	152 000	—
151	Kokand, USSR	1976(*E*)	152 000	—
152	Patiala, India	1971(*C*)	148 686	151 041
153	Kustanai, USSR	1976(*E*)	151 000	—
154	Urfa, Turkey	1973(*E*)	119 169	150 543
155	Paotow, China	1953(*C*)	149 400	—
156	Weifang, China	1953(*C*)	148 900	—
157	Ahmednagar, India	1971(*C*)	118 236	148 405
158	Gulbarga, India	1971(*C*)	145 588	—
159	Bukhara, USSR	1976(*E*)	144 000	—
160	Hamedan, Iran	1973(*E*)	144 000	—
161	Thanjavur, India	1971(*C*)	140 547	—
162	Mathura, India	1971(*C*)	132 028	140 150
163	Dhulia, India	1971(*C*)	137 129	—
164	Kurnool, India	1971(*C*)	136 710	—
165	Jhang, Pakistan	1972(*C*)	135 722	—
166	Najaf, Iraq	1965(*C*)	134 027	135 622
167	Bahawalpur, Pakistan	1972(*C*)	133 956	—
168	Firozabad, India	1971(*C*)	133 863	—
169	Nellore, India	1971(*C*)	133 590	—
170	Fergana, USSR	1976(*E*)	132 000	—
171	Guryev, USSR	1976(*E*)	131 000	—
172	Bandar, India	1971(*E*)	129 905	—
173	Chirchik, USSR	1976(*E*)	128 000	—
174	Dammam, Saudia Arabia	1974(*C*)	127 844	—
175	Ghaziabad, India	1971(*C*)	118 836	127 700
176	Eluru, India	1971(*C*)	127 023	—
177	Bellary, India	1971(*C*)	125 183	—
178	Anyang, China	1953(*C*)	124 900	—
179	Rohtak, India	1971(*C*)	124 755	—
180	Yaza, Iran	1973(*E*)	122 000	—
181	Leninabad, USSR	1976(*E*)	121 000	—
182	Sana, Yemen	1970(*E*)	120 000	—
183	Kumba Konam, India	1971(*C*)	113 130	119 655
184	Ratlam, India	1971(*C*)	106 666	119 247
185	Kolar Gold Fields, India	1971(*C*)	76 112	118 861
186	Mardan, Pakistan	1972(*C*)	115 218	—
187	Muzaffarnagar, India	1971(*C*)	114 783	—
188	Novokuibyshevsk, USSR	1976(*E*)	113 000	—

Table I.3.—*continued*

Rank	City/Country	Year and source E = Estimate C = Census	City proper	Urban agglomeration
189	Machilipatnam, India	1971(C)	112 612	—
190	Hilla, Iraq	1965(C)	84 704	111 338
191	Farrukhabad-Fategarh, India	1971(C)	102 768	110 835
192	Nadiad, India	1971(C)	108 269	—
193	Ining, China	1953(C)	108 200	—
194	Qazvin, Iran	1973(E)	107 000	—
195	Por-bander, India	1971(C)	96 881	106 727
196	Jalgaon, India	1971(C)	106 711	—
197	Hawalli, Kuwait	1970(C)	106 542	—
198	Sahiwal, Pakistan	1972(C)	106 213	—
199	Bhusawal, India	1971(C)	96 800	104 708
200	Bijapur, India	1971(C)	103 931	—
201	Tenali, India	1971(C)	102 937	—
202	Kasur, Pakistan	1972(C)	102 531	—
203	Okara, Pakistan	1972(C)	101 791	—
204	Huful, Saudi Arabia	1974(C)	101 271	—
205	Gujrat, Pakistan	1972(C)	100 581	—
206	Charekar, Afghanistan	1973(E)	100 443	—
207	Alwar, India	1971(C)	100 378	—
208	Kerman, India	1972(E)	100 000	—

Population figure derived from K. Davis (1969). Source of all other data: United Nations, Statistical Office (1977).

It includes highly urbanized countries such as Israel (81.9% urban in 1974), Kuwait (80.4%, 1970), Iraq (64.8%, 1976), the USSR (61.6%, 1976), Turkey (44.5%, 1976), and a number of less-highly urbanized nations such as Pakistan (25.5%, 1972), China (23.5%, 1970), and India (20.6%, 1974). Urbanism and urbanization have a long history in this region, dating back to some of the earliest sites of urban growth in Mesopotamia, Turkey, the Indus Valley, and China. Recent, rapid urban expansion has characterized the Middle East, and contemporary accounts of this phenomenon are numerous (Gulick, 1967; Lapidus, 1969; Berger, 1974; and Costello, 1977); Harris (1971) and Chen (1973) have respectively chronicled population growth and urbanization from the 1950s to 1970 in the Soviet Union and China.

Table I.1 shows 41 urban centres in Asia with over 500 000 inhabitants, 18 of which exceed a million, and six of which exceed two millions. Asian dryland cities in general show a smoother size–frequency distribution curve than those of Africa, a feature that may reflect a longer history of urbanization. Populations of the leading Asian dryland urban centres are listed in Table I.3. There are particularly large numbers of examples in India (62), the USSR (55), China (19), Pakistan (19), Iran (12), Turkey (10), Saudi Arabia (7), and Iraq (6) (Fig. I.3).

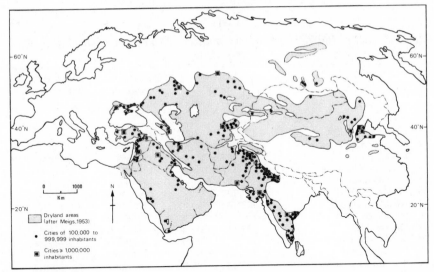

FIG. I.3 Drylands in Asia: cities with 100 000 or more inhabitants

4. Dryland cities in the Americas

The 94 dryland cities of North and South America, when ranked by total population, also show a smooth size–frequency progression, with a broad base of smaller urban settlements. In fact, the size–frequency relations (Table I.1) are typical of an integrated urban system within technologically advanced nations. Most of the American dryland countries are highly urbanized, such as Chile (78.7%, 1976), Venezuela (74.7%, 1976), USA (73.5%, 1970), Argentina (70.5%, 1970), Mexico (63.6%, 1976), and Peru (55.3%, 1974).

Table I.4 shows 22 dryland cities in the Americas with over 500 000 inhabitants, of which 10 had over a million, and five over two million inhabitants. Los Angeles, USA (7 millions) was the largest urban agglomeration, followed by Santiago, Chile (3.35 millions), and Lima, Peru (3.3 millions). The preponderance of dryland cities is clearly in Mexico and the southwestern USA: these two areas account for 64 of the cities over 100 000 and seven of the 'million' cities. The rapid urban growth in the desert areas of New Mexico, California, and Arizona in the past three decades has been an outstanding feature of the American metropolitan scene (Ullman, 1954; Borchert, 1967; Wilson, 1977; Langford-Smith, 1978). The rapid evolution of urban population in the drylands of Mexico during the twentieth century has recently been examined by MacGregor and Valverde (1975). Using data for 1970, they concluded that of a total national urban population of 22.5 millions, 7 millions or 31% live in the arid zone, while 101 (or 35%) of the country's 286 urban places are located in the arid zone. Table I.4 and Figs. I.4 and I.5 show that the other major countries in the Americas with dryland cities are Argentina (7), Peru (7), Chile (6), Venezuela (5), Canada (2), Bolivia (2), and Brazil (1).

TABLE I.4

Dryland cities of 100 000 or more inhabitants ranked by size of population: The Americas

Rank	City/Country	Year and source E = Estimate C = Census	City proper	Urban agglomeration
1	Los Angeles, USA	1975(E)/1970(C)	2 727 399	7 032 075
2	Santiago, Chile	1970(E)	3 273 600	3 350 680
3	Lima, Peru	1972(C)	2 833 609	3 302 523
4	Caracas, Venezuela	1971(C)	1 662 627	2 175 400
5	Guadalajara, Mexico	1976(E)	1 640 902	2 075 773
6	Monterrey, Mexico	1976(E)	1 090 226	1 725 013
7	Anaheim, USA	1975(E)/1970(C)	193 616	1 420 386
8	San Diego, USA	1975(E)/1970(C)	773 996	1 357 854
9	Denver, USA	1975(E)/1970(C)	484 531	1 227 529
10	San Bernardino, USA	1975(E)/1970(C)	102 076	1 143 146
11	Phoenix, USA	1975(E)/1970(C)	664 721	967 522
12	San Antonio, USA	1975(E)/1970(C)	773 248	864 014
13	Sacramento, USA	1975(E)/1970(C)	260 822	800 592
14	Córdoba, Argentina	1975(E)	781 565	790 508
15	La Paz, Bolivia	1976(C)	654 713	—
16	Maracaibo, Venezuela	1971(C)	651 574	—
17	Valparaiso, Chile	1970(E)	248 706	601 360
18	Salt Lake City, USA	1975(E)/1970(C)	169 917	557 635
19	Ciudad Juárez, Mexico	1976(E)	544 900	—
20	Tijuana, Mexico	1976(E)	411 643	535 535
21	Edmonton, Canada	1971(C)/1974(E)	438 150	529 000
22	León, Mexico	1976(E)	525 947	—
23	Mendoza, Argentina	1975(E)	118 568	470 896
24	Calgary, Canada	1971(C)/1974(E)	403 320	444 000
25	Fresno, USA	1975(E)/1970(C)	176 528	413 053
26	El Paso, USA	1975(E)/1970(C)	385 691	359 291
27	Oxnard, USA	1975(E)/1970(C)	71 225	376 430
28	Torreón, Mexico	1976(E)	256 995	372 888
29	San Miguel de Tucumán, Argentina	1975(E)	321 567	366 392
30	Chihuahua, Mexico	1976(E)	365 760	—
31	Tucson, USA	1975(E)/1970(C)	296 457	351 667
32	Mexicali, Mexico	1976(E)	345 943	—
33	Long Beach, USA	1975(E)	335 602	—
34	Bakersfield, USA	1970(C)	69 515	329 162
35	Lancaster, USA	1970(C)	57 690	319 693
36	Albuquerque, USA	1975(E)/1970(C)	279 401	315 774
37	Arequipa, Peru	1972(C)	302 316	—
38	Callao, Peru	1972(C)	296 721	—
39	San Luis Potosí, Mexico	1976(E)	292 345	—
40	Stockton, USA	1975(E)/1970(C)	117 600	290 208
41	Spokane, USA	1975(E)/1970(C)	173 698	287 487
42	Corpus Christi, USA	1975(E)/1970(C)	214 838	284 132
43	Las Vegas, USA	1975(E)/1970(C)	146 030	273 288
44	Hermosillo, Mexico	1976(E)	264 073	—
45	Culiacán, Mexico	1976(E)	262 504	—
46	Maracay, Venezuela	1971(C)	255 134	—

Table I.4—*continued*

Rank	City/Country	Year and source E = Estimate C = Census	City proper	Urban agglomeration
47	Salinas, USA	1970(*C*)	58 896	250 071
48	Mérida, Mexico	1976(*E*)	244 652	—
49	Trujillo, Peru	1972(*C*)	240 322	—
50	Viña del Mar, Chile	1970(*E*)	239 660	—
51	Colorado Springs, USA	1975(*E*)/1970(*C*)	179 584	235 972
52	Aguascalientes, Mexico	1976(*E*)	229 956	—
53	Saltillo, Mexico	1976(*E*)	222 087	—
54	San Juan, Argentina	1975(*E*)	112 582	217 514
55	Reynosa, Mexico	1976(*E*)	206 453	—
56	Nuevo Laredo, Mexico	1976(*E*)	203 739	—
57	Durango, Mexico	1976(*E*)	199 822	—
58	Modesto, USA	1970(*C*)	61 712	194 506
59	Chiclayo, Peru	1972(*C*)	187 809	—
60	Bahia Blanca, Argentina	1975(*E*)	—	182 158
61	McAllen, USA	1970(*C*)	37 636	181 535
62	Matamoros, Mexico	1976(*E*)	179 423	—
63	Lubbock, USA	1975(*E*)	163 525	179 295
64	Santa Ana, USA	1975(*E*)	177 304	—
65	Salta, Argentina	1975(*E*)/1970(*C*)	176 216	176 130
66	Campina Grande, Brazil	1970(*C*)	163 006	168 045
67	Ciudad Obregon, Mexico	1976(*E*)	161 319	—
68	Chimbote, Peru	1972(*C*)	159 045	—
69	Querétaro, Mexico	1976(*E*)	158 428	—
70	Antofagasta, Chile	1970(*E*)	154 800	—
71	Riverside, USA	1975(*E*)	150 612	—
72	Huntington Beach, USA	1975(*E*)	149 706	—
73	Irapuato, Mexico	1976(*E*)	145 254	—
74	Amarillo, USA	1975(*E*)/1970(*C*)	138 743	144 396
75	Brownsville, USA	1970(*C*)	52 522	140 368
76	Torrance, USA	1975(*E*)	139 776	—
77	Provo, USA	1970(*C*)	53 131	137 776
78	Glendale, USA	1975(*E*)	132 360	—
79	Ogden, USA	1970(*C*)	69 478	126 278
80	Piura, Peru	1972(*C*)	126 010	—
81	Oruro, Bolivia	1976(*C*)	124 091	—
82	Reno, USA	1970(*C*)	72 863	121 068
83	Norwalk, USA	1970(*C*)	79 118	120 099
84	Cumana, Venezuela	1971(*C*)	119 751	—
85	Garden Grove, USA	1975(*E*)	118 454	—
86	Pueblo, USA	1975(*E*)/1970(*C*)	105 312	118 238
87	Cabimas, Venezuela	1971(*C*)	118 037	—
88	Arica, Chile	1970(*E*)	117 020	—
89	Monclova, Mexico	1976(*E*)	115 707	—
90	Abilene, USA	1970(*C*)	89 653	113 959
91	Rancagua, Chile	1970(*E*)	112 710	—
92	Boise City, USA	1970(*C*)	74 990	112 230
93	Pasadena, USA	1975(*E*)	108 220	—
94	Santiago del Estero, Argentina	1975(*E*)/1970(*C*)	105 127	105 209

Source: United Nations, Statistical Office (1977).

(c) Growth of Dryland Cities

1. *Growth of dryland cities, 1950–70*

The preceding account provides a description of the present distribution and size of dryland cities. But these cities are changing, in common with cities elsewhere: in general they are growing rapidly in size and significance.

Several problems of data availability, reliability, and comparability arise in studies of urban growth. The main problem is that data for dryland city populations, derived from national estimates and censuses, are only available for a bewildering variety of dates, making comparisons difficult. Fortunately, tables in Davis (1969) provide data on urban growth that are standardized for 1950, 1960, and 1970, the figures having been projected either forwards or backwards to these dates from the nearest available census date or estimates. Davis's data only cover 225 of the 355 major dryland cities shown

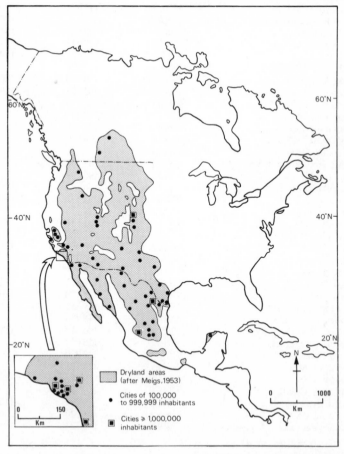

FIG. I.4 Drylands in North America: cities with 100 000 or more inhabitants

in Figs. I.2–I.5 (71.8%). The cities are shown in Tables I.5–I.7, ranked according to their estimated percentage population growth during the decade 1960–70, and where possible the 1950–60 growth figure is also included.

These tables reveal the very high estimated growth rates of dryland cities since 1950. From 1965 to 1975 world population grew at an annual rate of 1·9%, and in Africa it grew at 2.7%, in America, 2.0%, and in Asia 2.1%. In recent years world urban population growth has been approximately 3.2% p.a. The average annual growth rate of dryland cities is slightly higher than this at 3.93%. But the table also shows that many dryland cities were growing at much higher rates and in Tables I.5–I.7, figures as high as 16.0% p.a. are recorded, and rates above 5% p.a. are common. Thus during the decade 1960–70, 67.11% of dryland cities for which data are available showed growth rates of 3% p.a. or more, and 25.3% of them grew at more than 5% p.a. In

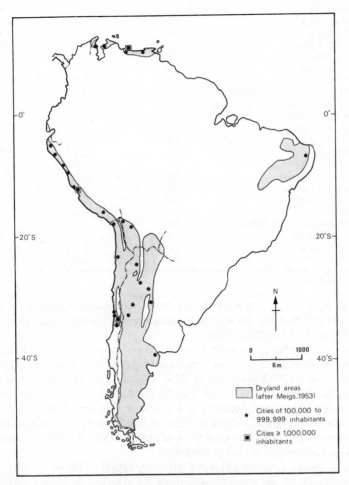

FIG. I.5 Drylands in South America: cities with 100 000 or more inhabitants

TABLE I.5

Growth rates for selected dryland cities of 100 000 or more inhabitants: Africa

Rank	City/Country	CP = City proper UA = Urban agglomeration	Per cent growth per annum 1950–60	Per cent growth per annum 1960–70
1	Accra, Ghana	UA	8.4	6.8
2	Bengazi, Libya	CP	—	6.8
3	Maputo, Mozambique	UA	—	6.2
4	Bamako, Mali	UA	—	5.6
5	Dakar, Senegal	UA	6.8	5.3
6	Kano, Nigeria	CP	5.0	5.0
7	Suez, Egypt	CP	5.0	4.7
8	Tripoli, Libya	CP	4.8	4.4
9	Meknès, Morocco	CP	2.6	4.2
10	Luanda, Angola	UA	4.7	4.1
11	Cairo, Egypt	UA	4.1	4.1
12	Faiyûm, Egypt	CP	—	4.1
13	Oujda, Morocco	CP	—	4.0
14	Tétouan, Morocco	CP	—	4.0
15	Fez, Morocco	CP	2.2	3.9
16	El Mahalla El Kûbra, Egypt	CP	3.3	3.7
17	Khartoum, Sudan	UA	3.4	3.5
18	Mansûra, Egypt	CP	3.0	3.5
19	Zagazig, Egypt	CP	—	3.5
20	Asyût, Egypt	CP	—	3.4
21	Tanta, Egypt	CP	2.1	3.3
22	Marrakech, Morocco	CP	1.4	3.2
23	Alexandria, Egypt	CP	3.9	3.1
24	Asmara, Ethiopia	CP	—	2.7
25	Port Said, Egypt	CP	2.5	2.5
26	Damanhûr	CP	—	2.4
27	Bloemfontein, South Africa	UA	3.3	1.9
28	Ismailia, Egypt	CP	—	1.3
29	Tunis, Tunisia	CP	0.5	0.8
30	Oran, Algeria	UA	3.1	−1.9

Source: Davis, K. (1969).

some oil-rich states, city growth has been particularly spectacular. In the United Arab Emirates, for example, annual growth rates range from 11.8 to 25.6% (Bials and Sinclair, 1980).

In Africa, capital cities seem to have grown most rapidly, reinforcing the position of primate cities. In Asia, rapid growth is apparent in certain dryland cities of Pakistan, the USSR, China, and the oil-rich states of Kuwait, Saudi Arabia, Iran, and Iraq. In the Americas, high rates of growth characterized the south-western US cities of Tucson, Phoenix, El Paso, Albuquerque, and the Californian cities of Sacramento, San Diego, and San Bernardino-Riverside. The growth of Mexican dryland urban areas such as Tijuana and Mexicali is also pronounced.

Clearly, many dryland cities are growing rapidly. Table I.8 emphasizes this conclusion by showing growth rates for selected dryland cities for the most recent ten-year period for which data are available. But there are

TABLE I.6

Growth rates for selected dryland cities of 100 000 or more inhabitants: Asia

Rank	City/Country	CP = City proper UA = Urban agglomeration	Per cent growth per annum 1950–60	Per cent growth per annum 1960–70
1	Lyallpur, Pakistan	UA	9.1	8.8
2	Kuwait City, Kuwait	UA	—	8.3
3	Huhehot, China	UA	7.9	8.1
4	Basra, Iraq	UA	7.7	7.9
5	Ulan Bator, Mongolia	CP	—	7.9
6	Paotow, China	UA	11.9	7.2
7	Shihkiachwang, China	UA	10.6	7.2
8	Kalgan, China	UA	7.2	7.2
9	Frunze, USSR	CP	7.6	7.0
10	Malegaon, India	UA	—	7.0
11	Taiyuan, China	UA	8.8	6.9
12	Ankara, Turkey	UA	8.4	6.8
13	Homs, Syria	CP	3.0	6.6
14	Mashhad, Iran	CP	2.4	6.6
15	Tselinograd, USSR	CP	—	6.5
16	Novocherkassk, USSR	CP	—	6.4
17	Multan, Pakistan	UA	6.6	6.3
18	Esfahan, Iran	CP	2.4	6.3
19	Jerusalem, Israel	UA	—	6.2
20	Hyderabad, Pakistan	UA	6.1	5.9
21	Riyadh, Saudi Arabia	CP	—	5.9
22	Karachi, Pakistan	UA	5.7	5.6
23	Shiraz, Iran	CP	2.4	5.6
24	Jeddah, Saudi Arabia	CP	—	5.6
25	Rawalpindi, Pakistan	UA	6.5	5.5
26	Ahvez, Iran	CP	2.5	5.5
27	Teheran, Iran	UA	—	5.4
28	Kirovabad, USSR	CP	—	5.3
29	Urumchi, China	UA	11.6	5.2
30	Sargodha, Pakistan	UA	—	5.2
31	Sining, China	UA	—	5.2
32	Gujranwala, Pakistan	UA	5.4	5.0
33	Kherson, USSR	CP	5.0	5.0
34	Tsinan, China	UA	3.5	5.0
35	Tzepo, China	UA	16.0	4.8
36	Aleppo, Syria	CP	3.1	4.8
37	Ining, China	UA	—	4.8
38	Lanchow, China	UA	12.2	4.7
39	Tsitsihar, China	UA	12.2	4.7
40	Dushanbe, USSR	CP	5.1	4.7
41	Ludhiana, India	UA	4.7	4.7
42	Jullundur, India	UA	4.6	4.7
43	Kurgan, USSR	CP	—	4.7
44	Ulyanovsk, USSR	CP	4.5	4.6
45	Amman, Jordan	CP	—	4.5
46	Chimkent, USSR	CP	4.6	4.4
47	Baghdad, Iraq	UA	4.2	4.4
48	Hama, Syria	CP	—	4.4
49	Aden, Yemen DR	UA	5.4	4.3

Table I.6—*continued*

Rank	City/Country	CP = City proper UA = Urban agglomeration	Per cent growth per annum 1950–60	Per cent growth per annum 1960–70
50	Sevastopol, USSR	CP	4.4	4.2
51	Ashkhabad, USSR	CP	4.3	4.2
52	Lahore, Pakistan	UA	4.3	4.2
53	Guntur, India	UA	4.1	4.2
54	Ust-Kamenogorsk, USSR	CP	4.4	4.1
55	Tashkent, USSR	CP	3.9	4.1
56	Mosul, Iraq	UA	3.6	4.1
57	Anyang, China	UA	2.3	4.1
58	Najaf, Iraq	UA	—	4.1
59	Rajkot, India	UA	4.0	4.0
60	Aktyubinsk, USSR	CP	—	4.0
61	Jammu, Jammu & Kashmir	CP	—	4.0
62	Kayseri, Turkey	CP	—	4.0
63	Kirkuk, Iraq	UA	—	4.0
64	Konya, Turkey	CP	—	4.0
65	Makhachkala, USSR	CP	—	4.0
66	Ahmednagar, India	UA	—	3.9
67	Peking, China	UA	9.8	3.8
68	Vijayawada, India	UA	3.6	3.8
69	Zhdanov, USSR	CP	3.9	3.7
70	Peshawar, Pakistan	UA	3.8	3.7
71	Agra, India	UA	3.6	3.7
72	Gaziantep, Turkey	CP	—	3.7
73	Omsk, USSR	CP	3.7	3.6
74	Baroda, India	UA	3.5	3.6
75	Ghom, Iran	CP	—	3.6
76	Tangshan, China	UA	5.4	3.5
77	Jamnagar, India	UA	3.6	3.5
78	Nikolaev, USSR	CP	3.6	3.5
79	Paoting, China	UA	4.6	3.4
80	Jaipur, India	UA	3.3	3.4
81	Kanpur, India	UA	3.2	3.4
82	Angarsk, USSR	CP	—	3.4
83	Semipalatinsk, USSR	CP	3.4	3.3
84	Ulan-Ude, USSR	CP	3.3	3.2
85	Weifang, China	UA	3.1	3.2
86	Andizhan, USSR	CP	—	3.2
87	Kabul, Afghanistan	UA	—	3.2
88	Mecca, Saudi Arabia	CP	4.7	3.1
89	Jhansi, India	UA	2.9	3.1
90	Delhi, India	UA	2.9	3.0
91	Sukkur, Pakistan	UA	—	3.0
92	Karaganda, USSR	CP	3.2	2.9
93	Samarkand, USSR	CP	3.0	2.9
94	Aligarh, India	UA	2.7	2.9
95	Astrakhan, USSR	CP	2.7	2.9
96	Nellore, India	UA	—	2.9
97	Rubtsovsk, USSR	CP	—	2.9
98	Volgograd, USSR	CP	2.9	2.8
99	Ahmedabad, India	UA	2.7	2.7
100	Bhavnagar, India	UA	2.5	2.7
101	Kandahar, Afghanistan	CP	—	2.7

Table I.6—*continued*

Rank	City/Country	CP = City proper UA = Urban agglomeration	Per cent growth per annum 1950–60	Per cent growth per annum 1960–70
102	Patiala, India	UA	—	2.7
103	Saratov, USSR	CP	2.7	2.6
104	Baku, USSR	UA	2.6	2.6
105	Kokand, USSR	CP	—	2.6
106	Tientsin, China	UA	4.8	2.5
107	Kuibyshev, USSR	CP	2.6	2.5
108	Amritsar, India	UA	2.0	2.5
109	Quetta, Pakistan	UA	—	2.5
110	Udaipur, India	UA	—	2.5
111	Damascus, Syria	CP	4.0	2.4
112	Taganrog, USSR	CP	2.4	2.4
113	Gwalior, India	UA	2.2	2.4
114	Saharanpur, India	UA	2.2	2.4
115	Eluru, India	UA	—	2.4
116	Tatung, China	UA	4.8	2.3
117	Simferopol, USSR	CP	2.3	2.3
118	Rajahmundry, India	UA	2.1	2.3
119	Meerut, India	UA	2.0	2.3
120	Jodhpur, India	UA	2.2	2.2
121	Salem, India	UA	2.1	2.2
122	Sholapur, India	UA	2.0	2.1
123	Kakinada, India	UA	—	2.2
124	Kerch, USSR	CP	—	2.2
125	Uralsk, USSR	CP	—	2.1
126	Odessa, USSR	CP	1.9	2.0
127	Ajmer, India	UA	1.6	1.9
128	Hyderabad, India	UA	1.4	1.9
129	Srinagar, Jammu & Kashmir	UA	1.9	1.8
130	Ambala, India	UA	1.8	1.8
131	Irkutsk, USSR	CP	1.8	1.8
132	Hamedan, Iran	CP	—	1.8
133	Abadan, Iran	CP	2.4	1.6
134	Bikaner, India	UA	1.5	1.6
135	Eskisehir, Turkey	CP	—	1.6
136	Syzran, USSR	CP	1.6	1.5
137	Trichurapalli, India	UA	1.3	1.4
138	Thanjavur, India	UA	1.0	1.3
139	Mathura, India	UA	1.0	1.1
140	Vellore, India	UA	1.7	1.0
141	Shakhty, USSR	CP	0.8	0.7
142	Chandigarh, India	UA	0.3	0.7
143	Sialkot, Pakistan	UA	0.6	0.6
144	Kolar Gold Fields, India	UA	− 0.8	− 0.2

Source: Davis, K. (1969).

important spatial variations in dryland city growth, and in the following section the 1960–70 figures are viewed in the context of location and size characteristics of major dryland cities.

TABLE I.7

Growth rates for selected dryland cities of 100 000 or more inhabitants: The Americas

Rank	City/Country	CP = City proper UA = Urban agglomeration	Per cent growth per annum 1950–60	Per cent growth per annum 1960–70
1	Tijuana, Mexico	CP	—	11.9
2	Tucson, USA	UA	—	11.6
3	Mexicali, Mexico	CP	—	9.6
4	San Bernardino-Riverside, USA	UA	10.8	8.0
5	Ciudad Juárez, Mexico	CP	7.9	7.3
6	Phoenix, USA	UA	9.8	7.2
7	Guadalajara, Mexico	CP	6.9	6.3
8	Calgary, Canada	UA	7.1	6.0
9	Sacramento, USA	UA	7.8	5.6
10	Monterrey, Mexico	CP	6.0	5.4
11	Caracas, Venezuela	UA	6.3	5.3
12	Maracaibo, Venezuela	CP	5.5	5.3
13	Chihuahua, Mexico	CP	—	5.3
14	Mendoza, Argentina	UA	5.2	5.2
15	Albuquerque, USA	UA	—	5.2
16	León, Mexico	CP	5.6	5.1
17	Lima, Callao, Peru	UA	4.8	5.1
18	El Paso, USA	UA	7.3	4.9
19	San Diego, USA	UA	6.8	4.9
20	Lubbock, USA	UA	—	4.8
21	Amarillo, USA	UA	—	4.5
22	San Juan, Argentina	UA	4.3	4.3
23	Colorado Springs, USA	UA	—	4.3
24	Salta, Argentina	UA	—	4.2
25	Maracay, Venezuela	CP	—	4.1
26	Edmonton, Canada	UA	7.0	4.0
27	Los Angeles-Long Beach, USA	UA	4.9	3.9
28	Arequipa, Peru	CP	—	3.9
29	Fresno, USA	UA	4.9	3.8
30	Denver, USA	UA	4.9	3.7
31	Salt Lake City, USA	UA	4.3	3.5
32	Bakersfield, USA	UA	—	3.5
33	San Antonio, USA	UA	3.6	3.3
34	Córdoba, Argentina	UA	3.3	3.3
35	Santiago, Chile	UA	4.1	3.1
36	Torreón, Mexico	CP	3.4	3.1
37	Aguascalientes, Mexico	CP	—	2.8
38	Corpus Christi, USA	UA	3.7	2.6
39	San Miguel de Tucumán, Argentina	UA	2.5	2.5
40	Campina Grande, Brazil	UA	—	2.4
41	Pueblo, USA	UA	—	2.4
42	La Paz, Bolivia	UA	2.9	2.3
43	Santiago del Estero, Argentina	UA	—	2.3
44	Viña del Mar, Chile	CP	—	2.3

Rank	City/Country	CP = City proper UA = Urban agglomeration	Per cent growth per annum 1950–60	Per cent growth per annum 1960–70
45	Stockton, USA	UA	2.3	2.1
46	Ogden, USA	UA	—	1.9
47	Mérida, Mexico	CP	1.7	1.7
48	Bahia Blanca, Argentina	UA	1.6	1.6
49	Valparaiso, Chile	CP	1.6	1.6
50	San Luis Potosí, Mexico	CP	2.3	1.5
51	Spokane, USA	UA	2.6	1.4

Source: Davis, K. (1969).

TABLE I.8

Percentage growth of selected dryland cities in the decade 1960/6–1970/6

Rank	City/Country	CP = City proper UA = Urban agglomeration	1960s population and date	1970s population and date	Growth during decade
1	Mérida, Mexico	CP	100 394(1966)	244 652(1976)	143.69
2	Bakersfield, USA	UA	141 763(1960)	329 162(1970)	132.19
3	Lima, Peru	UA	1 436 231(1962)	3 302 523(1972)	129.94
4	Culiacan, Mexico	CP	118 842(1966)	262 504(1976)	120.88
5	Ahvez, Iran	CP	155 054(1963)	302 000(1973)	94.77
6	Saltillo, Mexico	CP	121 996(1966)	222 087(1976)	82.04
7	León, Mexico	CP	290 634(1966)	525 947(1976)	80.97
8	Chihuahua, Mexico	CP	209 650(1966)	365 760(1976)	74.46
9	Teheran, Iran	CP	2 317 116(1963)	4 002 000(1973)	72.71
10	Hermosillo, Mexico	CP	154 987(1966)	264 073(1976)	70.36
11	Nuevo Laredo, Mexico	CP	123 449(1966)	203 739(1976)	65.04
12	San Luis Potosí, Mexico	CP	177 811(1966)	292 345(1976)	64.41
13	Durango, Mexico	CP	131 232(1966)	199 822(1976)	52.27
14	Aguascalientes, Mexico	CP	152 293(1966)	229 956(1976)	51.00
15	Guadalajara, Mexico	CP	1 105 930(1966)	1 640 902(1976)	48.37
16	Chimkent, USSR	CP	209 000(1966)	296 000(1976)	41.63
17	Tijuana, Mexico	CP	306 897(1966)	411 643(1976)	34.13
18	Phoenix, USA	CP	505 666(1965)	664 721(1975)	31.45
19	Ciudad Juárez, Mexico	CP	415 580(1966)	544 900(1976)	31.12
20	Angarsk, USSR	CP	179 000(1966)	231 000(1976)	29.05
21	Monterrey, Mexico	CP	849 677(1966)	1 090 226(1976)	28.31
22	Anaheim, USA	UA	1 110 000(1965)	1 420 386(1975)	27.96
23	Matamoros, Mexico	CP	141 341(1966)	179 423(1976)	26.94
24	Astrakhan, USSR	CP	361 000(1966)	458 000(1976)	26.87
25	Zhdanov, USSR	CP	373 000(1966)	467 000(1976)	25.20
26	Tucson, USA	CP	236 877(1965)	296 457(1975)	25.15
27	Kuibyshev, USSR	CP	969 000(1966)	1 186 000(1976)	22.39
28	Baku, USSR	UA	1 164 000(1966)	1 406 000(1976)	20.79
29	San Diego, USA	UA	1 145 000(1965)	1 357 854(1975)	18.59
30	Torréon, Mexico	CP	220 122(1966)	256 995(1976)	16.75
31	San Bernardino, USA	UA	1 033 000(1965)	1 143 146(1975)	10.66
32	Mexicali, Mexico	CP	317 041(1966)	345 943(1976)	9.12
33	San Antonio, USA	UA	807 000(1965)	864 014(1975)	7.06
34	Abadan, Iran	CP	302 189(1963)	312 000(1973)	3.25

Source: United Nations, Statistical Office (1967 and 1977).

2. Dryland city growth rates by continent and country

From 1960 to 1970, the estimated average annual growth rate of world dryland cities for which data are available was 3.93%, this being composed of a growth rate of 4.35% p.a. in the Americas, followed by 3.83% p.a. in Asian dryland cities, and 3.67% p.a. for African dryland cities. Average annual growth rates for 1960–70 were calculated for dryland cities grouped by country (Table I.9). Countries with particularly high rates of dryland urban growth include Libya, Mexico, Canada, China, Pakistan, Syria, Iraq, Saudi Arabia, Iran, and Venezuela.

3. Growth rates and city size

It has been argued frequently that larger cities tend to grow faster than smaller cities, and it is interesting to consider if this principle applies to dryland cities. Accordingly, average annual rates of growth 1960–70 were calculated for dryland cities grouped by population size. The results (Table I.10) show that for dryland cities taken as a whole, the highest average rate of growth was recorded by cities in the 500 000–1 000 000 category, followed by the category with over 1 000 000 inhabitants. Average annual growth rates for dryland cities of 250–500 000, and 100–250 000 are somewhat lower. On a continental basis, this overall pattern is repeated in African and Asian dryland cities, but in the Americas growth rates increased directly with city size.

TABLE I.9

Average annual growth rates during the decade 1960–70 for selected dryland cities grouped by country

	Country	No. of cities for which data available	Mean annual growth rate p.a. 1960–1970	Range	Standard deviation
1	Libya	2	5.60	4.4–6.8	—
2	Mexico	11	5.45	1.5–11.9	3.09
3	Canada	2	5.00	4.0–6.0	—
4	China	19	4.99	2.3–8.1	1.65
5	Iraq	5	4.90	4.0–7.9	1.50
6	Venezuela	3	4.90	4.1–5.3	0.56
7	Saudi Arabia	3	4.87	3.1–5.9	1.26
8	Pakistan	12	4.69	0.6–8.8	2.01
9	Syria	4	4.55	2.4–6.6	1.49
10	Iran	8	4.55	1.6–6.6	1.84
11	Peru	2	4.50	3.9–5.1	—
12	USA	21	4.43	1.4–11.6	2.26
13	Turkey	5	4.02	1.6–6.8	1.65
14	Morocco	5	3.86	3.2–4.2	0.34
15	USSR	39	3.51	0.7–7.0	1.35
16	Argentina	7	3.34	1.6–5.2	1.19
17	Egypt	12	3.30	1.3–4.7	0.86
18	Afghanistan	2	2.95	2.7–3.2	—
19	India	40	2.70	−0.2–7.0	1.26
20	Chile	3	2.33	1.6–3.1	0.61
	World drylands	225	3.93	−1.9–11.9	1.91

TABLE I.10

Average annual growth rates during the decade 1960–70 for selected dryland cities according to size and continental division

| City size | Continental divisions | | | | | | World drylands | |
| | Africa | | Americas | | Asia | | | |
	Mean	S.D.	Mean	S.D.	Mean	S.D.	Mean	S.D.
1 000 000	3.60	0.50	5.08	1.39	3.77	1.50	4.16	1.55
500 000–1 000 000	4.30	2.55	5.03	2.68	4.80	1.64	4.84	2.11
250 000–500 000	3.54	1.70	4.42	2.48	3.93	1.62	3.98	1.88
100 000–250 000	3.67	1.57	3.04	1.14	3.31	1.88	3.31	1.74
All dryland cities ⩽ 100 000	3.67	1.73	4.35	2.30	3.83	1.77	3.93	1.91

(d) Urban Growth and the Consumption of Land

The available data presented above show that there are many large urban agglomerations in drylands, and many of them are growing rapidly. Summation of the population totals in Tables I.2–I.4 suggests that 21.7 million people live in dryland cities with 100 000 or more inhabitants in Africa, 48.6 millions in the American drylands, and 93.7 millions in Asia. Kates *et al.* (1976) estimated the total population of drylands to be 628.4 millions (14% of the world total): it would thus seem that at least 26.1% of all dryland inhabitants live in urban agglomerations with populations of 100 000 or more.

Within the drylands, it is evident that urban growth and urbanization are proceeding in a wide variety of cultural, social, economic, and political contexts. From the point of view of environmental management, a range of types of urban development can be recognized. At one extreme, there is usually planned, technologically advanced urban development in societies such as those of the USA, the USSR, Israel, South Africa, and Australia. At the other extreme, there is development in technologically more poorly equipped societies, such as those in the Sudan, Algeria, Peru, India, and Pakistan. Between these extremes there is, for example, technologically advanced urban development in rapidly changing oil-based societies that have recently chosen to invest in urban growth; such countries as Kuwait, Saudi Arabia, Iran, and Iraq fall into this category.

Problems arising from urban development to some extent vary within these different contexts, but most dryland cities are united in their numerous common physical environmental problems and their need for planning controls and engineering expertise to help overcome them. If anything, the problems are most serious in those communities where growth is uncontrolled, and planning is primitive or absent; and it is in such areas that the rate of growth of urban populations is often so strikingly rapid.

The rapid growth of urban populations in less-developed countries may be regarded as the outcome of both rapid rates of rural-to-urban migration, and natural population increase. The combination of 'pre-industrial' fertility levels with 'post-industrial' mortality rates means that Third World cities are experiencing some of the highest rates of natural increase ever found in cities (Dwyer, 1975). At the same time, rural–urban migration is accelerating due mainly to the attraction of high urban wages and the collapse of traditional rural economies – the result is a tide of in-migration. Thus in Caracas, Venezuela, for example, it has been estimated that from 1960 to 1966 migrants comprised 50% of the total population increase (World Bank, 1972). Similarly, MacGregor and Valverde's (1975) study of Mexican cities showed that in Monterey 31.3% of the population are in-migrants, and the figure is 48% for Tijuana, 34% for Mexicali, and 32.9% for Nuevo Laredo. In addition, burgeoning urban populations in Third World dryland cities may lead to a dramatic increase in local water consumption and despoliation of agricultural land around cities, which, in turn, may generate further waves of rural migrants and yet more demands for water (e.g. Kates et al., 1976).

These forces have resulted in a situation in many Third World cities where urbanization is in advance of industrialization and employment provision, thereby replacing rural underemployment by urban unemployment, and the growing 'ruralization' of towns (Phillips, 1958; Abu-Lughod, 1961; Mountjoy, 1976). The most obvious tangible outcome has been the spontaneous explosion of squatter settlements in many Third World cities (Jones, 1964; Berry, 1973; Berry and Kasanda, 1977). The extent of these slums and uncontrolled settlements in several dryland cities is shown by data in Table I.11. The cities pose a multitude of problems for the urban community. Some of the physical environmental problems of such urban areas might be avoided or alleviated if methods used in planned, technologically advanced urban development in rapidly growing 'western' societies could be sensibly applied to them. The techniques and approaches used in this report are among those that may be of value.

Growth of urban populations in drylands, both in the recent past and in the future, undoubtedly involves the development of enormous areas of land, and this land is the main focus of attention in this volume. Data on land consumption are even more difficult to find than those on population, but two examples will suffice to illustrate the phenomenon. Figure I.6 shows the location of recorded subdivisions in Los Angeles, California, for successive periods from 1945 to 1959. The data are drawn from annual planning reports up to 1960 (after which time such data were no longer reported), and it should be emphasized that the recording of a subdivision does not necessarily indicate that urban development was completed (although this was usually the case). Nevertheless, Fig. I.6 clearly indicates the extent to which the massive urban explosion in the Los Angeles area has consumed enormous areas of land. Figure I.7A similarly illustrates the spatial expansion of Lima, Peru. Lima's growth was relatively slow until recently. In 1950, its population was 853 000; but today it is over three millions (Dwyer, 1975). This rapid contemporary growth strongly reflects the development of spontaneous settlements (barriadas) in and around the city since 1950 (Fig. I.7B). Thus while in 1956, squatters represented barely 10% of the total urban population (Deler, 1970), by 1969 they amounted to nearly 40% of the total (Table I.11),

TABLE I.11

Extent of slums and uncontrolled settlements in various dryland cities

City/Country	Year	City population (thousands)	uncontrolled settlement	
			Total population (thousands)	As percentage of city population
Dakar, Senegal	1969	500	150	30
Baghdad, Iraq	1965	1 745	500	29
Karachi, Pakistan	1964	2 280	752	33
Ankara, Turkey	1965	979	460	47
Santiago, Chile	1964	2 184	546	25
Lima, Peru	1969	2 800	1 000	36
Caracas, Venezuela	1964	1 590	556	35
Maracaibo, Venezuela	1966	559	280	50

Source: UN General Assembly (1970).

and Harris (1971) concluded that *barriadas* development probably accounted for over 80% of the total urban growth of Lima from 1960 to 1970. Los Angeles and Lima have both grown dramatically, but the contrast between the two landscapes of growth could scarcely be more dramatic. Undoubtedly, part of the contrast lies in differences in the planning and management of growth. If urban growth in drylands continues, as seems likely, the experience of planning and management attained so far is certain to be of enormous value in assisting future growth.

(e) The Control of Urban Growth

1. *Planned and unplanned urban growth*

Rapid urban growth in drylands has occurred in varied cultural contexts and, as a result, attitudes towards the phenomenon are quite diverse. At one extreme, planned, controlled development is considered essential in order to prevent environmental and social abuse, to husband resources, to control expenditure, and to ensure the creation of urban areas that are attractive and acceptable to a majority of their inhabitants. The USSR and the USA exemplify this view. At the other extreme, cities have grown in some dryland societies very rapidly by a process of almost spontaneous and uncontrolled expansion that has been generated largely by population growth and increasing employment opportunities. Such growth often reflects either a *laissez-faire* attitude or, more commonly, an inability – for political, economic, or other reasons – to control development. At its worst, uncontrolled growth gives rise to squatter settlements or shanty towns, as in many poorer Third-World countries, and it commonly creates serious social and environmental problems.

Central to this study is the view that planning, and in particular the use of environmental information in the planning process, can contribute fundamentally towards preventing some of the undesirable consequences of

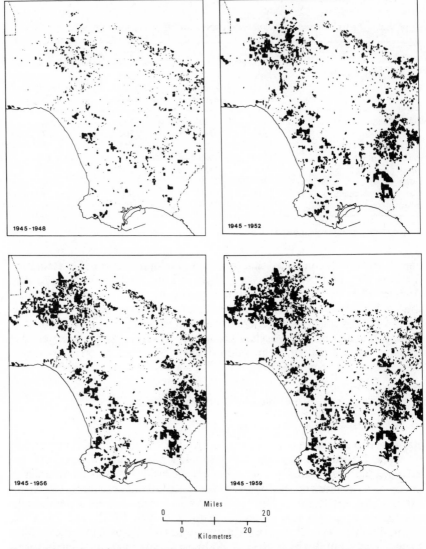

FIG. I.6 Location of recorded subdivisions in Los Angeles County, California from 1945 to 1959 (source: planning reports, City of Los Angeles, data collected by Cooke)

uncontrolled development; and that the experience in the use of environmental information in the planning process of dryland cities where development is controlled can be instructive to those responsible for the management of urban affairs in dryland cities that are at present without such control. For environmental information to be useful and effective, it is essential that it is presented at a pertinent time in the planning process.

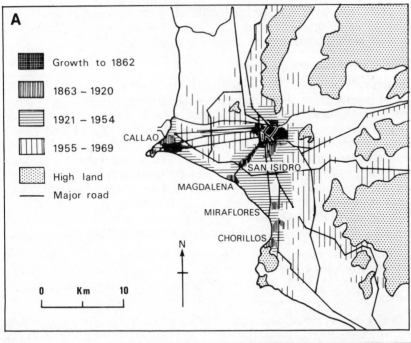

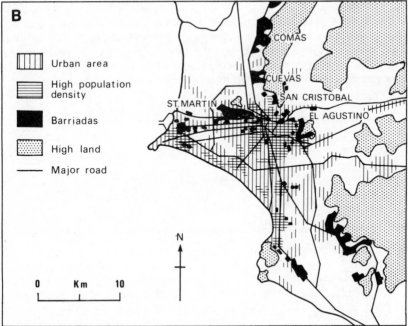

FIG. I.7 (A) The growth of Lima, Peru; (B) Lima, Peru: the development of *barriadas* since 1950 (after Dwyer, 1975)

Before proceeding to examine the specifically geomorphological problems of urban areas in drylands, it is therefore necessary to review briefly the structure of the planning process in urban areas.

2. *Structure of the planning process*

The literature on the philosophy of planning and the structure of the planning process is prodigious. It is sufficient to note here that the planning of urban areas normally involves several conceptually distinct phases of activity such as those shown in Fig. I.8A. Not all the phases in Fig. I.8 and I.9 are necessarily used everywhere, and the planning methods and procedures employed vary greatly from country to country and even within some countries. In addition, not all the phases are necessarily sequential – there may be considerable temporal overlap, and numerous 'feedback loops' commonly occur in what is fundamentally a continuous process.

There are five major phases in the planning process, which can be hierarchically arranged generally according to their spatial scale (Fig. I.8A).

(*i*) *National planning*, the highest order of planning, which is concerned with numerous objectives related to national resources and aspirations, and may be concerned only incidentally with urban development or environmental matters.

(*ii*) *Regional planning*, a complex field in its own right (e.g. Hall, 1974), commonly involves the broad outline plans for development at a sub-national level – for instance, at the scale of state planning in the USA or Australia. In small countries, such as some of the emirates of the Arabian Gulf, national and regional planning may be synonymous. Regional plans, like national plans, may be comprehensive, and perhaps only partly directed towards urban growth; or they may be focused on specific issues, such as industrial development or environmental hazards. An outstanding example of a

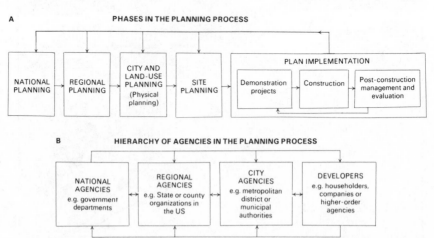

FIG. I.8 (A) Phases in the planning process (generalized); (B) Hierarchy of agencies in the planning process

A STAGES IN A PLANNING PHASE

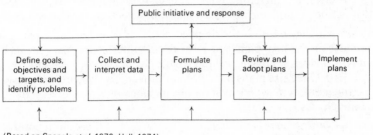

(Based on Spangle *et al*, 1976; Hall, 1974)

B CONTROLS OF CITY PLAN IMPLEMENTATION

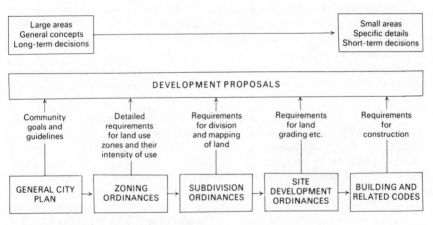

(Based on Mader and Crowder, 1969, with modifications)

FIG. I.9 Stages in a planning phase and controls of city plan implementation

planning document based on a specific set of problems at this level is the California Division of Mines and Geology's (1972) publication, *Urban Geology – Master Plan for California*; a comprehensive planning statement for part of the same area is that for Los Angeles County (1971).

(*iii*) *City and Land-use planning* usually determines the location of discrete areas for urban development and the location within them of different land-use activities (such as industry, housing estates, etc.). Essentially, it blocks out terrain for specific uses in the context of higher-order planning criteria and directives and the requirements of an integrated and growing urban system. Among numerous examples, the master plan for the new city of Suez, Egypt is typical (Egypt, Ministry of Housing and Reconstruction, 1976). Site selection is a major component of city planning. Selection procedures are well illustrated in the Libyan National Housing Corporation's housing design manual (Libyan Arab Republic, 1975). In this it is recommended that the

NHC submits to the *baladiyah* (municipality) its housing needs; the *baladiyah* responds by offering a range of sites; the NHC evaluates the sites in terms of specified criteria (including the 'site development characteristics' of landforms and ground-surface conditions) and selects the preferred locations.

(*iv*) *Site planning* requires the detailed planning of individual areas within the context of the city plan and local development controls, and includes layouts for factories, housing estates, roads etc. A typical site plan is shown in Fig. II.10. Once site plans have been approved, further progress relates to plan implementation which commonly includes one or more of the three following phases:

(*v*) *A demonstration project phase* may be the first of these; in it contractors or other developers build examples from proposed plans to illustrate for clients the reality they have envisaged and to permit more realistic appraisal of problems in the field so that modifications may be made to further developments in the light of experience, and potential problems can be better evaluated. An example is incorporated in the study of Suez City by Robert Matthew, Johnson-Marshall and Partners (Egypt, Ministry of Housing and Reconstruction, 1978).

(*vi*) *The construction phase*, in which development is implemented, is in many localities closely supervised and controlled through local regulations (see below).

(*vii*) *The post-construction management and evaluation phase* is a final but extremely important phase in the planning process because it requires continuous review and assessment of the effectiveness of earlier planning decisions, mainly by monitoring, and the feedback of experience into future plans: it is pre-eminently a phase of continuous reassessment.

A number of general observations on these phases in the planning process are pertinent to subsequent discussions. First, each planning phase characteristically includes a number of stages (Fig. I.9A) which, while they are generally sequential, are distinguished by continuous consultation, feedback, and overlap. Procedures vary locally, and in particular the extent of public involvement varies greatly between communities. Environmental information may be relevant at all stages.

Second, the phases in the planning process are usually associated with particular agencies of public and private administration in a generally hierarchical manner (Fig. I.8B). Thus, for example, national agencies are normally responsible for national planning, and municipal authorities generally control land-use and site planning. But the actual implementation of plans might be the responsibility of any agency in the hierarchy. For example, a factory might be developed by a national agency (e.g. a Department of Public Works) or by a local building contractor. In this generally hierarchical structure, there should be a continuous two-way flow of information. Higher-order agency planning will usually be a constraint on lower-order planning; at the same time, higher-order agencies may well find themselves involved in lower-order decision-making and their activities restricted by lower-order planning regulations.

Finally, implementation of city-level plans is normally accomplished through local guidelines and legal controls. A range of common controls is summarized in Figure I.9B. The city plan itself normally embodies community goals and a set of development guidelines. Detailed guidance on land-use areas, and on intensity of use and general requirements within designated areas may be achieved through zoning ordinances (in the USA), or their equivalent. Such ordinances not only designate and control use; they may also phase development (to ensure appropriate prior provision of services, for example) or provide for sequential land-use (to optimize the use of resources, for instance). Subdivision ordinances (or their equivalent) control the detailed design layout of a development area within a zone designated for a particular use. Site-development ordinances, which are usually incorporated in building regulations, control the way in which a site is engineered, and often incorporate requirements for slope development and environmental hazard avoidance. Building and related codes control the nature of construction.

Environmental information may be relevant and useful at all stages in the planning process. Such information is primarily of interest to the planner who ultimately controls development, and to the engineer, who is actually responsible for it, but it may also be of value to the developer, and public-interest groups. In general, the sooner the planner or other interested party is aware of, and appropriately informed about potential natural hazards and resources, the better he is able to anticipate areas of possible land-use conflict, to develop planning priorities, and recognize possible problems and solutions (e.g. W. Spangle and Associates, 1974). Thus environmental information may be required from the beginning of the planning process. The kind of environmental information that is relevant and useful and the form in which it is presented vary according to, *inter alia*, the responsibilities of the user, the scale of planning, and the stage in the planning process: it is essential that those who provide such information recognize this, their opportunities to help clients and their responsibilities to them.

(f) Summary

The world's drylands include 355 cities of over 100 000 inhabitants, together with a large number of smaller urban settlements. In recent years, and for a variety of reasons, the population of many of these urban areas has been growing rapidly, at rates of up to 16% a year. Such rapid growth has pushed urban limits into new territory – agricultural or pastoral land or, commonly, virgin desert. The voracious dryland towns and cities have consumed and are continuing to consume huge tracts of land.

This land, like all land, holds both potential resources and hazards. In order to benefit from the resources and avoid or alleviate the hazards, planning and control of development is desirable. The experience of some dryland countries in controlling urban growth may be of value to many countries without such experience. Central to this experience is the knowledge gained about the distinctive natural environment. The remainder of this volume explores the nature and value of one aspect of environmental knowledge, geomorphology, in the planning, management, and development of urban areas in drylands.

II

GEOMORPHOLOGY AND PLANNED URBAN DEVELOPMENT IN DRYLANDS

(a) Geomorphology: Some Preliminary Declarations

1. *Geomorphology, pure and applied*

Geomorphology is the study of landforms, and in particular of their nature, origin and evolution, processes of development, and material composition. It is pre-eminently a field science requiring detailed knowledge of the techniques available for studying landforms, terrestrial processes, and surface materials, and demanding a broad knowledge of several cognate disciplines, notably geology, climatology, meteorology, ecology, pedology, and hydrology.

Much work in geomorphology is of great potential value to man in his use of the physical environment (Cooke and Doornkamp, 1974; Coates, 1976b; Hails, 1977), and the application of geomorphological knowledge has increased in recent years, in harmony with growing public and political awareness of environmental problems and specifically because geomorphologists have recently come to give greater attention to those aspects of the subject of greatest practical application – the dynamic relations between landforms, materials, and contemporary processes.

Geomorphologists have, in the last two or three decades, increasingly directed their work to the solution of problems in environmental management, and their services have been increasingly sought by engineers, planners, and other environmental managers. This development in the subject has extended the required competence of the geomorphologist: in addition to his own techniques of analysis and a broad understanding of cognate sciences, he now requires an appreciation of the social, economic, and technical contexts in which his information may be relevant and in which it may be used. Above all, he must not only be able to acquire the information necessary to help solve problems in applied geomorphology, he must be able to present it in such a way that it is intelligible and useful to his client.

2. *Applied geomorphology, dryland and urban*

The broad field of applied geomorphology in drylands has recently been reviewed elsewhere (Cooke, 1977; Cooke, Goudie, and Doornkamp, 1978). Table II.1 summarizes the main areas of research activity in applied dryland geomorphology, and provides both general and detailed references for each major subject. Broadly, applied geomorphology in drylands is divided into two main categories. First, it is concerned with mapping and survey work in

TABLE II.1

Geomorphological contributions to engineering work in drylands: a summary

Geomorphology and resource appraisal

Terrain Evaluation	*General review references*	*Case studies in drylands*
Geomorphological surveys with potential value in applied work	Cooke & Doornkamp (1974); Demek (1972)	Neal (1965); Doornkamp *et al.* (1980)
Resource surveys with a geomorphological component	Mitchell (1973); Stewart (1968)	Christian *et al.* (1957)
Such surveys have included assessment of terrain in terms of:		
military conditions	Stewart (1968)	Howe *et al.* (1968)
trafficability, route planning and road construction	Stewart (1968)	Carter (1974); Perrin & Mitchell (1969, 1971)
surface materials (for aggregates, etc.)	Cooke & Doornkamp (1974)	Bull & Scott (1974); Doornkamp *et al.* (1980)
foundation engineering	Blyth & de Freitas (1974)	Holland & Stevenson (1963)
mineral deposits	UNESCO (1968)	Mueller (1968)
agricultural development	Mitchell (1973); Stewart (1968)	Prokopovich (1969); Doornkamp *et al.* (1980)
urban development and planning	Legget (1973)	Cleveland (1971)

Geomorphological aspects of engineering problems in drylands

Weathering processes	*General review references*	*Case studies in drylands*
salt weathering	Cooke & Warren (1973)	Minty (1965)
duricrusts	Goudie (1973)	Netterberg (1971)
soil development	Cooke & Warren (1973)	Ruhe (1967)

Fluvial processes	*General review references*	*Case studies in drylands*
sediment yield	USDA (1975)	Doornkamp & Tyson (1973)
soil erosion by water	Cooke & Doornkamp (1974);	
(a) contemporary	Hudson (1971);	Thornes (1976);
(b) historical	Butzer (1974)	Cooke & Reeves (1976)
water storage and flooding	Chow (1964)	Emmett (1974)
alluvial fans	Bull (1977)	Schick (1974a); Beaumont (1972)

Slope-stability processes	*General review references*	*Case studies in drylands*
landslides	Chandler (in Hails 1977)	Hayes (1971)
hydrocompaction	Lofgren (1969)	Lofgren (1969)
surface subsidence due to water abstraction	Cooke & Doornkamp (1974)	Bull (1964)
piping	Parker (1963)	Fletcher *et al.* (1954)
dessication phenomena	Cooke & Warren (1973)	Yaalon & Kalmar (1972)
water repellency	Qashu & Evans (1969)	Krammes & Osborn (no date)

Aeolian processes	*General review references*	*Case studies in drylands*
wind erosion of soil	Cooke & Doornkamp (1974)	Rapp (1974)
wind abrasion	Cooke & Warren (1973)	
aeolian deposition	Bagnold (1941)	
(a) bed forms	Cooke & Warren (1973)	
(b) aerosols	Cooke & Warren (1973)	Yaalon & Lomas (1970); Yaalon & Ganor (1968)
(c) loess	Schultz & Frye (1965)	Fookes & Best (1969)

Source: Cooke, 1977

terrain evaluation. Here its role is particularly important because landforms can provide an excellent basis in drylands for regional classification of terrain (see section (b) 1, below). Second, it deals with the identification, monitoring, and analysis of contemporary geomorphological processes – weathering, fluvial, slope-stability and, especially in drylands, aeolian activity. This work is self-evidently related to present-day conditions, and commonly these conditions present problems and provide hazards to the human use of drylands so that they are inevitably of interest to environmental managers (e.g. Table II.2). A fundamental problem here is that in order to evaluate geomorphological processes in drylands, the collection of appropriate data on the magnitude, frequency, and duration of events may take many years and most environmental managers have neither the time nor the experience to do the necessary collection and analysis. The geomorphologist may be able to help overcome this problem either by undertaking monitoring or by predicting the nature of processes on the basis of evidence of landforms and deposits.

Applied urban geomorphology is the study of landforms, and their related processes, materials, and hazards, in ways that are beneficial to planning, development, and management of urbanized areas or areas where urban growth is expected.

One important qualification must be made about urban geomorphology: different aspects of geomorphology have been studied in urban areas for many years – not only by geomorphologists, but also by engineers and others, some of whom may never have heard of the science. It is not, and never has been the exclusive preserve of the geomorphologist. For example, engineers in Los Angeles have studied the movement of sediment into, through, and out of the metropolis for decades, and the information they have collected has been used to predict, *inter alia*, the life span of reservoirs (e.g. Ruby, 1973). Nevertheless, as the field of urban geomorphology is one in which many different aspects of the environment are closely related and can be beneficially integrated, and is also one of rapidly growing knowledge requiring greater specialization among those studying it, it is scarcely surprising that planners and engineers are increasingly either receiving geomorphological training themselves or turning to specialist geomorphologists for help in tackling problems of a geomorphological nature.

TABLE II.2

Examples of geomorphological hazards in drylands

Flooding of valleys, fans and playas
Gullying
Hydrocompaction
Surface subsidence due to water abstraction
Sedimentation (fluvial)
Salt weathering
Piping
Landslides and related slope-failure phenomena
Sabkha inundation
Aeolian deposition, dune encroachment, dune reactivation, dust
Calcretization
Desiccation phenomena

Source: Cooke *et al.*, 1978.

Attention to urban geomorphology has increased in recent years in harmony with the growing recognition of the importance of the much broader, but closely related, fields of environmental and urban geology (e.g. McGill, 1964; Ass. Eng. Geol., 1965; Colorado Geological Survey, 1969; US Dept. Housing and Urban Development, 1969; Betz, 1975; Akili and Fletcher, 1978). Although some aspects of urban geomorphology have been considered in recent books, such as those by Coates (1971), Detwyler and Marcus (1972), Legget (1973), Cooke and Doornkamp (1974), and Leveson (1980), the rationale of the subject has not been clearly formulated. For this reason the objectives of urban geomorphology are explored in the following section.

(b) Objectives of Geomorphological Appraisal Before, During and After Urban Development

1. *Prior to urban development*

The planner's primary environmental requirement prior to urban development is a knowledge of the nature and disposition of natural resources and hazards: the need is for surveys that record and analyse these phenomena in order to facilitate initial planning decisions.

The principal objectives, therefore, of surveys designed to satisfy this requirement are the *identification* of the range of possible locations of resources and hazards or of possible locations suitable for particular activities (and, of course, the identification of locations unsuitable for such activities); to assist in the *selection* of suitable locations for particular activities within the range of possibilities; and to analyse conditions *within* selected locations, in the hope that the use of environmental resources will be more economic, beneficial, and efficient.

Within these broad objectives, the geomorphologist commonly has several aims: (i) to prevent urban growth from destroying or sterilizing valuable resources, especially aggregates: (ii) to identify and evaluate land and material resources required for development; (iii) to limit undesirable impact of urban development on geomorphological conditions; (iv) to predict the potential responses of ground surfaces to urban development (e.g. slope erosion and failure); and (v) to assess the potential impact of geomorphological hazards on the urban community. In addition, there lies behind these aims the need to provide intelligible and useful documents to planners and developers, and to demonstrate the value of the information to those who might benefit from it.

Geomorphologists commonly adopt one or more of three major approaches to appraisal prior to development. First, by far the most profitable and widely used approach is that of formally classifying and describing terrain features, through morphological or geomorphological mapping and/or the interpretation of air photographs or other remote sensing imagery. Second, analysis of process dynamics and landform change may be accomplished through, for example, the analysis of historical records (e.g. climatic and hydrologic data, and the evidence of topographical maps, aerial photographs, and satellite imagery). A third approach, less secure and less common, is to appraise one, poorly known situation by analogy with another similar but better-documented situation elsewhere. This approach is, of course, dependent on the availability of data from the analogous situations and it provides a strong argument for the collection of information in data banks such as that envisaged by the VIGIL network (Leopold, 1962) and in deserts by Beckett and others (1972; Mitchell *et al.*, 1979).

Before urban development, the geomorphological contribution is mainly of importance in providing surveys of direct use in themselves, of value as a source of derivative maps etc., and as a basis for more detailed subsequent surveys. The value of geomorphology as a basis for surveys prior to urban development in drylands rests on the premise that because landforms in drylands are normally displayed clearly in the field and on remote sensing imagery, they are an excellent basis for the spatial classification of terrain: the whole landscape of drylands is made up of landforms (where other environmental variables like soils and vegetation are commonly discontinuous) thus permitting comprehensive classification of terrain; and landforms can be classified relatively easily into a hierarchy of units (see Chapter III), thus providing a spatial framework that is potentially of value at a variety of planning scales. In addition, many environmental distributions, including to some extent sub-surface conditions, can be predicted on the basis of geomorphological information.

2. *During and after urban development*

During and after urban development the planner and urban manager normally require to know the effects of natural events and circumstances on the urban community, and the impact of urban development on the environment: their primary interest is in the environmental consequences of urban growth. Within this primary objective, planning and management aims may include (i) the minimization of environmental impact; (ii) the development of local, spatial, and temporal data bases from monitoring studies to facilitate the prediction of future changes; and (iii) the continuous modification of plans, management organizations, and procedures to ensure an harmonious environmental management.

The main aim of geomorphological work in this context is to monitor the dynamics of geomorphological systems with a view to predicting spatial and temporal changes in a way that allows the planner to respond effectively and in good time. Geomorphological approaches to appraisal during and after urban development are similar to those adopted prior to development, but their relative importance changes. Field monitoring is pre-eminent, whether it is on a global scale (such as Morales's (1977) studies of dust dispersion) or a city-wide scale (such as Schick's (1977) survey of sediment yield near Eilat, Israel). Monitoring often requires the establishment of fixed observation stations. In drylands the period of time required to collect sufficient data to allow reliable prediction is normally longer than in more humid areas because the recurrence interval of events of similar magnitude tends to be longer, so that useful results from monitoring studies are seldom generated quickly. For this reason, surrogate approaches that yield results more quickly are often adopted. Formal mapping and classification of *indices* of process activity rather than of the process itself provide one solution: such indices can often be derived from aerial photographs or fieldwork. Some types of hazard mapping (as illustrated in Chapters III and VI) exemplify this approach. And the prediction of change on the basis of temporal analogy (using historical data) or spatial analogy (using data from similar areas elsewhere) may be attempted.

The first two sections of this chapter have asserted that geomorphological studies can be of value in the planning and development of cities, both generally and in the specific context of drylands. The following section illustrates the use and value of geomorphological information in the planning and management of dryland cities by first examining through examples the relationships between management agencies and geomorphology, and, second, by reviewing examples of ways in which geomorphological information has helped in the various phases of the planning process outlined in Chapter I.

(c) The Use and Value of Geomorphological Information in Urban Areas of Drylands

1. *Urban management and geomorphology*

Each urban area in drylands has its own unique geomorphology and its own often distinctive ways of managing it within the local hierarchy of management organizations. The purpose of this section is briefly to introduce the relations between management and geomorphology by means of several matrix diagrams based on studies of selected dryland cities that fall mainly within the category of planned, technologically advanced communities in 'western' societies referred to in Chapter I. This category is selected because such cities reveal the most complex relations, and because many less technologically advanced cities not only have simpler relations but are in many cases developing management strategies that emulate those in the 'western' societies. The data for the matrix diagrams are derived from fieldwork, or from correspondence and discussions with authorities on the cities mentioned. The diagrams provide only a rudimentary description of the relations between geomorphology and management, but they do illustrate the existence of links, while at the same time hinting at the complexity of the links, and suggesting something of the scope for geomorphological contributions to urban management.

(*i*) *Los Angeles, California* (population 1975: 7032075) is a metropolitan area of great environmental and administrative complexity, and in some ways it epitomizes the technologically advanced urban area that has expanded very rapidly (Fig. I.6), and has encountered, in the process, a wide range of geomorphological problems. Los Angeles includes four major environments – the plains of the Mojave Desert, the mountain zone dominated by the San Gabriel and San Bernardino mountains, the intermontane valley plains (such as San Fernando Valley), and the Los Angeles coastal plain. The major geomorphological problems arise from the fact that settlement has expanded onto the plains (where it is vulnerable to floods and debris from the mountains), into the mountains (where it is vulnerable to a variety of slope failures), and along the coast (where slope-stability and coastal-process problems prevail). It is interesting to note that areas of predicted growth (County of Los Angeles Board of Supervisors, 1971) occur mainly in mountain areas and on the desert plain to the north. In addition, there are serious problems of subsidence due largely to oil removal (especially in the Long Beach area), a strong demand for good aggregates from the construction industry, and a concern to preserve and protect attractive, recreational land.

All these problems – some widespread, others localized; some severe, others slight (Fig. II.1A) – have generated responses from environmental managers. The responses are extremely complex – for many problems, all levels of the hierarchy of management agencies have some responsibility, whether it be for survey, assessment and planning, or for exploitation, repair and maintenance (Fig. II.1B). For example, in the case of surface stability problems, federal and state geologists and engineers, among others, have various responsibilities for survey, assessment and planning; the county and numerous cities within it have responsibilities for planning, control of construction, and repair and maintenance; and the private homeowner, the developer, private consultants, and insurance companies all have vested interests and distinctive roles to play. A selection of these relationships are examined in greater detail in Cooke (1977) and in section (c) 2 below.

(*ii*) In *Las Vegas, Nevada** (population 1975: 146 030) the hierarchy of management agencies is similar to that in Los Angeles but the geomorphological problems are somewhat different (Fig. II.2A). The Las Vegas region is dominated by mountains that are flanked by alluvial fans which encroach on a broad alluvial plain; runoff in the area is focused into a system of ephemeral stream channels and floodplains. Geomorphological problems in the region are more limited and more typical of drylands than those in Los Angeles – the dominant problems relate to flash flooding and sediment transport from the mountain areas across the alluvial plains, and to soil erosion by wind and water. Expansive soils (arising chiefly from the presence of clay minerals) and snow avalanching in the mountains are additional problems. Some problems are less serious now than previously: for example, debris flows are no longer a serious problem because urban development at the apices of alluvial fans has been restricted by local ordinance, and subsidence due to groundwater extraction has diminished as water is increasingly derived from the Colorado River system.

The diversity of management responses is greatest in the context of flash flooding and related problems, in which, for example, agencies at all levels in the hierarchy have responsibilities for both survey, planning, and assessment, and exploitation, repair, and maintenance (Fig. II.2B). Indeed, the problems are dealt with by such a multiplicity of agencies and affect so many different interests that conflicts and disagreements can arise – as revealed by the example of discharge estimation cited below (see Fig. II.8). Thus, these problems are central to much environmental litigation and such litigation, here as elsewhere, frequently involves geomorphological evidence. The next

* The help of Dr David Weide in preparing this section is gratefully acknowledged.

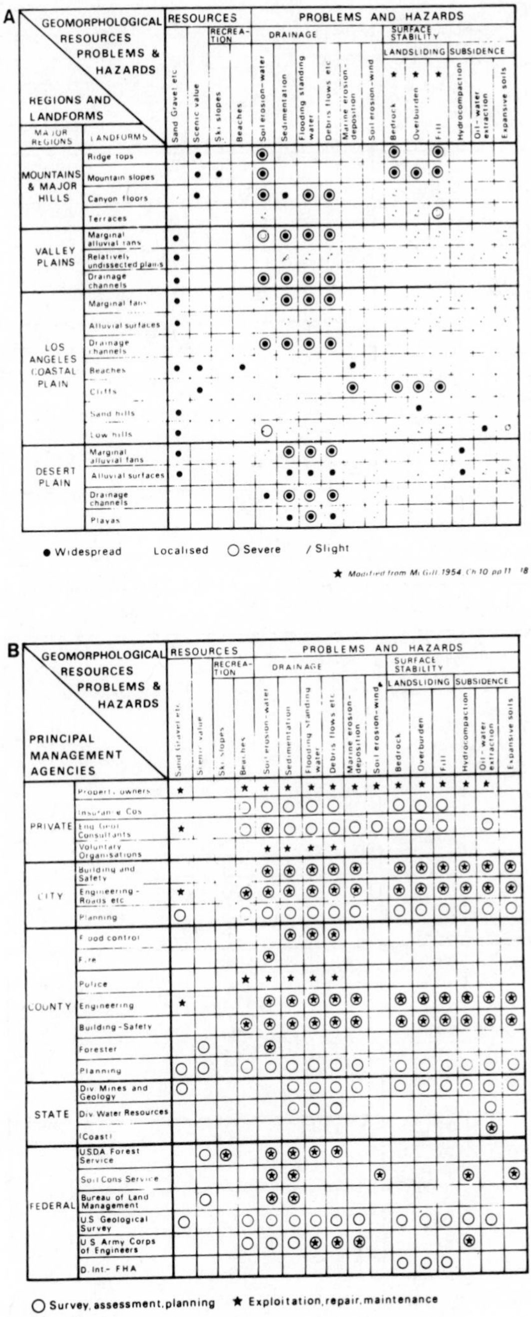

A

GEOMORPHOLOGICAL RESOURCES	RESOURCES				PROBLEMS AND HAZARDS											

RESOURCES
PROBLEMS & HAZARDS
REGIONS AND LANDFORMS

RECREATION · DRAINAGE · SURFACE STABILITY · LANDSLIDING · SUBSIDENCE

Sand Gravel etc · Scenic value · Ski slopes · Beaches · Soil erosion-water · Sedimentation · Flooding standing water · Debris flows etc · Marine erosion deposition · Soil erosion-wind · Bedrock · Overburden · Fill · Hydrocompaction · Oil-water extraction · Expansive soils

MAJOR REGIONS	LANDFORMS
MOUNTAINS & MAJOR HILLS	Ridge tops
	Mountain slopes
	Canyon floors
	Terraces
VALLEY PLAINS	Marginal alluvial fans
	Relatively undissected plains
	Drainage channels
LOS ANGELES COASTAL PLAIN	Marginal fans
	Alluvial surfaces
	Drainage channels
	Beaches
	Cliffs
	Sand hills
	Low hills
DESERT PLAIN	Marginal alluvial fans
	Alluvial surfaces
	Drainage channels
	Playas

● Widespread   Localised   ○ Severe   / Slight

★ Modified from M. Gill 1954 Ch 10 pp 11 18

B

GEOMORPHOLOGICAL RESOURCES	RESOURCES				PROBLEMS AND HAZARDS											

RESOURCES
PROBLEMS & HAZARDS
PRINCIPAL MANAGEMENT AGENCIES

RECREATION · DRAINAGE · SURFACE STABILITY · LANDSLIDING · SUBSIDENCE

Sand Gravel etc · Scenic value · Ski slopes · Beaches · Soil erosion-water · Sedimentation · Flooding standing · Debris flows etc · Marine erosion deposition · Soil erosion-wind · Bedrock · Overburden · Fill · Hydrocompaction · Oil-water extraction · Expansive soils

PRIVATE	Property owners
	Insurance Cos
	Eng Geol Consultants
	Voluntary Organisations
CITY	Building and Safety
	Engineering - Roads etc
	Planning
COUNTY	Flood control
	Fire
	Police
	Engineering
	Building - Safety
	Forester
	Planning
STATE	Div Mines and Geology
	Div Water Resources
	(Coast)
FEDERAL	USDA Forest Service
	Soil Cons Service
	Bureau of Land Management
	U S Geological Survey
	U S Army Corps of Engineers
	D Int- FHA

○ Survey, assessment, planning   ★ Exploitation, repair, maintenance

FIG. II.1 Relations between landforms, geomorphological resources, problems and hazards, and management agencies in Los Angeles, California

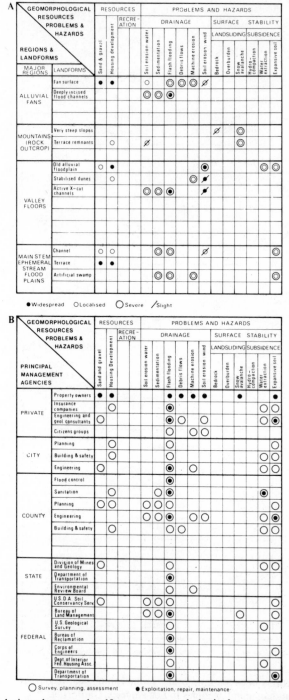

FIG. II.2 Relations between landforms, geomorphological resources, problems and hazards, and management agencies in Las Vegas

most serious set of problems relates to urban dust, mainly because so much of the land surface around and within the urban area (approximately 35% of the land surface within Las Vegas is undeveloped) has been disturbed and stripped of the superficial protection provided by vegetation, stone pavements, and surface crusts, often in anticipation of urban development, or by 'off-road vehicles', such as motor cycles (see 'machine erosion', in Fig. II.2).

(*iii*) Management agencies in *Phoenix, Arizona*** (population 1975: 664 721) have encountered geomorphological problems similar to those in Las Vegas during the course of recent, rapid urban development, but they have not all sought appropriate geomorphological advice. In this region, problems arise from alluvial-fan flooding, rockslides, ground-cracking associated with land subsidence due to groundwater withdrawal, and the occurrence of *calcrete* (a deposit of calcium carbonate) near to the surface; Péwé and others have prepared systematic maps of part of the area – showing, for instance, material resources, construction conditions, and geological hazards – and demonstrated the potential value of such information to planners and developers. Indeed, their maps have been used to help draft 'hillside' ordinances in Scottsdale, one of the cities in the region (e.g. Holway *et al.*, 1978; Péwé, 1978; Arizona Bureau of Geology and Mineral Technology, 1978). In general, management agencies in Phoenix and Las Vegas appear to lag behind similar agencies in Los Angeles in recognizing the potential of geological and geomorphological data in planning urban growth. But the three cities do show a very similar hierarchy of management agencies, extending from the individual to the state.

(*iv*) *N'Djamena, Tchad*† (population 1976: c.179 000), provides an African example, from a more arid region. Here, geomorphological problems relate mainly to the flow and use of water in the valleys of the Chazi and Logone which meet in the area of urban development (Fig. II.3). In this relatively under-developed area, it is interesting to note that the hierarchy of management agencies, while lacking some of the local complexity of major dryland cities, is supplemented by an international level.

(*v*) *Elsewhere in technologically advanced, 'western' dryland cities*, the pattern of management–geomorphology relations is generally similar, although it varies greatly in detail. Thus, many of the new and expanding towns in Israel‡ have encountered geomorphological problems, to the extent that they are increasingly anticipated by planning agencies, and geomorphologists and others are beginning to tackle them. Table II.3 shows for some urban areas the associated landforms and geomorphological problems and also the professional affiliations of the consultants who have studied them: geomorphologists, for example, have contributed significantly to studies of Tiberias, Beer Sheva, and Eilat. The most important point is that geomorphological problems are now recognized as being relevant at different levels of the management hierarchy, and professional help in solving them is increasingly sought.

* Thanks are due to Dr Troy Péwé for providing data on which this paragraph is based.

† Information kindly provided by Dr C. Bardinet, Montrouge, France.

‡ Dr Moshe Inbar (Haifa, Israel) kindly helped in the preparation of this paragraph.

PRINCIPAL MANAGEMENT AGENCIES	GEOMORPHOLOGICAL RESOURCES / PROBLEMS & HAZARDS	RESOURCES		RECREATION		DRAINAGE						SURFACE STABILITY — LANDSLIDING		SURFACE STABILITY — SUBSIDENCE	
		Sand and gravel	Forage water	Water skiing	River borders	Urban flooding	Rural flooding	Soil erosion	Afforestation	Irrigation	Navigation	Embankments	River bank failure	Buildings	Land
PRIVATE	Water Service S.T.E.E.		●												
	Building Firms	●												●	
	Public works enterprises	●												●	●
	Market garden Co-operatives					●									
	Clubs			●	●										
CITY OF N'DJAMENA	Hydrological Service		◉										○		
	Urban Commission					○						○	○	○	○
	Mayors office					○						○	○	○	○
	Land Register					○							○	○	○
STATE SERVICES IN N'DJAMENA	Geology and Mines	○													○
	Water and Forests								◉						
	Public Works	◉				◉	◉				◉	●	◉	◉	◉
INTER-NATIONAL SERVICES IN N'DJAMENA	Agence de Bassin C.B.L.T.						◉	○	◉	○	○				
	U.N.D.P.- O.N.U.						◉	○	○	○					
	O.R.S.T.O.M.						○								

○ Survey, assessment, planning   ● Exploitation, repair, maintenance

FIG. II.3 Relations between landforms, geomorphological resources, problems and hazards, and management agencies in N'Djamena, Tchad

Despite the fact that so much of Australia is arid, all the largest cities are located outside the dryland boundary shown in Fig. I.1. Nevertheless, some of these cities do suffer from typically dryland geomorphological problems and, of course, many towns and cities within the dryland zone also suffer from such problems. Of the largest cities, Adelaide (population 1974: 809 482) is closest to the drylands, and its principal problems relate to the fact that it is developed on over-consolidated Pleistocene clays that swell due to wetting as a consequence of urbanization. As a result, gilgai (mounds) and sink-hole development pose serious problems (e.g. Cox, 1970; Stapledon, 1970). By way of contrast, in the small town of Alice Springs (population 1977: 14 149), in the heart of the Australian drylands (Fig. II.4)* problems of resource use, wind and water erosion, and flooding are the concern of a similar administrative hierarchy to that encountered in, for example, the USA,

* Professor J. A. Mabbutt (Sydney, Australia) kindly provided the data for this example.

TABLE II.3

Geomorphological problems in urban areas of Israel

Urban area	Major landforms	Geomorphological problems	Problems studied by:
Kiryat Shmone	Mountains	Landslides, rockfalls	Engineering geologists
Tiberias	Mountains lakeshore	Flooding, soil erosion, shore erosion	Geomorphologists Hydrologists
Meron	Mountains	karst cavities	Engineers
Haifa	Mountains	landslides, rockfall flooding, shore erosion-deposition	Engineering geologists Marine engineers
Tel Aviv	Coastal plain	flooding, river deposition, cliff erosion, shore erosion	Hydrologists Engineers
Natanya	Coastal plain	cliff erosion, shore processes-deposition, sand movement in beach and platform	Engineering geologists
Hadera	Coastal plain	sand movement	Marine geologists
Ashdod	Coastal plain	sand movement, sediment transport	Marine geologists
Kiryat Gat	Southern (loess) badlands	flooding, gullying, bank erosion	Soil scientists Hydrologists Geomorphologists
Beer Sheva	Desert plains	flooding, dust storms	Hydrologists Soil scientists
Eilat	Desert plains	flooding, marine erosion and deposition	Geomorphologists Hydrologists Marine biologists
Bat Yam-Rishon	Coastal plain	coastal cliff erosion, marine erosion and deposition	Engineers

despite the fact that the urban area is not large and the geomorphological problems are not severe.

The preceding examples illustrate, in a very general way, some of the complex relations that exist between geomorphology and management agencies in a selection of dryland cities. In some cities, geomorphological studies may now play only a minor role, but in others their role is already substantial and increasing. What is clear, however, is that there is enormous scope for geomorphological contributions to the work of management agencies responsible for the geomorphological problems at all levels in the administrative hierarchy. The following section provides specific examples which illustrate some aspects of the value of geomorphology at the various stages of the urban planning process in drylands discussed in Chapter I.

2. *Examples of the value of geomorphological information in the planning process*

(*i*) *Geomorphology and regional planning.* At the scale of regional planning, geomorphological mapping and related research can often help in classifying

A

GEOMORPHOLOGICAL RESOURCES PROBLEMS & HAZARDS / REGIONS & LANDFORMS		RESOURCES		RECRE-ATION		PROBLEMS AND HAZARDS			
MAJOR REGIONS	LANDFORMS	Sand and Gravel	Ground Water	Tourism	Local Recreation	Flash Flooding	Dust Haze	Wind Erosion	Sheet Erosion
MAC-DONNELL RANGES	Intermontane Plain				○		○		
	River Channel and adjacent Flood Plain	○	○	○		●	○		
	Hill Slopes	○		○					
PIEDMONT PLAIN	Alluvial Slopes		●				●	●	●

● Widespread   ○ Localised

B

GEOMORPHOLOGICAL RESOURCES PROBLEMS & HAZARDS / PRINCIPAL MANAGEMENT AGENCIES		RESOURCES		RECRE-ATION		PROBLEMS AND HAZARDS			
		Sand and Gravel	Ground Water	Tourism	Local Recreation	Flash Flooding	Dust Haze	Wind Erosion	Sheet Erosion
PRIVATE	Companies	●		●					
	Citizens Groups				●				
CITY	Town council			○	◉	◉	●		
NORTHERN TERRITORY	Animal Industry Bureau						○	○	○
	Water Resources		◉			◉			
	Geological Branch	○	◉						
	Reserves Board			◉				○	
FEDERAL	Department of Construction	◉		○		◉			●
	C.S.I.R.O.						○	○	○

● Exploitation, repair, maintenance   ○ Survey, Planning, Assessment

FIG. II.4 Relations between landforms, geomorphological resources, problems and hazards, and management agencies in Alice Springs, Australia

terrain and locating resources. An example related to terrain classification at a regional scale is provided by a survey of geomorphological hazards in the central Luni Basin of the Rajasthan Desert (India) (Singh, 1977). The survey produced a hazard map (Fig. II.5) which classifies the terrain according to the intensity (slight, moderate, severe, and very severe) of wind, water, and salinity hazards. This map provides a fundamental, albeit qualitative, document for the planning of agricultural land-use, reclamation, and the development of urban settlements in the region.

An example of the use of geomorphological evidence in the context of resource location at a regional scale is provided by a survey in Bahrain. Here, the formulation of an explanation of the distribution of surface sands – a distribution first determined by geomorphological mapping – shows how the study of landform evolution may be of value in the search for aggregate resources. Lack of sand for the construction industry is a serious problem in Bahrain. Detailed field mapping for the Bahrain Surface Materials Resources Survey (Bahrain, Ministry of Works, Power and Water, 1976; Doornkamp *et al.*, 1979) revealed a fundamental regional division of the island into two, along a line trending north-west–south-east, the trend of the prevailing 'Shamal' wind (Fig. II.6). To the north-east of the line, quartz sand is largely absent, but there is silt and coastal carbonate sand. To the south-west, in addition to silt, there are wind-blown quartz sand deposits in various forms and locations, together with widespread wind-abrasion features. The boundary, the 'sand line', if projected north-west runs closely parallel to the Saudi Arabian coast in an area at present characterized in part by submerged dune fields. Clearly the aeolian quartz sand could not be blown on to Bahrain today. All the evidence suggests that it originated in the dune fields of Saudi Arabia and was propelled across the Gulf of Salwa by the Pleistocene

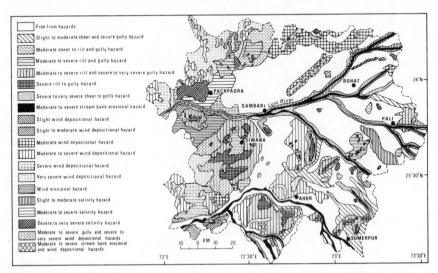

FIG. II.5 A map of geomorphological hazards in the central Luni Basin of the Rajasthan Desert (India) (after Singh, 1977)

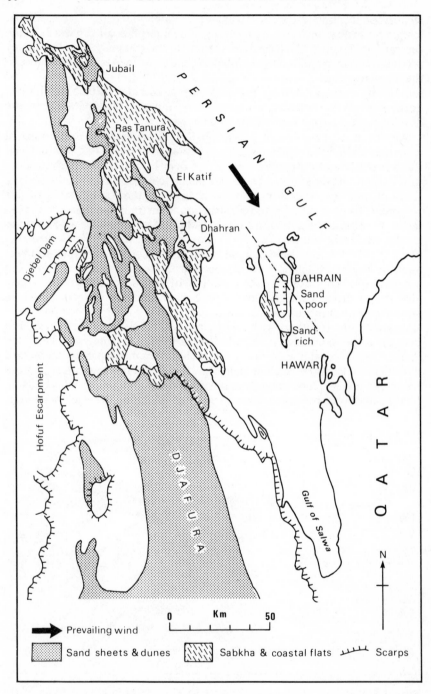

FIG. II.6 The regional setting of Bahrain

equivalent of the Shamal in one or more periods of low sea-level prior to 5000 BP when the Gulf was dry; the supply of sand would have been cut off when the Gulf was flooded by the post-glacial marine transgression. This hypothesis not only explains the regional distribution of aeolian quartz sand on the island; it also indicates where new extensive deposits of sand are most likely to be found: beneath the sea separating Saudi Arabia and Bahrain. At the same time, the survey successfully identified several areas of aggregate resources on the island, south-west of the 'sand line' (see Fig. III.16).

(*ii*) *Geomorphology in city planning.* If geomorphological surveys are carried out sufficiently early in the planning of a new city, and their results are digested into planning decisions, some of the potential dissonance between environmental conditions and the city plan can be avoided. Figure II.7 is a hypothetical illustration of this advantage, based on actual problems encountered in a variety of dryland cities. The first city plan (**B**) has an attractive spatial geometry, but it would have encountered several problems – building over scarce aggregates; hydro-compaction; flooding, sedimentation and erosion on roads crossing alluvial-fan channels; blowing sand and salt weathering. In the revised city plan (**C**), which attempts to accommodate the implications of the geomorphological map (**A**), most of these problems are either avoided or sensibly controlled, thus saving resources, time, and money.

At the scale of a whole city, a common problem is that the management of a single, natural unit, such as a drainage basin, is divided between several administrative organizations, with resulting duplication or dispersion of effort, and perhaps competition and conflict. This is a problem that can be avoided if environmental criteria are used intelligently in the initial formulation of responsibilities of different authorities – although it would be naïve to assume that other factors are not usually more important in such formulations! Figure II.8 shows an example, based on data from a city in the south-western USA, of the ways in which flood control of land drained by two ephemeral washes that cross the city is the responsibility of several quite different agencies whose interests are not necessarily accordant. Channel or culvert capacity was estimated at five locations along one of the washes as follows (D. Weide, personal communication, 1978):

A: 283 cumec   (natural channel)
B: 56 cumec   (Federal specifications)
C: 1.4 cumec   (1905 railroad culvert)
D: 1.4 and 14 cumec   (no specifications)
E: 140 cumec   (channelized)

Such extreme variations of estimated capacity clearly create potential flooding problems if and when a large runoff event occurs; the diverse responses to flow reflect, in part, the arbitrary administrative subdivision of the catchment, the difficulties of predicting ephemeral runoff, the improvement of predictions over time, and different bases of prediction. Clearly some form of rationalized and integrated channel management would be advantageous.

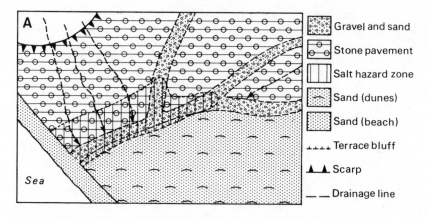

Legend for A:
- Gravel and sand
- Stone pavement
- Salt hazard zone
- Sand (dunes)
- Sand (beach)
- Terrace bluff
- Scarp
- Drainage line

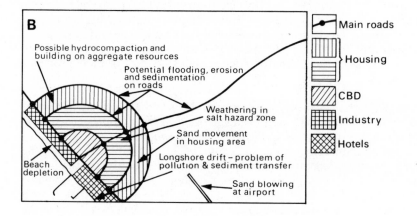

Legend for B:
- Main roads
- Housing
- CBD
- Industry
- Hotels

Labels in B:
- Possible hydrocompaction and building on aggregate resources
- Potential flooding, erosion and sedimentation on roads
- Weathering in salt hazard zone
- Sand movement in housing area
- Longshore drift – problem of pollution & sediment transfer
- Sand blowing at airport
- Beach depletion

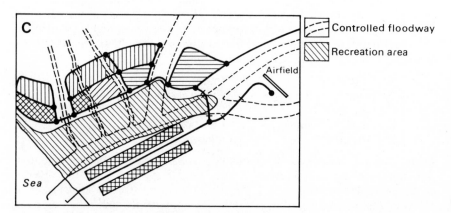

Legend for C:
- Controlled floodway
- Recreation area

Labels in C:
- Airfield
- Sea

FIG. II.7 Hypothetical illustration of some of the ways in which environmental problems inherent in a town plan (B) may be overcome by redesigning the plan (C) in the light of geomorphological evidence (A)

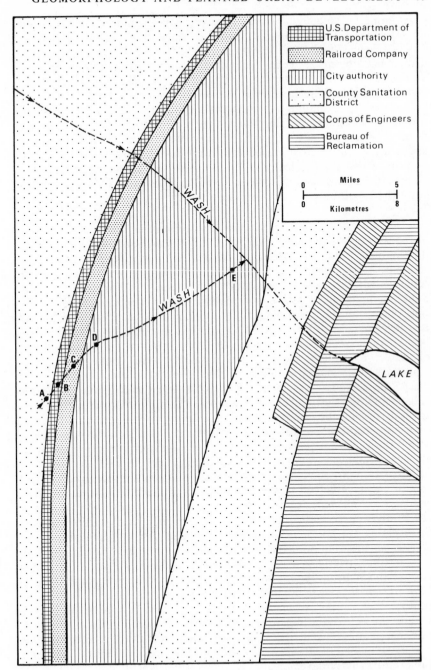

FIG. II.8 Hypothetical illustration of the ways in which flood control of land drained by ephemeral washes that cross a dryland city may be the responsibility of different agencies. For explanation of A-E, see text. Based on a USA example

Administrative boundaries rarely accord with geomorphological or other environmental boundaries, but there are exceptions. In Los Angeles, a County Flood Control District was established in 1914, largely in response to serious flooding in that year. The boundary of the district fully enclosed the potentially floodable region surrounding the city *at that time* (Figure II.9A). But since 1914, there have been two major changes: the urban area has grown rapidly, extending now into the north of the county, where there is now a serious real 'desert' flood hazard (Figure II.9B); and engineering work since 1914 has alleviated flood hazards in the original flood district in the south of the county. Thus the location of the flood hazard has moved, but the administrative district has not: a change in the boundaries of the flood control district would seem to be required, in accordance with the urbanized watersheds within which flooding is contained.

(*iii*) *Geomorphology and site planning and development.* An excellent example of the way in which geomorphological advice can beneficially modify site-development plans is provided by Mader and Crowder (1969). In 1956 a residential development was proposed in the hill country of the growing settlement of Portola Valley, south of San Francisco, California (Fig. II.10A). Subsequent geological and geomorphological studies related to the formulation of the town's general plan and zoning and subdivision ordinances revealed the nature and extent of potential slope instability in the proposed development area. A 'relative slope stability' map (Fig. II.10B) showed areas of stable, potentially moving, and moving ground, and located major landslides. This map, together with other relevant information formed the basis for a new plan, in which (Fig. II.10C) houses were clustered on stable ridge crests, and the number of lots was only slightly lower than the maximum permitted under the general formula relating lot size to average slope; even more important, about 15 house sites and some roads on the original plan were removed from actively moving ground, and considerably more house sites were removed from potentially moving ground.

There are numerous instances where specific geomorphological information might have helped to avoid problems and made the site-development process more efficient. It is very common in drylands, for instance, to develop the gently sloping plains that extend between mountains and basin floors. Such plains are normally underlain by graded alluvium and present few serious construction problems. But in some locations the alluvium on plains may be thin, overlying a relatively smooth bedrock surface, or pediment (Cooke and Warren, 1973; Mabbutt, 1977). Such a pediment may be a mixed blessing – on the one hand it might provide sound foundations for buildings; on the other it may considerably increase the cost of trench digging and foundation excavation: in either case, it is very important to be aware of the presence of near-surface pediments and to anticipate their consequences. The development of Desert Knolls Manor in the Mojave Desert, California, for example, required extensive and expensive blasting to create service trenches through a quartz-monzonite pediment (Cooke and Mason, 1973); on a much larger scale, the new city of Al Ain (Abu Dhabi) will be built, in part, over thinly veneered limestone pediments.

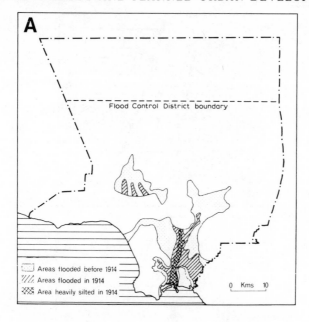

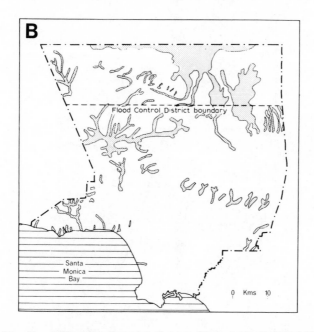

Fig. II.9 Flooding in Los Angeles: actual area of inundation in 1914 (A); potential area of inundation in 1970 (B). The County Flood Control District, established in 1914, is to the south of the boundary shown. (Sources: Los Angeles County Board of Supervisors, 1915 and 1971)

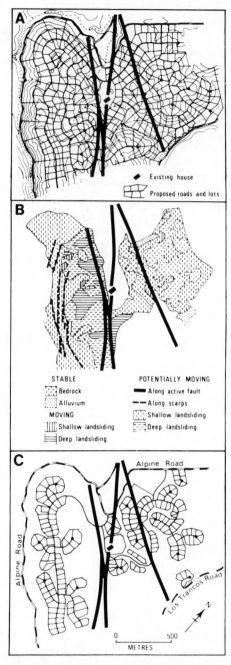

FIG. II.10 Portola Valley, California: example of the way in which a site development plan (A) can be beneficially modified (C) in the light of geomorphological evidence (B) (after Mader and Crowder, 1956)

Detailed site investigation is more properly the concern of the soil mechanics engineer and the engineering geologist than the geomorphologist, but the latter can provide the general framework within which sub-surface investigations and 'materials' sampling may proceed. One of the most attractive advantages of using a geomorphological basis to such work, which is usually quite cheap to commission, is that it can save money. For example, the number of boreholes required to sample adequately an area using a conventional grid system may be considerably reduced without loss of information if a stratified random sampling design is used that is based on a geomorphological map in which each unit is assumed to be reasonably homogeneous (Fig. II.11). What is true of a borehole programme is almost certainly also true of a 'trial pit' programme for sampling surface sediments.

(*iv*) *Geomorphology in the formulation of planning ordinances and building codes*. If a geomorphological problem is recognized, and is to be avoided or satisfactorily managed, it must be translated into effective planning regulations. Among many examples of such translations, one of the most elaborate and successful is to be found in the controls of hillside development in the Los Angeles metropolitan area.

The County of Los Angeles includes 77 cities, all of which have building codes that control development (Fig. II.12). Many of these cities include potentially unstable hillslopes (Cooke, 1977) and they have attempted to incorporate into their building codes measures designed to control urban development on slopes and prevent slope failure and erosion, through, for example, the designation of hazard areas, where special regulations apply, the imposition of maximum limits to 'cut' and 'fill' slopes, specifications for benching and terracing slopes at regular intervals, provisions for slope drainage and vegetation cover, fill compaction controls, and requirements for geological surveys. There is no single administrative response to the common geomorphological problems, however: as Fig. II.12 shows, some cities have adopted the so-called Uniform Building Code, some have modified it to suit themselves, and others have formulated their own codes. Detailed examination of these codes reveals further diversity – especially in the type and number of control measures adopted and in their specific provisions. The causes of such diversity are complex, but fundamental to them is the fact that community perception of the problem is spatially variable and develops over time. The political conflict, in which the building codes represent the tangible compromise, is essentially between developers concerned to maximize their opportunities and the community exercising what it perceives to be its social responsibility: resolution of the conflict inevitably varies from place to place.

The city of Los Angeles illustrates well the recent history of community response to slope problems. Hillslope development was first seriously controlled in the city with a 1952 ordinance. This ordinance, and most of the numerous subsequent amendments to it, were provoked by serious winter storms and the damage they caused to the community (e.g. 1952, 1958, 1963, 1969, 1978, and 1980). Part of the community's response to the hazard was to commission – either formally through the codes or by special contract –

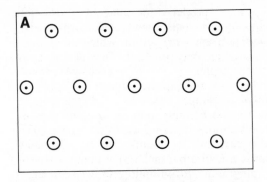

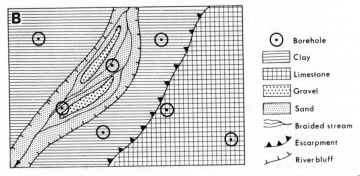

FIG. II.11 The use of geomorphological mapping to economize on sampling costs: (A) the reticulate grid sample design commonly used by engineers; (B) a stratified sampling design based on a geomorphological map of the same area

research by geologists into the problems of slope stability with the result that the scientific data available for community legislation have been much improved.

The storms of 1969 provided an excellent opportunity to assess the effectiveness of successive hillside development ordinances in Los Angeles City. Figure II.13 shows the location of a 20% random sample of slope failures reported to and acted upon by the city's Building and Safety Department after the 1969 storms; each location is classified according to its age – before the 1952 ordinance, after the 1952 ordinance but before the more restrictive 1963 ordinance, and after the 1963 ordinance. Figure II.13 shows several interesting features. First, a preponderance of the reported failures were at sites developed before 1952 or between 1952 and 1963 – the number of failures as an approximate percentage of the estimated number of sites developed was pre-1952, 10.4; 1952–63, 1.3; and post-1963, 0.15 (Slosson, 1969). Further analysis of the storm and slope-failure damage reports reveals that the extent and cost of damage have also decreased with time (Table II.4). Second, nearly all reported damage was confined to the 'hillside area', where special planning and building regulations have been introduced to control slope failure: clearly, even though the 'hillside area' is delimited by cultural boundaries (Palmer, 1976), it effectively defines the hazard zone.

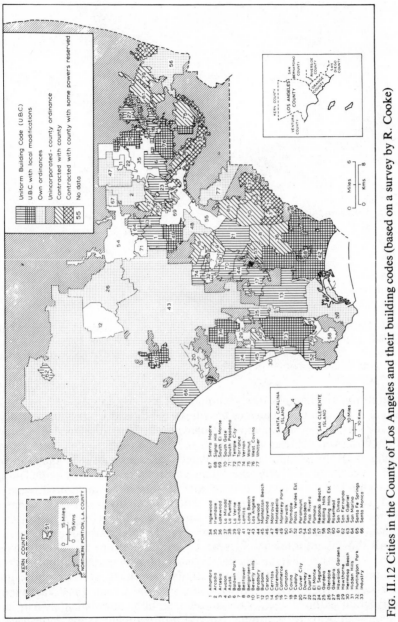

FIG. II.12 Cities in the County of Los Angeles and their building codes (based on a survey by R. Cooke)

(v) Geormorphology and post-construction management. The monitoring of geomorphological processes and landform changes after construction can play an important role in helping environmental managers to influence future planning policy and alleviate or avoid environmental hazards. Ruby's (1973) study of sediment-yield trends in the Los Angeles River catchment is a

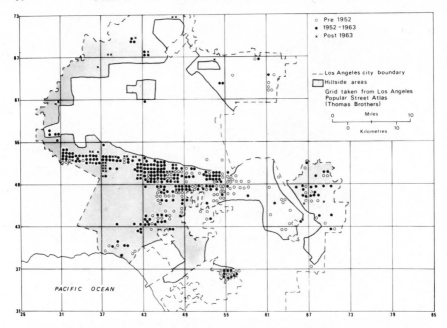

FIG. II.13 Location of buildings, classified according to age, within Los Angeles City affected by slope failure and slope erosion during the storms of 1969: based on an analysis of a 20% random sample of reports by the Building and Safety Department of the City of Los Angeles by R. Cooke

straightforward example. The accumulation of sediment in debris basins at the mouths of mountain canyons provides a rough measure of sediment yield and allows the performance of smaller check dams (designed to restrict sediment movement from the canyons on to urbanized alluvial plains) to be evaluated. Ruby compared accumulated sediment yields in one canyon (Dunsmore) before and after check-dam construction with the regional norm (the accumulated sediment yield per km² for the whole monitored region – the 'Los Angeles River Watershed'). There are many problems associated with the data and their interpretation, but regression lines (Fig. II.14) relating sediment yield in Dunsmore Canyon to the regional norm for 30 years of records show that for the first 21 years the canyon performed similarly to the regional watershed, that during the ten years following construction of check dams sediment yield from the canyon relatively declined, and that the decline has become progressively less over time. In the period since check-dam construction sediment yield was reduced overall by approximately half, but there now appears to be a trend back to pre-treatment yields, indicating a decline in the trap efficiency of the check dams. Clearly an alternative strategy for sediment control is required.

Before geomorphological problems of dryland urban development are reviewed systematically, two further introductory themes require examination, and these are considered in the next two sections: the availability of

TABLE II.4

The effect of land use management on landslide damage in Los Angeles during the wet winter of 1969

Pre-1952	1952–62*	1963 to Present*
No grading codes, no soils engineering, no engineering geology	Semi-adequate grading code, soils engineering required, very limited geology but no status and no responsibility	New modern grading codes; soils engineering and engineering geology required during design; soils engineering and engineering geology required during construction; Design Engineer, Soils Engineer, and Engineering Geologist all assume legal responsibility
Approx. 10 000 sites constructed	Approx. 27 000 sites constructed	Approx. 11 000 sites constructed
Approx $3 300 000 damage	Approx. $2 767 000 damage	Approx. $182 400 damage†
Approx. 1 040 sites damaged	Approx. 350 sites damaged	Approx. 17 sites damaged
An average of $330 per site for the total number produced	An average of $100 per site for the total produced	An average of $7.00 per site for the total produced
$3 300 000	$2 767 000	$80 000
Sites 10 000	Sites 27 000	Sites 11 000
Predictable failure percentage: 10.4%	Predictable failure percentage: 1.3%	Predictable failure percentage: 0.15%
1 040 damaged	350 damaged	17 damaged
10 000 total sites	37 000 total sites	11 000 total sites

* It should be noted that the storms of 1952, 1957–8, 1962, 1965, and 1969 all produced similar total losses associated with similar destructive storms.
† Over $100 000 of the $182 000 was incurred on projects where grading was in operation and no residences were involved, thus less than $80 000 occurred on sites constructed since 1963.
Source: Slosson, 1969.

information relevant to geomorphological studies in urban development of drylands, and the integration of geomorphological data into the whole assemblage of environmental data relevant to planning decisions.

(d) Information relevant to Geomorphological Studies in Urban Development of Drylands

1. *Problems of obtaining data*

Geomorphological studies, in drylands as elsewhere, normally require at their outset an investigation of all extant and relevant data sources, and the acquisition of potentially useful topographic maps and remote sensing imagery. A critical problem in this initial phase is the practical one of data retrieval. The search for background, basic data can be time-consuming and frustrating, and it frequently depends on the establishment of good personal contacts and the exercise of considerable discretion.

While it is dangerous to suggest that any one approach to data retrieval should be adopted prior to field investigation, most searches can be subdivided into stages, each of which is concerned with particular types of information. Figure II.15 summarizes these for an individual intending to work on an applied geomorphological problem in a dryland country from a

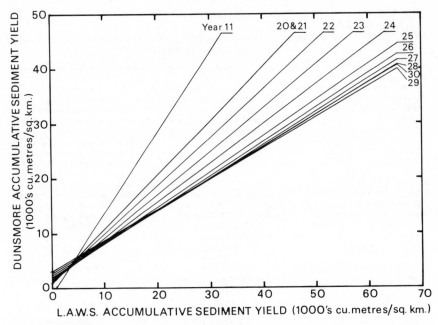

FIG. II.14 Sediment yield from Dunsmore Canyon compared with that from the whole Los Angeles Watershed (LAWS), over a 30 year period. Check dams were introduced at year eleven (after Ruby, 1973)

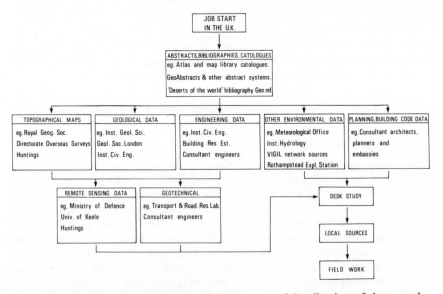

FIG. II.15 Idealized scheme for the efficient, sequential collection of data on the natural environment prior to a geomorphological survey in a dryland region (from a base in the UK) (modified from Eastaff et al., 1978)

base in the United Kingdom. The main categories of data are discussed in section 2, below.

Within the 'local sources' category in Fig. II.15 data are likely to be required from national, regional and local authorities within the country. In a typical Middle Eastern state, for example, the search probably needs to extend into several government departments (e.g. those responsible for lands, public works, water supply, defence, housing and town planning, transport, agriculture, and industry) in order to discover and to obtain relevant material; further enquiries will certainly be required at the municipality level for building codes, street plans etc., it they exist.

Difficulties likely to be encountered in obtaining basic yet essentially background and preliminary information are numerous, and undoubtedly familiar to those with experience of such searches. The problem of discovering the existence of information and then determining its location is usually perplexing. Even in the United States, where public access to documentation is particularly free, this can be a serious problem because relevant data may be generated by over a dozen federal agencies, hundreds of state, county, and city authorities, and a host of private organizations (e.g. Spangle, W. and Associates, 1976, p. 125). Commonly the material is not catalogued or referenced so that access to it is not straightforward for an individual from outside the organization concerned. In many countries the view is widely held that reports should be confidential (even if they are not officially so), especially if they relate to environmental resources: this notion is not unrelated in some dryland countries to the cloak of secrecy surrounding oil-industry data. What is true of public agencies is even more true of private companies (such as consulting engineers), many of whom may be competitors. In addition, there is the bureaucratic habit of allowing reports to collect dust or to disappear beneath a rising tide of paper, especially if the documents did not justify politically based action at the time of their receipt, or if their contents are unintelligible to the administrators who handle them. Finally, many reports are not published and are produced only in limited numbers, and copies may be difficult to locate for this reason. Such difficulties are, of course, almost always only effectively overcome through the establishment of personal contacts.

2. *Background data sources relating to geomorphological studies of urban development in drylands*

The background data required at the outset of geomorphological studies in drylands are varied (e.g. Fig. II.15); some of them are briefly reviewed below.

(*i*) *Abstracts, bibliographies, and similar secondary data sources* have been listed by Cooke (1977) and, in a Middle Eastern context, by Eastaff *et al.* (1978). *GeoAbstracts* (Norwich, UK) and *Geotechnical Abstracts* are sources of pertinent abstracts, and the office of Arid Land Studies at the University of Arizona (Tucson) is a less regular source; reference retrieval systems, such as GEOREF are of occasional value, as are the bibliographies and indexes of most important geological and geographical societies. An impressive feature of a search for geomorphological literature on drylands, especially in the former colonial realms of the French, German, and British, is the frequent occurrence of pertinent, if ancient, studies, of even the remotest areas.

(*ii*) *Maps* are required both as a source of relevant data and as planimetric bases for the plotting of field information. International indexes to map availability are few, Winch's (1976) guide being the best known. More detailed national guides are often available through military and survey agencies. Those provided by the US Geological Survey, and the annotated indexes of aerial photographs and maps of topography and natural resources compiled by member countries of the Organization of American States, are useful examples. Many Middle Eastern countries have surveying and mapping departments, although some of their map production is done under contract by specialist firms such as Huntings Surveys Ltd and Clyde Surveys Ltd.

Maps showing the results of systematic surveys – such as those of geology, soils, and land-use potential – are also valuable background documents to some geomorphological studies. International examples of potential value at an early stage of a study include several maps produced by UN agencies, such as UNESCO's survey of *Water Balance of Lakes and Reservoirs of the World* (*Nature and Resources*, 1974), FAO and UNESCO's *Soil Map of the World* (Dudal and Batisse, 1978), and UNESCO's *Geological Atlas of the World* (Marçais, 1977). UNESCO's encouragement of international vegetation mapping schemes (UNESCO, 1973) and recommendations for the compilation of national atlases of natural conditions and resources (Salishchev, 1972) also promise to yield useful documents. Generalizations concerning systematic maps are difficult at a national scale because subjects covered and availability are so varied. Major sources usually include government agencies and universities and special reports produced by consultants.

(*iii*) *Aerial photographs and remote sensing imagery.* Aerial photographs are a fundamental requirement for most geomorphological fieldwork in drylands, at all levels of enquiry (see chapters III and VII). Not only are they often more valuable than topographic maps for recording field data, they now form the basis for the preparation of many topographic maps. Unfortunately, aerial photographs can be sensitive military documents (because of the great detail of surface characteristics that can be discerned) and are therefore not always easily obtained. Sources of aerial photographs at scales up to 1:50 000 (the smallest scale suitable for detailed fieldwork) come from such excellent archives as those held by former colonial powers (e.g. the coverage held by the Ministry of Defence and the University of Keele in the UK), national agencies in the relevant countries, and the collections of international photomapping companies. Invariably, political and military permission is required before photographs can be obtained and used. On occasion it may be essential to have aerial photography taken, especially for a particular project.

Much more freely available, at least for the time being, is the imagery derived from LANDSAT satellites. This imagery, which is still fairly cheap, is not only available for desert areas, but also for those same areas at different periods of time (e.g. Bardinet *et al.*, 1978). With its four channels (0.5–0.6 μm, 0.6–0.7 μm, 0.7–0.8 μm and 0.8–1.1 μm), and an effective spatial resolution of 80–150 m, LANDSAT 1 imagery is potentially useful for much reconnaissance work (e.g. Simonett, 1976; Robinove, 1979). Satellite imagery is likely to become increasingly valuable as enhancement techniques improve, as the resolution of imagery from new satellites improves (Return Beam Vidicon, RBV, imagery from LANDSAT 3, for example, has an effective spatial resolution of approximately 40 m) (Allan, 1978), and as new forms of imagery (such as that from the Heat Capacity Mapping Mission, the space shuttle, and meteorological satellites) become available.

(*iv*) *Engineering, geological, and other environmental data.* The search for relevant data invariably passes into the shadowy area of unpublished sources, including geological, meteorological, and hydrological information, and data contained in unpublished consultants reports. There have been some attempts to make accessible on an international basis some valuable data. For example, the UK Meteorological Office publishes summary tables of temperature, relative humidity and precipitation for the world (HMSO, London), and during the International Hydrological Decade an effort was made to record formally and store accessibly hydrological information through the VIGIL network (Leopold, 1962). But such data are of only limited value to specific projects, and local unpublished data, if available, are also required. For example, in the study of sand movement, wind observations from local airports are often useful.

The diversity of local unpublished data that may be available for a particular area or topic may be considerably greater than might be expected. The following examples illustrate this point and the facts that the data are often scattered, provided to different authorities, and subject to all the problems discussed in section (d) 1 above.

Example 1: Aggregate availability in Kuwait.
   Surveys of direct relevance to this subject which, like the aggregate reserves, are far from exhausted, include:
W. Fuchs, T. E. Gattinger, and H. F. Holzer, 1968, *Explanatory text to the synoptic geologic map of Kuwait* (Geological Survey of Austria).
GeoServices, 1970, *Geological survey and raw material evaluation of superficial formations* (Ministry of Commerce and Industry, Kuwait).
Kuwait Institute of Scientific Research, 1978, *Proposal for a geological and geophysical exploration for gravel and sand in northern Kuwait.*
J. C. Sproule and Associates Ltd, 1974, *Sources of construction materials in the State of Kuwait. Summary report* (Ministry of Commerce and Industry, Kuwait).

Example 2: Surface materials survey of Bahrain.
   Surveys of direct relevance to a project on this subject undertaken by the authors (Bahrain, Ministry of Works, Power and Water, 1976; Doornkamp *et al.*, 1980) include:

Bahrain Petroleum Company, 1974, 'Report on the geology of Bahrain with hydrology and economic aspects'. Unpublished company report.

Engineering and Resources Consultants, 1971, 'Strengthening of the Department of Agriculture services in Bahrain'. Unpublished report to the Government of Bahrain.

Geophoto Services, ?1971, 'Report on the photogeological map of Bahrain'. Unpublished report, BAPCO.

Italconsult, 1971, 'Water and agricultural studies in Bahrain'. Unpublished report to the Government of Saudi Arabia.

Messrs Sandberg, 1974, 'A photogeological interpretation of Bahrain, adjacent islands and near-shore shallows for materials resources and engineering purposes'. Unpublished report to the Government of Bahrain.

Messrs Sandberg, 1975, 'A review of materials aspects of geological and resources studies of Bahrain 1908–74'. Unpublished report to the Government of Bahrain.

E. P. Wright and M. Ayub, 1975, *Groundwater abstraction and irrigation in Bahrain – interim report*. Department of Agriculture, Bahrain.

(e) Integration of Geomorphological Data into the Assemblage of Environmental Data Relevant to Urban Development

1. *Relations between geomorphology and other scientific information of value to planners of urban areas in drylands*

Geomorphological information forms but one part of the body of environmental data that may be of value to urban planners and engineers in drylands. Many environmental attributes of interest are both closely linked and highly interdependent (Fig. II.16). Table II.5 outlines the basic environmental information on regional setting and site that ideally should be available at various stages in the planning process. Data on the regional setting is primarily relevant in assessing hazards and resources and in choosing locations for development; data on site conditions related to decisions about specific developments (e.g. the foundation capacity of soils). Some of the information (†, Table II.5) requires long-term monitoring or, failing that, informed prediction; some predictions (or estimates of what has happened in the recent past and of what is going on at present) may be based on geomorphological criteria, such as some of those listed for hydrology, and in coastal, mountain, and volcanic areas (Table II.5*).

Despite the links between environmental attributes, information is often highly compartmentalized within the boundaries of specialist disciplines, and it is clear that some of it can only be collected by those with training and experience in particular aspects of environmental assessment – geologists, geohydrologists, and soil mechanics engineers, for instance. The geomorphologists' main distinctive contribution is likely to be through the assessment of topography, surface materials, and processes of landscape development. But precisely because of the close ties between the distinct environmental data sets, co-operation between specialists is not only desirable but essential; equally, such co-operation is likely to be both profitable and cordial because the data sets are distinct but complementary. The co-operation is most

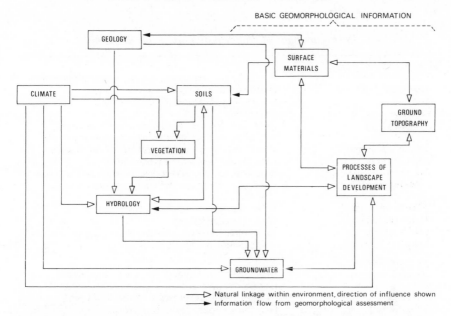

FIG. II.16 General relationships between types of environmental information and the planning process

efficiently and cheaply accomplished if data are collected in a sensible, sequential manner, so that each new increment is added in a meaningful way to existing information.

In this context, it can be argued that geomorphological surveys can provide a useful first stage in the environmental assessment, not only because geomorphology is a fundamental basis for urban development, but also because studies of geology, soils, hydrology, etc., can all benefit from, and be facilitated by access to geomorphological surveys. Figure II.17, for example, shows a sequence of assessment of environmental data in which information from one survey might efficiently feed into the next, beginning with the geomorphological survey.

2. Geomorphology and environmental impact assessment

(*i*) *Impact of urban developments on the environment.* The previous section makes clear that geomorphological information is linked both philosophically and practically to other environmental information of value to planners. In this context it is important to record that demands are increasing from planners for wide-ranging environmental reports (of which geomorphology forms a part) prior to making and implementing planning decisions. The most important new requirement is not so much for resource surveys – for that demand has existed for many years, and much of this book is concerned with geomorphological aspects of such surveys – but for studies to evaluate the impact a proposed development is likely to have on the physical environment.

TABLE II.5

Basic environmental information for urban planning

For the regional setting	
Climate*	Precipitation, temperature, winds, evapotranspiration
Geology‡	Stratigraphy, lithology, structure, minerals
Soils‡	Soil types, land capability for agriculture
Vegetation	Ecosystems
Hydrology*†‡	River flow characteristics, surface water supplies
Groundwater	Available water supplies, excessively high watertables
Ground topography ¶	Slopes (amount and direction of steepness) slope form
Surface materials¶	Bedrock outcrops, sediment and soil types
Processes of landscape development¶	e.g. Slope stability, soil erosion, weathering of soils and bedrock, fluvial activity, wind processes, sediment removal and transport
(Coastal areas*†‡	Tides, flooding, erosion, sedimentation).
(Mountain areas†‡	Avalanches, landslides)
(Earthquake zones	Susceptibility to severe earthquakes)
(Volcanic areas†‡	Susceptibility to volcanic activity)
For the urban site	
Geology‡	Depth to bedrock foundations, bedrock properties
Soils‡	Engineering properties of soils (incl. salt content)
Hydrology*†	Channel flow, sedimentation/erosion, flooding
Groundwater‡	Depth to watertable, salinity of groundwater
Ground topography¶	Slopes (amount and direction of steepness) slope form
Surface materials¶	Bedrock outcrops, sediment and soil types
Processes of landscape development¶	e.g. Slope stability, soil erosion, weathering of soils and bedrock, fluvial activity, wind processes, sediment removal and transport

* = requires long-term monitoring
† = sensible estimates possible from geomorphological assessment
‡ = invaluable basic data available to specialists from a primary geomorphological survey
§ = more work could be done by specialists but geomorphological assessment probably adequate for the urban planning stage
¶ = substance of a geomorphological assessment

Although many planning authorities have required environmental impact assessments for years, recent legislation has enforced and codified the requirement in several countries, providing a substantial stimulus to the development of methods for assessment and for integrating environmental data. The most important new law was the National Environmental Policy Act of 1969 (NEPA), passed by the US Congress, and this was followed by similar measures in many states, and in other countries such as Australia and Israel (e.g. Ditton and Goodale, 1972). Critical to NEPA was the requirement that environmental impact statements (EIS) should be prepared on federal proposals that could significantly affect the quality of the human environment.

To assess potential environmental impact is extraordinarily difficult because it involves predicting complex responses on the basis of what is usually woefully inadequate scientific data. However, political necessity has prompted several attempts to develop standardized techniques of assessment in order to streamline production of reports (in the USA for example, over 4000 EISs were produced by federal agencies in the first three-and-a-half

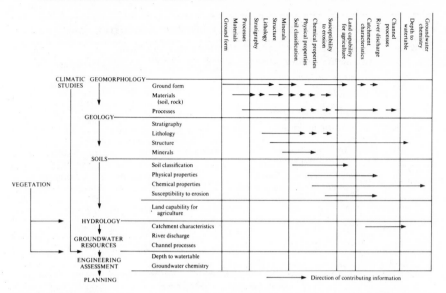

F<small>IG</small>. II.17 Sequence of environmental data collection based on an initial geomorphological survey

years after NEPA was passed), to facilitate comparisons, and to simplify the preparation and presentation of complex problems.

The range of techniques can be divided conveniently into three (Skutsch and Flowerdew, 1976): (*a*) techniques designed to identify or aid in the identification of probable impacts of alternative projects (impact identification); (*b*) techniques designed to assess or quantify the probable impacts of alternative projects (impact evaluation) and (*c*) techniques designed to compare impacts of alternative projects and select the least damaging overall (impact comparison). These techniques, which it is not appropriate to review in detail here, all allow the inclusion of geomorphological data and provide frameworks for integrating environmental data in the context of planning problems.

(*ii*) *Impact of environment on development.* This theme has much in common with the previous topic and the techniques of environmental impact assessment mentioned there could well be appropriate here, for clearly cause and effect are intimately related in man's relations with his environment. But emphasis in this field commonly rests mainly on the assessment of natural hazards at different scales. Thus, a major problem of environmental planning at a regional scale is to establish priorities. In this context, assessment of the potential relative impact of environmental hazards is particularly important, and such impacts will vary both spatially and temporally.

The California Division of Mines and Geology (1973) devised a preliminary method for assessing loss-reduction programmes related to urban development in hazard areas at a regional scale. The location and degrees of severity of several hazards (including geomorphological hazards) were mapped for

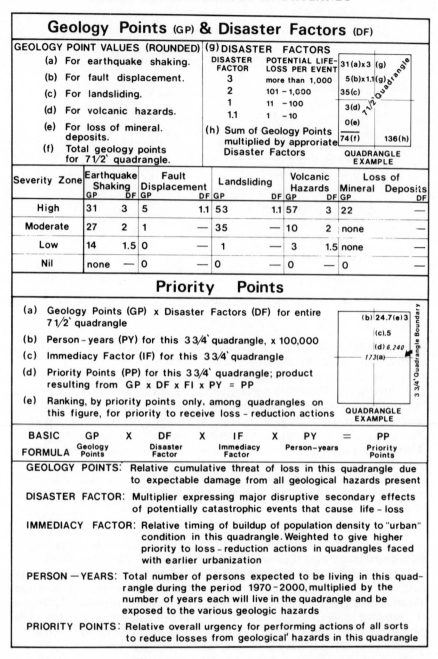

Geology Points (GP) & Disaster Factors (DF)

GEOLOGY POINT VALUES (ROUNDED)

(a) For earthquake shaking.

(b) For fault displacement.

(c) For landsliding.

(d) For volcanic hazards.

(e) For loss of mineral. deposits.

(f) Total geology points for 7 1/2' quadrangle.

(g) DISASTER FACTORS

DISASTER FACTOR	POTENTIAL LIFE-LOSS PER EVENT
3	more than 1,000
2	101 – 1,000
1	11 – 100
1.1	1 – 10

(h) Sum of Geology Points multiplied by appropriate Disaster Factors

QUADRANGLE EXAMPLE:
31(a)x3 (g)
5(b)x1.1(g)
35(c)
3(d)
0(e)
74(f)   136(h)
7 1/2' Quadrangle

Severity Zone	Earthquake Shaking		Fault Displacement		Landsliding		Volcanic Hazards		Loss of Mineral Deposits	
	GP	DF	GP	DF	GP	DF	GP	DF	GP	DF
High	31	3	5	1.1	53	1.1	57	3	22	—
Moderate	27	2	1	—	35	—	10	2	none	—
Low	14	1.5	0	—	1	—	3	1.5	none	—
Nil	none	—	0	—	0	—	0	—	0	

Priority Points

(a) Geology Points (GP) x Disaster Factors (DF) for entire 7 1/2' quadrangle

(b) Person–years (PY) for this 3 3/4' quadrangle, x 100,000

(c) Immediacy Factor (IF) for this 3 3/4' quadrangle

(d) Priority Points (PP) for this 3 3/4' quadrangle; product resulting from GP x DF x FI x PY = PP

(e) Ranking, by priority points only, among quadrangles on this figure, for priority to receive loss – reduction actions

QUADRANGLE EXAMPLE:
(b) 24.7(e)3
(c).5
(d) 6,240
173(a)
3 3/4' Quadrangle Boundary

BASIC FORMULA	GP Geology Points	X	DF Disaster Factor	X	IF Immediacy Factor	X	PY Person–years	=	PP Priority Points

GEOLOGY POINTS: Relative cumulative threat of loss in this quadrangle due to expectable damage from all geological hazards present

DISASTER FACTOR: Multiplier expressing major disruptive secondary effects of potentially catastrophic events that cause life – loss

IMMEDIACY FACTOR: Relative timing of buildup of population density to "urban" condition in this quadrangle. Weighted to give higher priority to loss – reduction actions in quadrangles faced with earlier urbanization

PERSON — YEARS: Total number of persons expected to be living in this quadrangle during the period 1970 – 2000, multiplied by the number of years each will live in the quadrangle and be exposed to the various geologic hazards

PRIORITY POINTS: Relative overall urgency for performing actions of all sorts to reduce losses from geological' hazards in this quadrangle

FIG. II.18 Summary of the procedure for assessing loss-reduction programme priorities related to urban development in hazard areas, by the California Division of Mines and Geology (1973)

California and then digitized on a rectangular grid in which each cell corresponds to a $7\frac{1}{2}'$ quadrangle (Fig. II.18). Each hazard was then given a numerical rating in each quadrangle indicative of its 'dollar loss potential' for destruction of life and property for each degree of severity using, as a standard basis for estimation, a hypothetical 'urban unit' of 3 000 population and $90 m total value. A numerical value (expressed as Geology Points, GP) was assigned that represents expectable annual loss to any resident in each severity zone for each problem. The value then becomes a relative weighting factor for that type of problem for any quadrangle in the indicated severity zone. The sum of the GPs for a quadrangle expresses the average dollar loss that the average resident might expect to suffer per annum from all the problems considered. Then, for each quadrangle, GPs are multiplied by a Disaster Factor (DF), a measure of predicted population (Person-years, PY), and estimates of the timing of population growth (Immediacy Factor, IF) to identify Priority Points (PP). In this way spatial priorities for field surveys and other loss-reduction actions can be established.

This method clearly provides a way of integrating regional assessments of hazards; it is only one of several methods (e.g. W. Spangle and Associates, 1974), all of which need to face the central problem of judging the relative importance of quite different hazards. Such methods are not appropriate for the larger-scale study of one city; at this scale, the assessment of hazards requires different techniques such as those described in subsequent chapters.

(f) Summary

Geomorphology, the study of land forms (their nature, origin, and evolution, processes of development and material composition) has a valuable role to play in the planning, development, and management of urban areas in drylands, especially in evaluating terrain prior to urban development, and in monitoring changes during and after urban development. Responsibility for studying and managing geomorphological resources, hazards, and other problems may rest with a variety of agencies within local hierarchies of management organizations. The type of information required will depend chiefly on the local geomorphology and the responsibilities and attitudes of the relevant agencies. It will also vary with the phase of planning: regional planning, city planning, site planning and development, and post-construction management all require information that can be provided by the geomorphologist. In such circumstances, he must invariably be able to use a wide range of relevant data sources, some of which may be difficult to obtain. Commonly, the geomorphological information must be integrated into a broader assemblage of environmental information that is useful to planners: among the available methods, it is suggested that those in which geomorphological surveys provide a first stage for environmental assessment may be particularly valuable.

III

SYSTEMATIC MAPPING OF GEOMORPHOLOGY

(a) Introduction

Fundamental to most geomorphological contributions to urban development in drylands is the systematic recording of geomorphological data in the form of maps. The primary purpose of such maps is to show at an appropriate scale and in an intelligible way the distribution of landforms, surface and near-surface materials, and the processes that sculpture the landforms and modify the materials. The usefulness of the maps in urban development is largely dependent on the abstraction of data from them of terrain information required in the planning process. This chapter examines briefly the nature of the main mapping methods (details of techniques have been omitted because they have been discussed extensively elsewhere) and presents some of the results of geomorphological mapping programmes undertaken in drylands at a variety of planning scales.

There are essentially two alternative methods of terrain assessment applicable to dryland studies in which geomorphology plays a central role: land-systems surveys and geomorphological mapping *sensu stricto*. The former involve making a regional subdivision based on the assemblage of landforms, while the latter requires the mapping of individual geomorphological features. Land-systems mapping has already proved to be a very efficient method for the rapid classification of the character of large areas of land, and has been used widely in land resources assessment projects around the world. Geomorphological mapping, on the other hand, tends to be more useful at the scales of site and city investigation, and is thus particularly suited to the assessment of terrain conditions in an area of proposed urban development. The methods of portraying terrain information differ between the two approaches. But the two techniques are not mutually exclusive, and in some situations both can profitably be used together.

(b) Land Systems Surveys

1. *Philosophy and background*

The starting-point for all land-systems surveys is the recognition of regions within each of which there is a specific assemblage of types. Each region is internally homogeneous with respect to its landforms but differs from neighbouring regions. Thus an area of sand dunes is distinctly different from one composed of dry lake floors, which is, in its turn, distinct from an area of alluvial fans. The pattern of landforms in any one area is called a *land system*, and this may vary in size from a few square kilometres to several hundred square kilometres.

Each land system is capable of further subdivision – the dune field is composed of individual dunes, and so on. These smaller areas are called *land units* (or *land facets* in some studies). Each land unit is composed of individual slopes (e.g. the individual side-slopes of a seif dune), and these are known as *land elements,* the smallest component in the hierarchy of units of the land system (Fig. III.1). As indicated in the previous chapter, the hierarchical classification in this way is a feature of substantial importance.

Published maps based on land-systems surveys are either of land-systems boundaries or, in more detailed studies, of land-unit boundaries. The maps do no more than show the location and extent of areas. Data on the areas are tabulated separately in an accompanying report that will also include either schematic block diagrams and ground and/or aerial photographs of each mapped unit to enhance the visual impression of terrain characteristics. Figure III.2 shows a subdivision of the Kyalami land system near Johannesburg and some of the data relevant to its land units. Although the data provided vary with the goals of the survey, the land-systems approach is underlain by the fundamental assumption that the land unit is a meaningful component of the land surface in terms of landform, bedrock, soils, and sometimes vegetation or site drainage. In effect, therefore, the land system and its component units can provide a convenient data store for terrain information, and they have been used as a basis for data banks (Brink *et al.,* 1966, 1968; Grant and Lodwick, 1968; Beckett and Webster, 1971; Ollier, 1977), especially in South Africa (Brink and Partridge, 1967; Kantey, 1971; National Institute for Road Research, 1971, 1976).

Fuller reviews of land systems surveys are provided by Christian (1957), Mabbutt and Stewart (1963), Christian and Stewart (1968), Thomas (1969, 1976), Webster and Beckett (1970), Lawrance (1972), Mitchell (1973), Cooke and Doornkamp (1974), and Ollier (1977). Russian views are reflected by Solentsev (1962) and Isachenko (1973). The main organizations using the land-systems approach in drylands are listed in Table III.1.

Aerial photographs usually provide the basis for identifying land systems (e.g. Webster, 1963; Bawden, 1967; Ollier *et al.,* 1967; Webster and Beckett, 1970), and the preliminary definition is followed by field checking and sampling (usually by a team of scientists). In recent years the method has been developed with attempts at the quantification of terrain characteristics (e.g. Speight, 1968; Haantjens, 1968; King, 1970), at the identification of units from morphometric land-surface data (Speight, 1974, 1976, 1977), and at the use of LANDSAT imagery for defining land systems (e.g. Mitchell and Howard, 1978).

To be of most value in urban development, land-systems maps and related data need to be produced at a relatively large scale and modified for use in engineering investigations and planning (Aitchison and Grant, 1967, 1968; Brink and Partridge, 1967; Brink *et al.,* 1968; Dowling, 1968; Grant, 1968, 1972; National Institute for Road Research, 1971). In South Africa, for instance, the method is used as the basis for a scheme of engineering soil-mapping in which assessment of mapping units includes a strong geomor-

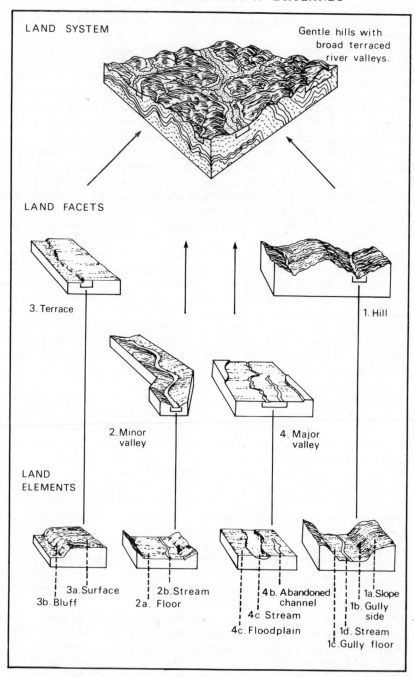

FIG. III.1 Diagram to show the relationship between land system, land facet, and land element (after Lawrance, 1972)

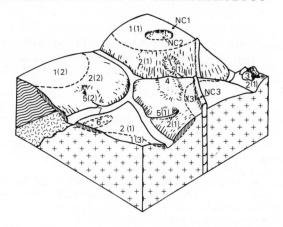

The engineering properties of some of the soils in the Kyalami Land System

Unit	Variant	Form	Soils and hydrology	Soil constants			
				Statistic	Liquid limit	Plasticity index	Linear shrinkage
1	1	Hillcrest, 0–2°, width 450–1800 m	Residual sandy clay with collapsing grain structure on granite (9–23 m). Ferrallitic soil. Above groundwater influence except at depth	X̄ S V	44.1 11.0 0.25	22.6 10.88 0.48	9.78 4.27 0.44
2	1	Convex side slope 2–12°, width 450–910 m	Hillwash of silty sand derived from granite (0.3–0.9 m) on granite schist or basic metamorphic rocks. Occasionally saturated	X̄ S V	17.3 8.98 0.52	4.90 3.70 0.76	2.04 1.04 0.51
7	—	Alluvial flood-plain 1–3°, width 45–230 m	Expansive alluvial clays and sands (3–6 m) on granite schists and basic metamorphic rocks. Mineral hydromorphic soil. High watertable (data for black alluvium)	X̄ S V	58.5 14.7 0.25	37.1 13.67 0.37	13.7 10.31 0.75

X̄ = mean,   S = standard deviation,   V = coefficient of variation

FIG. III.2 Schematic block diagram of the Kyalami area, near Johannesburg, South Africa, showing its subdivision into land units, together with the data on engineering properties of soils in the land system (after Brink *et al.*, 1970)

phological component (Kantey and Williams, 1962; Brink and Williams, 1968; Kantey and Mountain, 1968). For example, 'transported soils' occur in many South African landscapes, and their occurrence and types are closely related to their geomorphological position and the processes that transported them (Fig. III.3; see also Fig. VI.4). With engineering purposes in mind, Haantjens (1968) has suggested that the land-systems map should be the basis for the preparation of four maps of potential value in planning; these should show (i) engineering materials, (ii) ruggedness and relief, (iii) drainage status and flood hazard, and (iv) soil permeability, all of which have a strong geomorphological basis.

Land-systems mapping, therefore, is potentially useful in the urban development of drylands as a means of classifying land into units of relatively homogeneous character; the units can serve as a basis for data storage, and the data can be both precise and often quantified. The method does not,

TABLE III.1

Principal organizations using a land-systems approach in deserts, with comments on the geomorphological information used

1. CSIRO Division of Land Research and Regional Survey (Australia)
Example. Stewart *et al*. (1970)
Geomorphology. General qualitative comments on relief and drainage of the survey area, and generalized accounts of the major types of country (with quantitative estimates, at times, of slope characteristics). More recent studies include more precise geomorphological information. This example includes a geomorphological map and a descriptive chapter on geomorphology.

2. Directorate of Overseas Surveys, Land Resources Division (UK)
Example. DOS, Land Resources Division, 1968
Geomorphology. Introductory qualitative remarks on geomorphology; the characteristics of land systems (grouped into larger, topographically based *land regions*) are each described with brief comments on physiography, and illustrated by block diagrams.

3. Military Engineering Experimental Establishment (UK)
Example. Perrin and Mitchell (1969)
Geomorphology. Abstract land systems include simple, quantified statements of such variables as altitude and relief, and general, qualitative comments on landforms. Abstract facet and clump descriptions include approximate slope limits, adjectival classification of relief amplitude, and brief qualitative descriptions of topography.

4. Transport and Road Research Laboratory (UK)
Example. Dowling (1968)
Geomorphology. Descriptions are generalized (much as in the CSIRO surveys), but geomorphological data are embodied in the evaluation of engineering properties. Land systems were used effectively in the pre-design stage of roadworks, and features sampled in land systems included particle-size distribution, Atterberg limits, and soil strength.

5. National Institute for Road Research (South Africa)
Example. NIRR (1971)
Geomorphology. Descriptions and indexing of *land patterns* ($\simeq$ land systems) includes a classification of terrain according to its *erosion cycle* (after King, 1963) and its *physiographic type* (after Hammond, 1954; Wallace, 1955). Erosion cycles are significant in engineering terms ('residual soils may confidently be expected to be deep below old planed erosion surfaces, while transported soils would be deep below young planed depositional surfaces'; NIRR, 1971, p. 15), and so is physiographic type ('in relation to general accessibility and likely degree of cut and fill'; NIRR, 1971, p. 15). In addition, the description of facets within land patterns includes notes on their shape, inclination, length and altitude.

6. CSIRO, Division of Applied Geomechanics (Australia)
Example. Grant (1972)
Geomorphology. The *terrain component* ($\simeq$ sub-facet) is defined on the basis of precisely defined slope characteristics. *Terrain units* (= facet) fall into classes, and each class has a characteristic association of slopes, and a consistent local relief amplitude. *Terrain patterns* (= land systems) are classified numerically in terms of relief amplitude and general physiography. In addition, the summary of terrain patterns includes a block diagram, and the description of a typical drainage net, stream frequency, and drainage type. This type of land-systems survey includes more precise geomorphological data than the other examples.

Source: Cooke (1977).

however, provide a dynamic view of the environment (Thomas, 1976), so that, for example, processes that transgress unit boundaries may be underestimated or even misunderstood. Geomorphological mapping, on the other hand, does allow both a more detailed and a more dynamic assessment of the landscape.

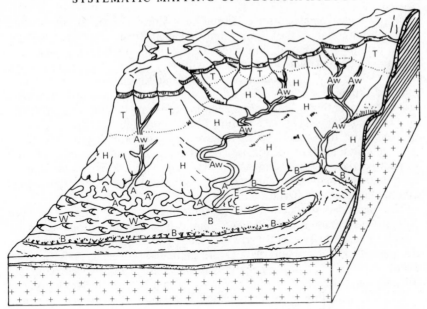

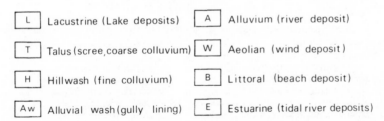

L	Lacustrine (Lake deposits)	A	Alluvium (river deposit)
T	Talus (scree, coarse colluvium)	W	Aeolian (wind deposit)
H	Hillwash (fine colluvium)	B	Littoral (beach deposit)
Aw	Alluvial wash (gully lining)	E	Estuarine (tidal river deposits)

FIG. III.3 Idealized block diagram showing transported soils in relation to their landform setting (after Brink and Williams, 1964)

(c) Morphological and Geomorphological Mapping

1. *Philosophy and background*

Mapping surface form is fundamental to all geomorphological mapping, and it can be accomplished in a variety of ways. Contours, the method used almost universally to show landform on topographic maps, are unfortunately of only limited value in geomorphological work because they reveal little about what happens between contours and their accuracy is conditioned by the manner of survey, scale, and contour interval; furthermore, contour maps are difficult and expensive to compile, and many areas still do not have them.

Morphological mapping (see section 2 below) provides an alternative that does not share some of these limitations (Savigear, 1956, 1961, 1965; Waters, 1958; Bridges and Doornkamp, 1963; Curtis *et al.,* 1965). In engineering or planning decisions related to urban growth, it is not only important to know about the form of the ground, but also about the nature of the features mapped, their material composition, and the environmental processes associated with them. Thus forms need to be interpreted and classified, and *geomorphological mapping* (see section 3 below) provides an efficient method for achieving this goal. The chief aim of a geomorphological map is to provide a statement of the location of different forms and to indicate their respective age, origin, dimension, and material composition. Thus a morphological map shows the shape of the surface, and a geomorphological map includes an interpretation of that shape.

The practical applications of geomorphological mapping have been demonstrated in many environments, mainly in temperate Europe (Klimaszewski, 1956, 1961; Tricart, 1959, 1961, 1966; Fränzle, 1966; Verstappen and Zuidam, 1968; Verstappen, 1970; Doornkamp, 1971; Demek, 1972; Panizza, 1978). Some of the main applications in the context of urban development are listed in Table III.2. A similar list has been produced by Brunsden *et al.* (1975) for evaluating route alignments. In drylands, the authors have used detailed geomorphological mapping in projects related to urban development over a period of years, chiefly in Bahrain (Bahrain, Ministry of Works,

TABLE III.2

Some applications of geomorphological mapping in urban development

From General Geomorphological Maps	
1 General ground components	e.g. plains, valleys, dune fields
2 Data on slope steepness	e.g. shown in steepness classes
3 Data on natural hazards	e.g. landslides, dune migration, flooding
4 Data on existing man-made features	e.g. burial mounds, canals
5 Data on mineral/aggregate resources	e.g. clays, sands and gravels.
6 Data on neotectonic movements	e.g. signs of tectonic activity (fault scarps, etc.).
From Special Geomorphological Maps	
1 Slope steepness	e.g. as it will affect construction methods and costs
2 Slope stability	e.g. isolation of sites already unstable and those with a potential for failure
3 Soil erosion sites	e.g. sites liable to erosion in storms, and sites where sediment will accumulate
4 Sites liable to flooding	e.g. in relation to storms of differing magnitudes
5 Karst areas liable to subsidence	e.g. sites liable to collapse into caverns
6 Superficial materials	e.g. their origin, nature and predicted thickness
7 Watertable	e.g. location of springs, salinity conditions
8 Foundation conditions	e.g. variations in conditions for footings, aggressiveness of saline soils

Source: based on Demek (1972).

Power and Water, 1976), Dubai (Halcrow Middle East, 1977a), Ras al Khaimah (Halcrow Middle East, 1977b), and Egypt (Egypt, Ministry of Housing and Reconstruction, 1978).

With different geomorphologists working on a variety of problems in contrasting terrain conditions, it is scarcely surprising that several mapping systems have come into use (e.g. Geographical Research Institute of the Hungarian Academy of Sciences, 1963; Tricart, 1965; Verstappen and Zuidam, 1968; International Geographical Union Subcommission, 1968; Demek, 1972; Cooke and Doornkamp, 1974). They often use similar symbols, but differ in emphasis and detail (see, for example, discussion in Gellert and Scholz, 1964; Gilewska, 1967; and Dorsser and Salome, 1973); and attempts to produce unified systems (Demek, 1972; Demek and Embleton, 1978) do not seem to have been widely adopted.

2. *Morphological mapping*

Morphological mapping is among the most accurate methods for producing slope maps. It is based on the recognition in the field or on aerial photographs of breaks and changes of slope that delimit areas of uniform slope characteristics (i.e. rectilinear, concave, or convex slopes). Figure III.4 summarizes the method (see Savigear, 1965 for further details). Maps of this sort can be valuable in slope management and in planning land use because many activities cannot be undertaken satisfactorily on slopes above certain critical inclinations (Table III.3; Crofts, 1973). In addition, the method is most suitable for producing maps in the scale range 1 : 5 000 – 1 : 50 000, the range of greatest value in urban development.

3. *Comprehensive geomorphological mapping*

Although morphological mapping may in some instances be the first stage in geomorphological mapping, the recognition of breaks and changes of slope, the identification of slope units, and their classification may all be absorbed into the process of comprehensive geomorphological mapping. But the mapping of surface materials is more formal. The materials fall into two basic groups – solid rock or superficial deposits. All the deposits are *soils* to the engineer. The distinction between fresh and weathered bedrock is fundamental, and the boundary between the two (the *weathering front*) normally separates materials with quite different physical properties, and different influences on the retention, passage, and chemical composition of water. Mapping bedrock lithology and structure in the field is described in standard geology texts (e.g. Gilluly *et al.,* 1968; Longwell *et al.,* 1969), as is the mapping of geology from aerial photographs (Lueder, 1959; American Society of Photogrammetry, 1960; Miller, 1961; and Allum, 1966).

Many distinctive properties of both bedrock and lithology can be mapped in the field and recorded on a geomorphological map. Bedrock lithology can be recorded (e.g. granite, limestone, etc.) and structures such as fold axes and faults can be indicated. For soils (in the engineering sense) it is useful to display not only the spatial distribution of different types, but also the different grades of rock decomposition or types of transported soil. Such mapping, however, can be heavily dependent on available exposures, or on

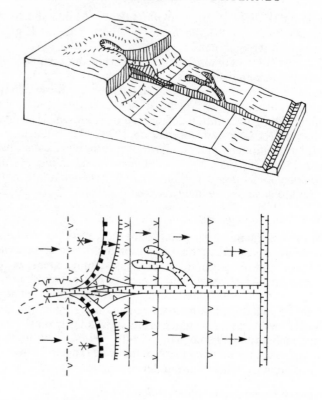

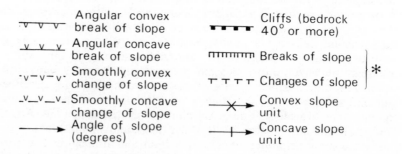

MORPHOLOGICAL   MAPPING   SYMBOLS

ᴠ ᴠ ᴠ	Angular convex break of slope	
ᴠ ᴠ ᴠ	Angular concave break of slope	
ᴠ–ᴠ–ᴠ	Smoothly convex change of slope	
ᴠ_ᴠ_ᴠ	Smoothly concave change of slope	
——→	Angle of slope (degrees)	

Cliffs (bedrock 40° or more)

Breaks of slope

Changes of slope

Convex slope unit

Concave slope unit

∗

∗ Convex and concave too close together to allow use of
separate symbols

FIG. III.4 A morphological mapping system (after Cooke and Doornkamp, 1974)

TABLE III.3

Critical slope steepness for specified activities

Steepness per cent	Critical for
1	International airport runways
2	Main-line passenger and freight rail transport
	Maximum for loaded commercial vehicles without speed reduction
	Local aerodrome runways
	Free ploughing and cultivation
	Below 2% – flooding and drainage problems in site development
4	Major roads
5	Agricultural machinery for weeding, seeding
	Soil erosion begins to become a problem
	Land development (constructional) difficult above 5%
8	Housing, roads
	Excessive slope for general development
	Intensive camp and picnic areas
9	Absolute maximum for railways
10	Heavy agricultural machinery
	Large-scale industrial site development
15	Standard wheeled tractor
20	Two-way ploughing
	Combine harvesting
	Housing-site development
25	Loading trailers
	Recreational paths and trails

Source: Crofts (1973).

the correct interpretation of tonal boundaries on aerial photographs. Many soil types can, however, be adequately mapped from their association with specific landforms identified from aerial photographs (see Table III.4). Such soil information is directly relevant to assessing the availability of local construction materials, to recognizing foundation hazards, and to problems of soil mechanics. The close coincidence of soil and landform boundaries is a characteristic of many drylands and, as a result, the geomorphological map may provide a preliminary substitute for a soils map until a comprehensive soil survey can be undertaken.

Of the several comprehensive mapping systems in use, that which provides the *basis* of most of the surveys carried out by the authors is described in Cooke and Doornkamp (1974) and has proved to be both flexible and durable. It comprises a simple classification of the features into bedrock lithology, geological structure, and landforms related to structure (drawn in purple); superficial sediments (undifferentiated, in black); features of volcanic origin (red); slope stability features (brown); man-made features (black); permafrost, glacial, and periglacial features (light blue); aeolian features (yellow); coastal features (green); fluvial phenomena (dark blue), and limestone features (orange). This classification and its associated mapping key has to be modified and extended to meet the specific needs of any particular survey. Figure III.5, for example, shows how the general

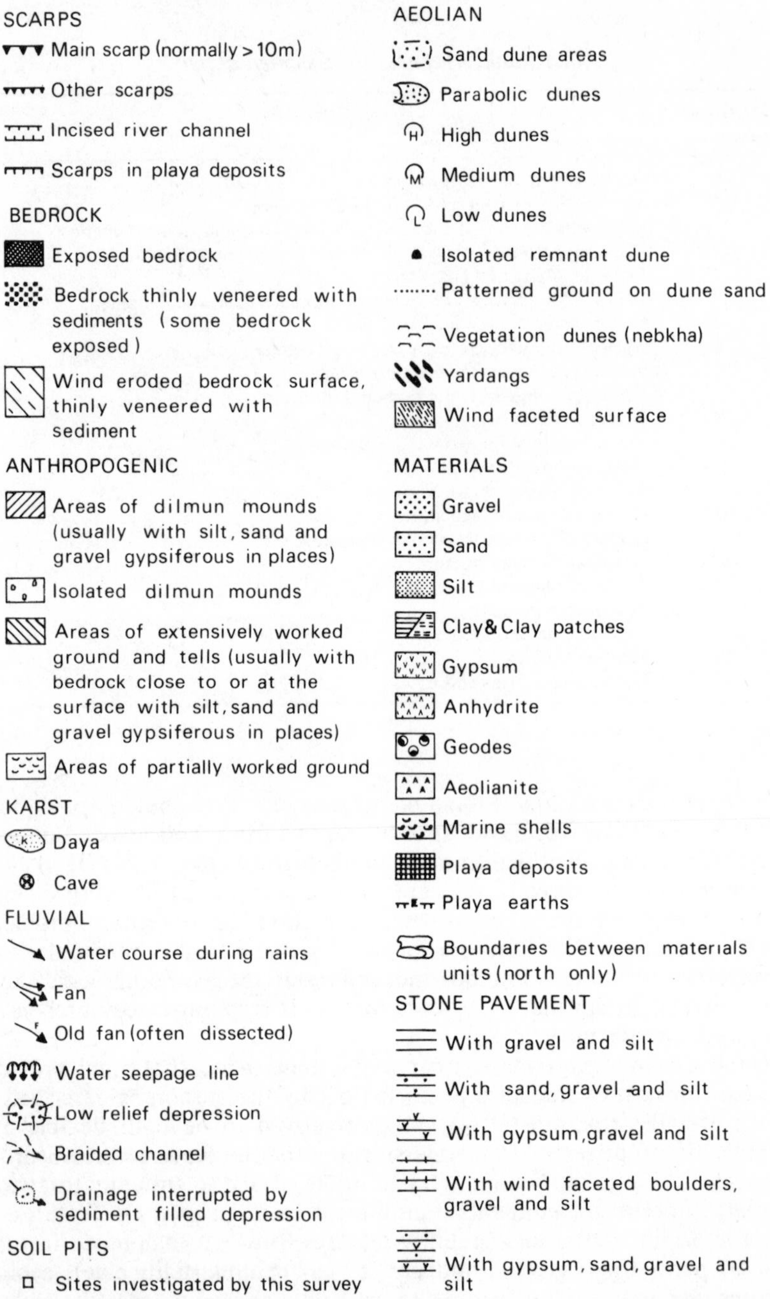

SCARPS

▼▼▼ Main scarp (normally > 10m)

▼▼▼▼ Other scarps

Incised river channel

Scarps in playa deposits

BEDROCK

Exposed bedrock

Bedrock thinly veneered with sediments (some bedrock exposed)

Wind eroded bedrock surface, thinly veneered with sediment

ANTHROPOGENIC

Areas of dilmun mounds (usually with silt, sand and gravel gypsiferous in places)

Isolated dilmun mounds

Areas of extensively worked ground and tells (usually with bedrock close to or at the surface with silt, sand and gravel gypsiferous in places)

Areas of partially worked ground

KARST

Daya

⊗ Cave

FLUVIAL

Water course during rains

Fan

Old fan (often dissected)

Water seepage line

Low relief depression

Braided channel

Drainage interrupted by sediment filled depression

SOIL PITS

□ Sites investigated by this survey

AEOLIAN

Sand dune areas

Parabolic dunes

High dunes

Medium dunes

Low dunes

● Isolated remnant dune

········ Patterned ground on dune sand

Vegetation dunes (nebkha)

Yardangs

Wind faceted surface

MATERIALS

Gravel

Sand

Silt

Clay & Clay patches

Gypsum

Anhydrite

Geodes

Aeolianite

Marine shells

Playa deposits

Playa earths

Boundaries between materials units (north only)

STONE PAVEMENT

With gravel and silt

With sand, gravel and silt

With gypsum, gravel and silt

With wind faceted boulders, gravel and silt

With gypsum, sand, gravel and silt

FIG. III.5 Geomorphological mapping legend used in the drylands of Bahrain (Bahrain, Ministry of Works, Power and Water, 1976; Doornkamp *et al.*, 1979). Fig. III.13 shows a portion of the map

scheme was adapted to accommodate the conditions found in Bahrain (Bahrain, Ministry of Works, Power and Water, 1976; Doornkamp *et al.*, 1980).

In addition to the geomorphological mapping systems mentioned above, geomorphological variables feature prominently in engineering geology mapping (Anon, 1972; UNESCO, 1976), mapping soil degradation (Mitchell and Howard, 1978), and in systems of hydrogeological mapping (UNESCO, 1970) – again revealing the common interest in, and the importance of, geomorphological features to scientists concerned with the earth's surface, and the crucial role geomorphological data can play in other surveys. For example, the background information necessary for a hydrogeological map is that of bedrock and superficial materials and river characteristics, all of which are identified on a geomorphological map. The nature of surface materials influences groundwater storage characteristics; the definition of watersheds is relevant to project planning; variations in soil properties are often closely coincident with the landforms identified on the geomorphological map. With a little additional fieldwork a geomorphological map can soon be converted into one showing hydrogeology.

In brief, what is required most from a geomorphological mapping system is flexibility to adapt to the landscape being mapped, the scale of mapping, and the problems posed by any given urban development.

4. *Other methods of recording geomorphological data*

Although geomorphological mapping has to date been the most widely used technique for recording land-surface characteristics, other methods are also in use. For example, a relief profile across an area of special interest may provide a good expression of the ground undulations. Such profiles are additionally useful if information concerning either soils, bedrock, or drainage conditions are plotted on them. They are particularly valuable for the analysis of slope stability in landslide studies.

Another useful method for recording geomorphological site information is a well-designed pro-forma. This acts not only as an easily filed data source, but also as a check list for the field investigator enabling him to make sure that he has examined all the characteristics of the site that have a bearing on the problem in hand. Such information normally includes details on slope form and steepness, position on the hillside or along the valley floor, soils, bedrock, drainage, landform processes, landform genesis, and so on. Special attention is normally given to site hazards or particular resources that may affect the development project.

5. *Stages in a geomorphological mapping programme*

A geomorphological mapping programme generally follows a series of well-defined stages (Fig. III.6). It is not always possible to obtain completely satisfactory results at each stage (e.g. aerial photographs may only exist at an inappropriate scale). The consequences of inadequate conditions at any one stage tend to have repercussions at each following stage, leading to inefficiency, inaccuracy, or increased costs. Rarely are all conditions fully

Fig. III.6

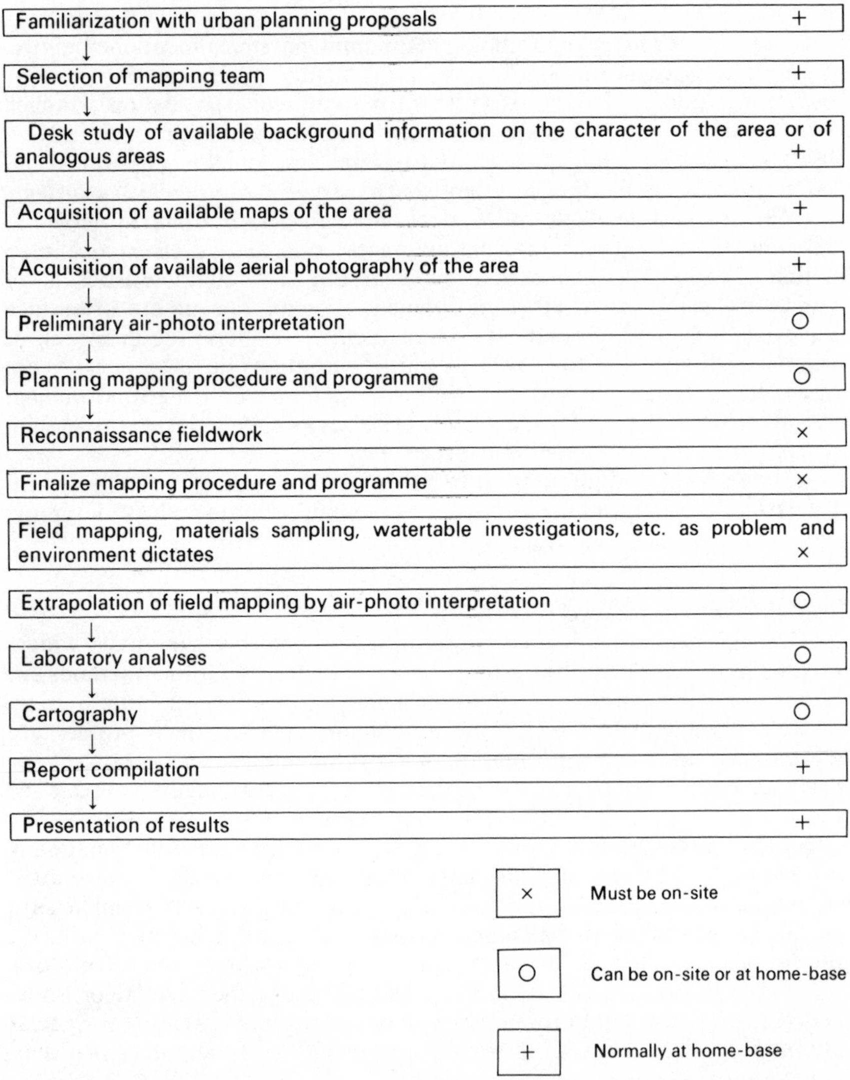

Familiarization with urban planning proposals	+

↓

Selection of mapping team	+

↓

Desk study of available background information on the character of the area or of analogous areas	+

↓

Acquisition of available maps of the area	+

↓

Acquisition of available aerial photography of the area	+

↓

Preliminary air-photo interpretation	O

↓

Planning mapping procedure and programme	O

↓

Reconnaissance fieldwork	×

Finalize mapping procedure and programme	×

Field mapping, materials sampling, watertable investigations, etc. as problem and environment dictates	×

Extrapolation of field mapping by air-photo interpretation	O

↓

Laboratory analyses	O

↓

Cartography	O

↓

Report compilation	+

↓

Presentation of results	+

×	Must be on-site
O	Can be on-site or at home-base
+	Normally at home-base

FIG. III.6 Stages in a geomorphological mapping programme for urban expansion in drylands

satisfied, especially in dryland surveys. On occasions, the ingenuity of the mapper can help to overcome some of the problems.

(i) Familiarization with urban planning proposals. Although the task of the geomorphologist in a survey for urban development is to provide basic environmental data, he is often faced with the decision as to whether or not

certain aspects of the terrain should be included in the survey. Relevance can only be judged in the light of the purpose of the survey and, even more important, the client's wishes: full briefing by the client is therefore essential. In addition, part of the geomorphologist's task is to abstract data from the original survey pertinent to particular problems or stages of development: clearly, planning of the survey must take its subsequent uses into account, reinforcing the need for a proper briefing

(ii) Selection of the mapping team. Under most of the circumstances experienced by the authors, geomorphological data are required quickly, and the environments to be mapped are usually complex or large, or both. Thus a mapping team is essential, especially in deserts where climatic conditions can be extremely arduous (remembering that much of the work is done on foot), and where it is impractical or unsafe for the mapping to be done alone. A minimum team of four is usually desirable, augmented by additional staff if large areas are to be mapped, great speed is required, or additional skills are necessary.

(iii) Desk study. The availability of background information for a geomorphological survey is often extensive (see Chapter II, section (d) above). A study of available data often helps to focus attention on the features most likely to be of critical importance in an understanding of the area to be examined; it also contributes to formulating the mapping legend. Comparative studies may provide a basis for estimating rates of process operation. The desk study also allows the general setting of a location to be established, especially with respect to climatic, terrain, and geological conditions (e.g. by reference to the work of Perrin and Mitchell, 1969, 1971; see section (d) below for an example).

(iv) Map and aerial photograph aquisition. These data sources have been discussed above (Chapter II). Of the two types, aerial photographs are undoubtedly the most important. If there is sufficient time and resources, one of the first technical steps in a project will be to commission an air-photo survey, and the production from them of ortho-photo maps and plans carrying contours obtained by a combination of photogrammetric methods and some primary ground control. Such maps not only provide a data source for the geomorphologist, but are also invaluable for field mapping and for control during map compilation. In addition, they can be invaluable to both the planner and the engineer at every subsequent stage in the urban development programme.

(v) Air-photo interpretation. Since the camera cannot distinguish the relevant from the irrelevant, it is essential that a preliminary interpretation of the aerial photographs is carried out by geomorphologists before fieldwork begins in order to gain an insight into the character of the terrain. Stereoscopic examination is highly desirable, and techniques of air-photo interpretation are well documented (e.g. Tator, 1958; Lueder, 1959; Verstappen, 1959, 1977; American Society of Photogrammetry, 1960; Miller, 1961, 1968; Fezer, 1971 (which contains a useful bibliography)). Some studies are specifically concerned with air-photo analysis in drylands (e.g. Smith,

1969; Ehlen, 1976). The interpreter is, of course, dependent largely on the elements of photographic pattern, as exemplified in Table III.4. The stages in the geomorphological survey when aerial photographs are particularly valuable are listed in Table III.5. Aerial photographs can be obtained in a variety of forms. The most common is panchromatic (black-and-white) prints (9" × 9" – 22.86 cm × 22.86 cm), but monochromatic transparencies and colour (true and false colour) prints may also be obtainable. Increasingly, other forms of imagery, such as infra-red and side-scanning radar images, may be available, each of which have particular applications and potential; but none is as generally useful as panchromatic or colour prints. Satellite imagery is as yet of unproven value except as a means of regional extrapolation of large landforms for the purposes listed in Table III.5. Its value will increase with improved resolution when it may replace aerial photography in all but stereoscopic analysis. Spectral bands of particular value in drylands include 400 – 525 nm and 600 – 675 nm for sand deserts, 600 – 675 nm and 750 – 850 nm for cultural areas, and 400 – 500 nm and 620 – 680 nm for piedmont slopes.

(vi) Planning mapping procedure. Preliminary air-photo analysis usually allows sensible planning of the nature and amount of mapping to be done. During this stage it is important to decide equipment requirements and the logistic support needed. Given that fieldwork in deserts can be very arduous, especially in high temperatures, equipment should be kept to a minimum. Logistic support normally required includes accommodation, suitable vehicles and, if possible, manual labour.

(vii) Reconnaissance fieldwork should be undertaken by a member of the team, whenever possible, in order to meet other personnel linked to the development project, ensure that all the required logistic support is available, check the feasibility of the mapping programme, and check the appropriateness of the mapping legend devised from air-photo interpretation. In this way adjustments to the programme can be made before the whole team is committed to being in the field area. The need for a reconnaissance visit is not always appreciated; but it can in fact do much to reduce costs and ensure the success of the geomorphological mapping programme.

(viii) Field mapping comprises systematic recording of all pertinent geomorphological characteristics of the area, and it is only successful if it involves close inspection of the terrain, usually be walking across it – vehicular travel may be useful in large, homogeneous areas, but it can easily lead to the overlooking of significant facts. Experience suggests that mapping in pairs is the most efficient way of covering the ground, with one person concentrating on maintaining correct location and plotting observations on the base map or aerial photographs, the second investigating specific features, and exposures and making field notes, and both sharing ideas and developing working hypotheses. Field information is transferred to a master map, preferably each evening, until the whole area has been covered. If additional and specific information is required for a particular survey (e.g. depth of water table, evidence of recent dune movement, etc.), the general survey

Elements in an aerial-photograph image used in soil identification

	Elements of form			Elements of tone and texture		Profile identification of mapping units	Potential as source of construction materials
Topographic form	Drainage form	Erosional form	Vegetation	Land-use	Surface material		
plains and gentle slopes	poorly developed and non-integrated	U-section donga erosion along cattle tracks	scrub and bush	pasture	dark grey tone from red soil colour	W/C/S	Aeolian clayey sand (W) is a possible source of binder
plains and gentle slopes	poorly developed and non-integrated	U-section donga erosion along cattle tracks	scrub and bush	pasture	dolerite weathering spheroids visible under high magnification	W/D_R	Aeolian clayey sand (W) is a possible source of binder
gentle slopes	well-integrated sub-parallel	sheetwash and flattened V-section gulley erosion	thornbush along gulleys	partly cultivated	white tone; light grey in poorly drained depressions	H/C/S	Not suitable for upper layers, base or sub-base because of high clay content of hillwash derived from dolerite and shale
steep slopes	radial	rill erosion in shale where talus thin or absent	dense thornbush	—	light grey where talus cover thin, grading into dark grey towards foot slopes	T_D/S	Suitable sub-base material may be located within the talus deposit
closed depressions	nil	small gulleys	absent	—	dark grey tone	V/W/C/S	Unsuitable for any construction material because of clay content
plains and gentle slopes	—	sheetwash	scattered thornbush	—	mottled owing to changes in moisture content and soil texture	H/W/C/S	Generally unsuitable for construction because of variable nature
low-lying plain	—	slight gulley development	absent	partly cultivated	mottled owing to changes in moisture content and soil texture	H/S	Generally unsuitable because of plastic properties
partly rounded hillocks	joint-controlled	joint-controlled	scanty	—	light grey tone	D_R	Weathered dolerite below the residual soil may provide possible source of base or sub-base material which may require lime stabilization
hilltops	joint-controlled	joint-controlled	scanty	—	light grey tone	D	Source of hard aggregate for concrete or surfacing stone after crushing; or as crusher-run for base-course
ridge-like erosion remnants	co-linear	rill erosion	scattered bush	—	light grey tone	S	Might be used for sub-base where indurated, i.e. near dolerite intrusions

Source: Brink and Williams (1964).

TABLE III.5

The uses of aerial photographs in geomorphological surveys of drylands for urban development

---

A. Before fieldwork
    1. General familiarization with ground conditions
    2. Tentative interpretation of geomorphological features
    3 First draft of mapping legend
    4 Planning site access and survey routes
    5 Planning team mapping responsibilities
    6 Production of ortho-photo maps*

B. During fieldwork
    1 Use as base maps for field survey
    2 Identification of the causes for variations in photographic tones
    3 Identification and investigation of apparently anomalous areas

C. After fieldwork
    1 Extrapolation of field mapping into areas not visited (because inaccessible or remote)
    2 Extraction (by photogrammetry) of dimensions and attributes thought to be significant for the project

---

* carried out by an outside agency

invariably enhances the understanding of such information which can often be collected during the course of the geomorphological survey.

(ix) Extrapolation by air-photo interpretation of mapped features into unmapped areas may be both possible and desirable, especially because it can increase the rate at which field mapping (and field checking) can be accomplished. Occasionally a late decision may require a post-fieldwork desk study in which mapped boundaries are extended into adjacent areas, using aerial photographs only.

(x) Laboratory analyses are normally required of rock, soil, and water samples collected during fieldwork to establish (or confirm) such things as field identification, mineralogical content, grading curve characteristics, and rock durability. In each case, the tests tend to serve a twofold purpose: they make map information more specific and precise, and they illuminate particular problems of interest later in the development programme.

(xi) Cartography. In terms of production, a neat manuscript map often lags behind the field programme. Indeed, the final geomorphological map may not be completed until much later. Unfortunately, the client frequently requires derivative maps quickly – maps that he understands and can use – so that the later stages in the investigation programme are not held up. Time therefore has to be spent in abstracting such maps from the field maps even before the completed geomorphological map can be tidily presented. Derivative maps may include those showing local resources (e.g. aggregates for use in construction) and hazards (e.g. flooding).

(xii) Report compilation is necessary in order to present information that may not be suitable for cartographic presentation. Such information (e.g. borehole and trial-pit records, laboratory data), together with descriptions

and explanations of the map, must be recorded in the report. In addition, the mapping programme frequently leads to ideas and recommendations about the relationship between development plans and the landscape, and this information is often eagerly sought from the report by the client.

(xiii) Presentation of results should be given as a verbal report to *all* who might be concerned with using the data in the later stages of a development project. This verbal report should be accompanied by displays of the maps, colour photographs of the features discussed, and all the derivative maps. Interpretation of the maps and suggestions as to how the data should be used must be provided. This is, in effect, a de-briefing session. In practice, geomorphological surveys usually provide more data than can be absorbed by the planner or engineer in a single de-briefing session. The critical next step, therefore, is to ensure a continued close liaison between the geomorphologists and those concerned with subsequent stages of the development programme, so that all of the available data can be used as efficiently and effectively as possible.

(d) Geomorphological Mapping at Different Scales

1. *Introduction*

The preceding sections have indicated the need to think and work at a variety of different scales, ranging from 1 : 5 000 to 1 : 2 m, for effective problem solving. Also, it is often appropriate to adopt different scales of investigation at different stages in the planning process. Thus at the scale of national planning, surveys (maps) at 1 : 100 000 – 1 : 1 m are often appropriate but site-specific studies involve an intensive survey with plans normally in the scale range 1 : 2 500 – 1 : 10 000. There is a danger here; features mapped at smaller scales still have to be correctly identified, and this usually requires field assessments that may be quite detailed. Thus a distinction has to be made between the scale of the final map and the detailed work that may be required to produce that map. In some cases the over-view presentation may only be possible by generalization from larger-scale studies; in other cases it may be possible to establish generalizations before embarking on large-scale studies.

Geomorphological surveys and mapping have been carried out in various parts of the world's drylands at each of the survey levels shown in Fig. III.7. The following review looks at some of these and at some of their implications for urban planning in drylands.

2. *National planning scales (normally c. 1 : 100 000–1 : 1 m)*

National landform maps exist for a few countries (e.g. France, UK, Hungary, Poland, Switzerland, Papua New Guinea) but there is none for any dryland state except Bahrain, which has a comprehensive survey at the unusually large scale of 1 : 10 000 (Bahrain, Ministry of Works, Power and Water, 1976; Doornkamp *et al.,* 1980). The national survey of Hungary has

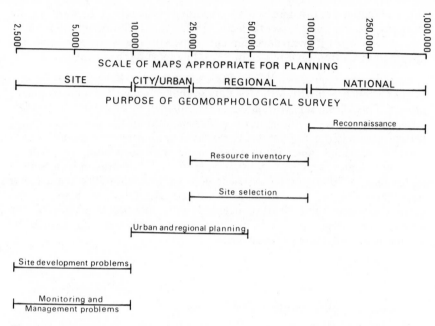

FIG. III.7 Relations between scale and purposes of geomorphological mapping

a map of 1 : 1 m (Peçsi, 1970) that is a good example of what might be achieved in many dryland countries. On the published map a fundamental distinction is drawn, by skilful use of colours, between different types of mountains, hills, and plains. Important details are then provided within each of these on minor landforms classified in general either by the processes that produced them or by the materials of which they are composed. The map is the culmination of several years of research: it is a synoptic summary of accumulated detail and experience.

Most drylands have no such history of geomorphological investigations, and so the approach – at least at present – has to be different. National maps have to be compiled by beginning with the general and working towards the particular. Small-scale examples are those of Mitchell and Perrin (1967) and Perrin and Mitchell (1969, 1971), Glennie (1970), and Mitchell and Howard (1978). In these cases the studies and maps were not limited by national boundaries, and both were based on the interpretation of aerial or satellite imagery with some ground control. Mitchell and Perrin (1967, and Perrin and Mitchell, 1969, 1971, and Mitchell, 1973) subdivided the arid lands into separate *land systems* (see section (b)), each of which is internally relatively homogeneous with respect to its landform characteristics, and maps showing land system boundaries were produced at scales ranging from 1 : 250 000 to 1 : 1 m. A subdivision of the land systems was achieved by recognizing and describing assemblages of smaller units mappable at a scale of 1 : 100 000 and called *land facets*. At such a scale, each land facet contains internal

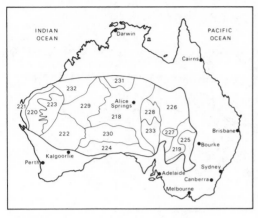

MOUNTAINS
218   Crystalline and metamorphic
HILLS
219   Undulating country–foliated rocks
PIEDMONTS
220   Erosional–crystalline
221   Depositional–mainly acidic
PLATEAUX
222   Slightly dissected–crystalline or
       metamorphic
224   Slightly dissected–bare limestone
223   Moderately dissected–crystalline or
       metamorphic
PLAINS
225   With inselbergs–sandy
226   Without inselbergs–sandy
AEOLIAN AREAS
227–232 Ergs
ENCLOSED DEPRESSIONS
233   Tectonic or fold–internal drainage and salt
       lakes

FIG. III.8 Physiographic regions in arid Australia (after Perrin and Mitchell, 1969)

variability not shown by the mapping, but at the level of national planning this is generally unimportant. Figure III.8 shows a modified extract for the drylands of Australia from Perrin and Mitchell (1971). This scale of mapping, while useful to national-scale planning, is of little direct value to urban planning, but it can form the basis for more detailed classification of terrain. For example, land-systems mapping in the Alice Springs area (Perry *et al.*, 1962) has divided the area into 88 land systems based on geomorphological divisions. Figure III.9 shows an example of the description of ground conditions, soil, and bedrock accompanying one such land system. Glennie (1970), on the other hand, in his study of the Oman Peninsula, identified specific landforms (e.g. coastal sabkhas, sand-dune areas, etc.) and thus more closely approaches, at a national scale, the mapping system that might be used at the stages of city or site planning (Fig. III.10). Despite these studies, the data available at the national scale are still grossly inadequate for most countries and will remain so until further geomorphological surveys are commissioned.

The land-systems approach, as modified for engineering purposes (Aitchison and Grant, 1967), has been applied by Grant (1972) to the Melbourne area of Australia specifically as a basis for urban and rural planning. This work originated from preliminary studies within the Town and Country Planning Board, Victoria, on the problem of land classification for urban growth. The published map (Grant, 1972) is at a scale of 1 : 250 000, and provides the regional setting for any proposed urban developments within an area of about 130 km × 110 km. The area is subdivided into major *provinces,* defined by bedrock geology, which are further subdivided on the map into *terrain patterns* defined by landforms. These are the units for which topography, soil, land use, and vegetation are described in the published memoir. It is essentially a catalogue, a data store, without a feeling for the dynamics of the environment, but nevertheless it is a helpful compendium of environmental information for planning purposes.

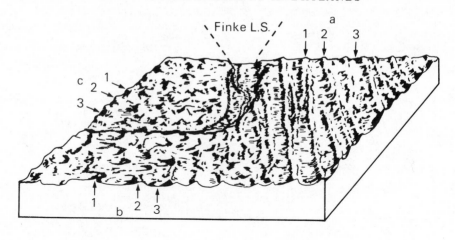

SIMPSON LAND SYSTEM (37 800 SQ MILES)

Spinifex-covered sand dunes   mainly in the southern half of the area.

Geology. Quaternary aeolian sand. Overlying Pre-Cambrian metamorphic and igneous rocks and sedimentar rocks of Proterozoic to Quarternary age.

Geomorphology. Depositional surfaces: dune fields of three main types; (*a*) parallel linear dunes in the Simpso Desert and in the west of the area, up to 70 ft high, $\frac{1}{4}$–$\frac{3}{4}$ mile apart, and orientated NNW.–SSE. or westerly with flat, mainly sandy swales; (*b*) reticulate dunes up to 40 ft high, made up of braiding sand ridges c connected smaller dunes, and typical of the desert margins; (*c*) irregular or aligned short dunes.

Water Resources. Groundwater prospects are extremely variable because of the wide variety of underlyin rocks. Sub-artesian water is available from depths of 300 to 1000 ft in the Simpson Desert. Elsewhere variabl supplies of good-quality to saline water are available at varying depths from alluvium, sandstone, limeston and fractured metamorphic aquifers. There is little prospect of obtaining supplies of ground water on th Missionary Plain and to the north of Deep Well.

Climate. Comparable climatic stations are Charlotte Waters, Henbury, Hermannsburg, Alice Springs, Te Tree Well, and Barrow Creek.

Unit	Area	Land Form	Soil	Plant Community
1	Small	Dune crests: hummocky surfaces locally unstable and eroding attaining up to 20 ft above unit 2; slopes generally above 25%, with small slip-faces exceeding 60% on eastern and northern sides	Red sands (3*f*)	*Zygochloa paradoxa* (desert cane grass)
2	Large	Dune flanks: smooth, stable slopes, up to 300 yd long and attaining 20% on the west and south, up to 150 yd long and attaining 35% on the east and north		Sparse shrubs and low trees, or mino *Casuarina decaisneana* (desert oak) ove *Triodia basedowii* (spinifex), minor *Plec trachne schinzii* (spinifex), *Aristida brownian (kerosene grass), or *Eragrostis eriopod* (woollybutt)
3	Medium	Swales: flat or concave surfaces up to 400 yd wide; mainly sandy, with small clay pans, minor calcrete exposures, few drainage channels, and minor alluvial flats associated with through-going drainage	Mainly red clayey sands (3*d*), locally red earths (4*d*), texture-contrast soils (7*e*), calcareous earths (6*c*), and shallow sandy soils (3*b*) over soft calcareous and/or gypsiferous deposits ('kopi')	Mainly as unit 2. Minor variable includin *A. aneura* (mulga). *E. microtheca* (coolibah) or sparse low trees over *Bassia* spp., *Arthro nemum* spp. (samphire), or *Atriplex nummu laria* (old-man saltbush)

FIG. III.9 Extract from the land systems report on the Alice Springs area, Australia, showing summary of data for one land system (after Perry *et al.*, 1962)

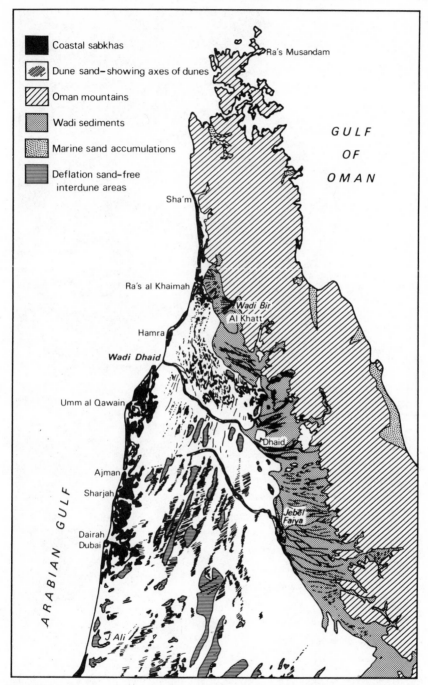

FIG. III.10 Geomorphology of the northern part of the Oman Peninsula (after Glennie, 1970)

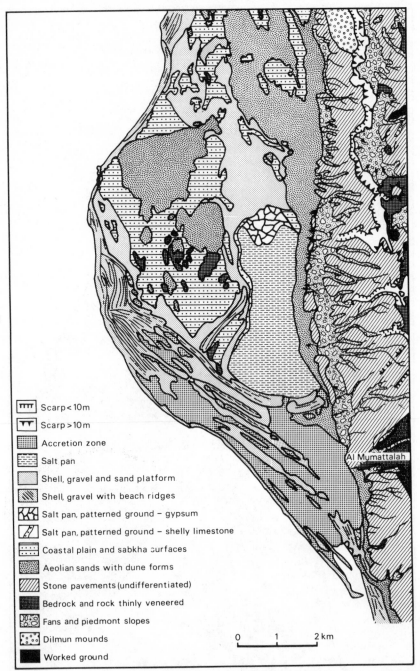

Scarp < 10 m

Scarp > 10 m

Accretion zone

Salt pan

Shell, gravel and sand platform

Shell, gravel with beach ridges

Salt pan, patterned ground – gypsum

Salt pan, patterned ground – shelly limestone

Coastal plain and sabkha surfaces

Aeolian sands with dune forms

Stone pavements (undifferentiated)

Bedrock and rock thinly veneered

Fans and piedmont slopes

Dilmun mounds

Worked ground

Al Mumattalah

0   1   2 km

FIG. III.11 Extract from the geomorphology and superficial materials map of Bahrain at a scale of 1:50 000 (after Bahrain, Ministry of Works, Power and Water, 1976 and Doornkamp *et al.*, 1980)

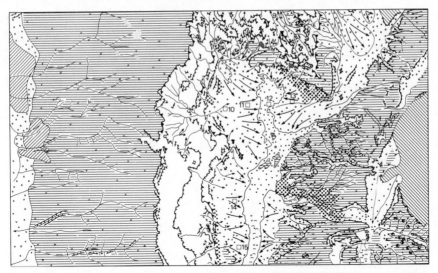

FIG. III.12 Extract from the geomorphology and superficial materials map of Bahrain at a scale of 1:10 000. For key, see Fig. III.5 (after Bahrain, Ministry of Works, Power and Water, 1976 and Doornkamp *et al.*, 1979) □ = soil pit sites

3. *Regional planning scales (1 : 25 000–1 : 100 000)*

The national survey of Bahrain, published at 1 : 50 000 (Bahrain, Ministry of Works, Power and Water, 1976; Doornkamp *et al.*, 1980), falls into the range of scales appropriate for regional planning, and exemplifies the type of geomorphological information that can be shown at such scales (Fig. III.11). The original field survey was carried out at a scale of 1 : 10 000 and the detail so obtained was reduced in compiling the 1 : 50 000 map. The map (Fig. III.11) identifies separate sedimentary types (and hence different materials) of the coastal area. And it clearly shows some areas that are entirely unsuitable for construction purposes (e.g. salt pan, sabkha surfaces), other areas that are possibly potential sources of fine aggregates for use in building (e.g. aeolian sands, fans, and piedmont slopes) and worthy of further specific investigations, as well as bedrock areas that have a greater attraction as sites with good foundation conditions for building development. In short, such a map is a first stage in delimiting sites suitable (or indeed completely unsuitable) at the regional planning scale for various types of land-use, and thus helps in the planning of an urban development programme (Brunsden *et al.*, 1979). Further details of the character of the different areas would normally occur in the accompanying reports.

Regional surveys with maps produced in the range 1 : 25 000–1 : 100 000 have also employed the land-systems approach, especially when land units are identified. The surveys by the department of Geography, Durham University, in the Arabian Gulf and Oman provide excellent examples (Fisher and Bowen-Jones, 1974). Another such survey is that of the Kyalami

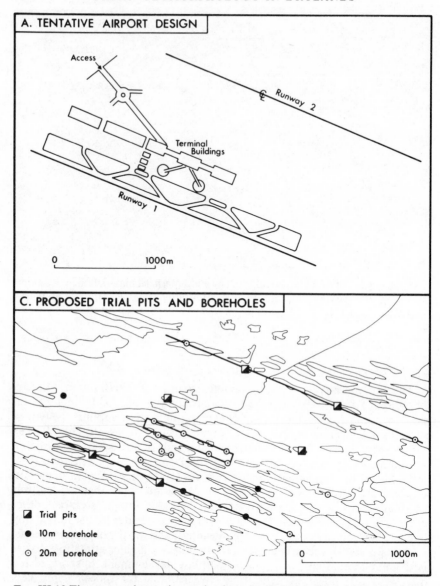

Fig. III.13 The proposed new airport site, between Dubai and Jebel Ali. (A) Airport plan; (B) Geomorphology of the site; (C) Proposed trial pits and boreholes; (D) Legends for (B) (after Halcrow Middle East, 1977a and Doornkamp *et al.*, 1979)

area near Johannesburg (National Institute for Road Research 1971; see also Fig. III.2), which was designed to act as basis for soil classification for engineering purposes. Similar surveys have been carried out elsewhere in South Africa (e.g. for a trunk-road construction programme in the drylands of the northern Cape Province, South Africa – Brink and Williams, 1964).

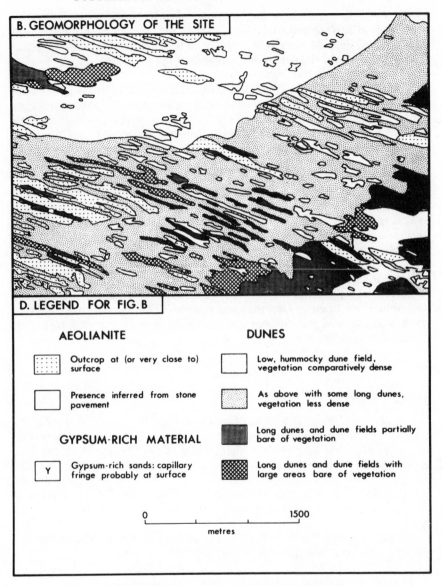

B. GEOMORPHOLOGY OF THE SITE

D. LEGEND FOR FIG. B

AEOLIANITE

Outcrop at (or very close to) surface

Presence inferred from stone pavement

GYPSUM-RICH MATERIAL

Y Gypsum-rich sands: capillary fringe probably at surface

DUNES

Low, hummocky dune field, vegetation comparatively dense

As above with some long dunes, vegetation less dense

Long dunes and dune fields partially bare of vegetation

Long dunes and dune fields with large areas bare of vegetation

0 —————————————— 1500
metres

4. City and land-use planning scales (1 : 10 000–1 : 25 000)

As emphasized previously, the physical character of a site is a particularly important consideration when decisions have to be made about site choice and layout for urban development. Land-unit maps can be compiled within the scale range 1 : 10 000–1 : 25 000, and can provide a useful basis for storing data of planning or engineering significance. A geomorphological mapping approach, however, provides a much better record of the dynamic elements of the landscape (e.g. mobile or unstable dune faces) and of variations in surface materials. Figure III.12 provides an example at a scale of 1 : 10 000, derived from field mapping in Bahrain (for key, see Fig. III.5). Maps prepared for the urban development of the Suez area, Egypt, provide additional examples (see section (e) below).

Vallejo (1977) also described how geomorphological mapping at this scale can form an integral part of an engineering geology investigation for urban planning. His map of part of Tenerife, which was originally produced at a scale of 1 : 25 000, stresses surface morphology and areas of instability, and isolates a few specific landforms.

5. *Site planning scales (1 : 2 500–1 : 10 000)*

Within areas of urban development, some sites may present special problems that require detailed geomorphological and, probably, geological mapping. A large landslide area provides such an example: here, in addition to the mapping mentioned, investigations by other specialists, such as a specialist in soil mechanics, may be essential. Normally, it may be more appropriate for the geomorphologist to give way at this stage to an engineering geologist with specialist training in the requirements of the site engineer and who can, for instance, better provide the geotechnical mapping techniques described by Dearman and Fookes (1974). Nevertheless, detailed morphological and geomorphological mapping may be a useful adjunct to other surveys at this scale.

(e) Derivative Maps

1. *Introduction*

The geomorphological map is, of course, primarily a device for portraying the geomorphology of an area, and its compilation ensures the representation of form, material, and process characteristics for the whole of the area examined. But the map may be complex, and the legend not readily understood by planners and engineers. This means that to be of real value the geomorphological map has to be used as a basis for the production of derivative maps, with each map showing pertinent information on a specific topic.

Such abstraction has been done by the authors in several dryland studies, and the following examples are taken from these studies.

2. *Planning geotechnical investigations at a new airport site.*

During discussions in Dubai over the building of a new international airport, two alternative sites dominated the proposals. One site lay between Dubai and Jebel Ali (Fig. III.13A); little was known about it except that it was a dune field that was apparently sufficiently well-vegetated for the dunes to be regarded as stable. Geomorphological mapping of the site (Fig. III.13B) showed the dunes to be very variable in character, with some of them potentially highly mobile (Halcrow Middle East, 1977a; Doornkamp *et al.,* 1979). In addition, other landforms occurred, notably inter-dune areas of low relief composed of aeolianite (a calcareous siliceous sandstone of old dune sands, often eroded by the wind), and some highly gypsiferous sands in low ground on the eastern border of the area. With the proposed plan for the airport layout in mind (Figure III.13A), it was possible from the geomorphological evidence to suggest a pattern of boreholes and trial pits for a site investigation (Fig. III.13C) that would provide an optimum amount of information about sub-surface conditions with a minimum of excavation. By comparing the proposed position of engineering structures with the geomorphology, it was thus possible to match the engineering needs for the information with the variability of the terrain. The proposed locations for site investigations on the derivative map (Fig. III.13C) are what the engineer requires – they are of more value to him than the original geomorphological map. At a later stage, when sub-surface information is available, the engineer can consult with the geomorphologist in order to interpret the site-investigation data and to determine the lateral extent of site conditions in the light of the geomorphological map.

3. *Hazard of migrating dunes at a new airport site*

The second proposed site for a new airport in Dubai lay south of Jebel Ali (Fig. III.14A) (Halcrow Middle East, 1977a; Doornkamp *et al.,* 1979). The geomorphological features of this site were similar to those at the first site, but occurred in different proportions. During the geomorphological survey, it became clear that some of the dunes within the airport site were mobile (Fig. III.14B). On the basis of detailed field examination, dune characteristics were recognized that could be related to the degree of mobility of the dunes and could also be recognized on the available 1 : 15 000 aerial photography. From these characteristics, hazard classes were defined (see Table VII.3) in which the high, angular fresh dunes without vegetation were recognized as being potentially or actively mobile. These were mapped on the windward side of the airport site as dunes in hazard category I (Fig. III.14C). In terms of stabilization work, these dunes could be completely removed where they are shallow or, where they are larger, could be extensively stabilized (Fig. III.14C). At the other extreme the broad, rounded, low and well-vegetated dunes provide a low hazard threat (category III) and hence only require moderate stabilization works (category III, Fig. III.14C). Thus the mapping of the dunes according to their form and density of vegetation allows the boundaries between dune types (mapped as potential hazard classes) to be used as boundaries between types, or classes of dune-stabilization work. In this way, the engineering can be directly related to the variations in the dune field that have a bearing on the operation of the airport.

4. *The expansion of Suez City, Egypt*

A geomorphological survey of the area to be occupied by the expansion of the city of Suez was undertaken by the authors, together with Dr P. R. Bush in 1977 as part of the Suez area sub-surface investigation for Sir William Halcrow and Partners (Egypt, Ministry of Housing and Reconstruction, 1978). Before the survey began, a master plan existed (Egypt, Ministry of Housing and Reconstruction, 1976) that suggested a layout of residential,

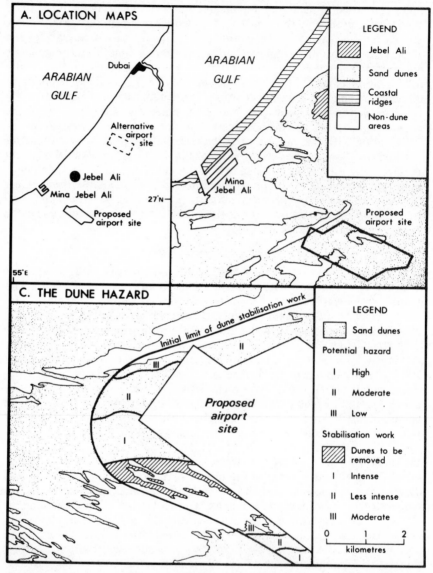

FIG. III.14 The second proposed airport site, south of Jebel Ali, Dubai (A) Location maps; (B) Details of dune morphology; (C) The dune hazard; (D) Legend and orientation diagrams (after Halcrow Middle East, 1977a, and Doornkamp *et al.*, 1979)

business, administrative, and recreational areas, but the planning of its details awaited more specific information about ground conditions. A subsurface investigation by means of trial pits and boreholes was being undertaken. Geomorphological mapping was carried out in order to (i) define

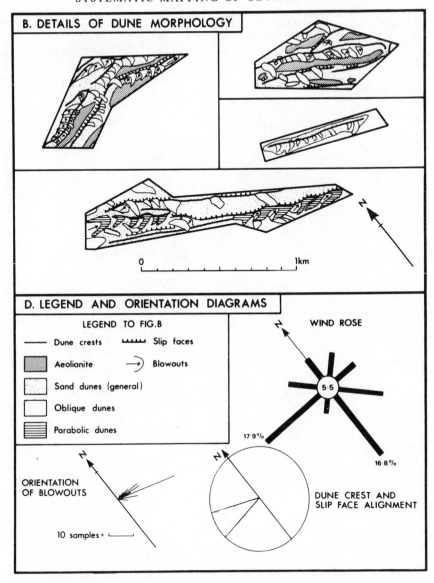

the lateral extent of soil and bedrock materials found in trial pits and boreholes, (ii) classify these materials according to their genesis as revealed by geomorphological and sedimentological analysis, (iii) assess the hazard to the site of periodic flooding, and (iv) assess groundwater conditions in the context of the salt-weathering hazard to foundations. The results of the survey, which involved 100 man-days of fieldwork, were: (i) synoptic maps at a scale of 1 : 25 000 of geomorphology, surface materials, drainage and flood hazard, and salt hazard to foundations (with written commentary), (ii)

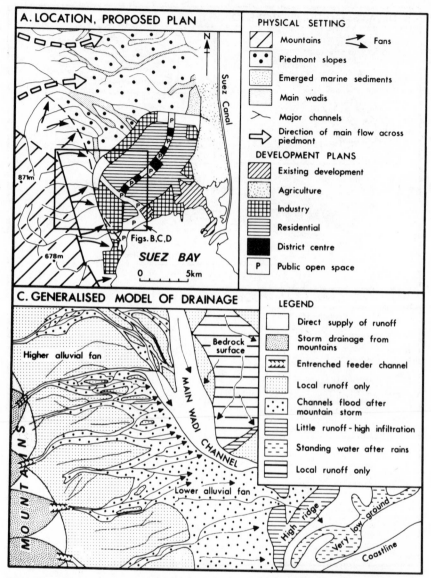

Fig. III.15 Geomorphological surveys of the Suez area, Egypt: flood hazard mapping (A) Location and proposed urban development; (B) Geomorphology; (C) Generalized model of drainage conditions; (D) Potential flood hazard (after Egypt, Ministry of Housing and Reconstruction, 1978 and Doornkamp *et al.*, 1979)

a report on the natural and artificial exposures of soil and bedrock examined, and (iii) an account of the mapped conditions for each planned development zone within the proposed city area. Some of the results of this study are

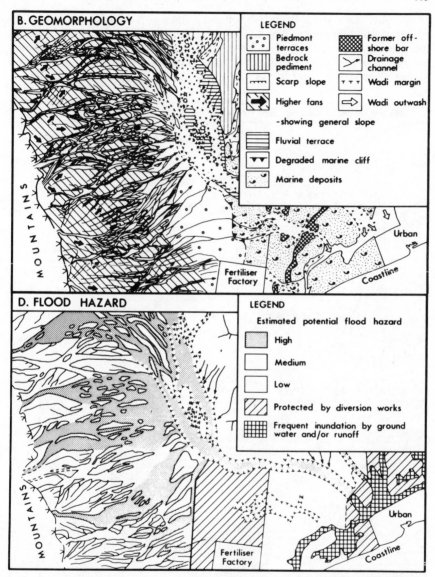

B. GEOMORPHOLOGY

LEGEND

Piedmont terraces
Bedrock pediment
Scarp slope
Higher fans
  -showing general slope
Fluvial terrace
Degraded marine cliff
Marine deposits

Former off-shore bar
Drainage channel
Wadi margin
Wadi outwash

Urban
Fertiliser Factory
Coastline
MOUNTAINS

D. FLOOD HAZARD

LEGEND

Estimated potential flood hazard

High
Medium
Low
Protected by diversion works
Frequent inundation by ground water and/or runoff

Urban
Fertiliser Factory
Coastline
MOUNTAINS

described elsewhere in this volume (Chapters V and VI); the examples used here illustrate the way in which an assessment of potential flood hazard was made and how borehole and trial-pit data were extrapolated to provide a map of bedrock and superficial materials. It should be emphasized, however, that the survey was based on geomorphological mapping and that each specific study (e.g. salt hazard) was undertaken by the mapping team *at the same time* as the geomorphological map was being prepared.

In the flood-hazard study it was essential not only to assess the geomorphology of the site itself but also that of its general situation. Flooding in the Suez area arises from two sources – high groundwater at the coastal margins, and from storm flow along wadi channels. In the latter case, storms are generated by rainfall over Jebel Ataka which reaches the lower ground around Suez along valleys that drain directly eastwards from the mountains across alluvial fans, and by way of northward drainage towards the Suez–

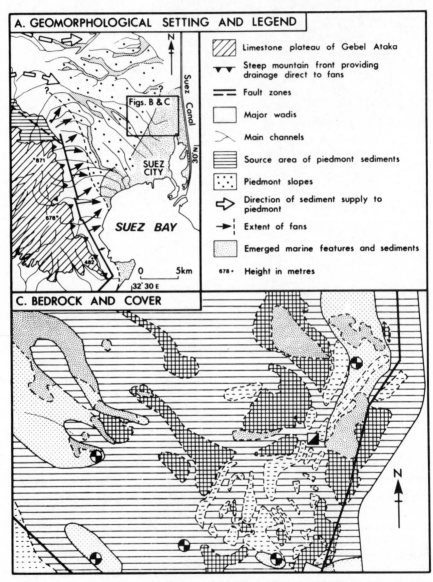

FIG. III.16 Geomorphological surveys of the Suez area, Egypt; application to mapping bedrock and superficial sediments (after Egypt, Ministry of Housing and Reconstruction, 1978 and Doornkamp *et al.*, 1979)

Cairo road, and thence eastwards into the area across piedmont slopes (Fig. III.15). Figure III.15C shows a generalized model of natural drainage conditions in terms of the landforms mapped during the geomorphological survey (cf. Fig. III.15B); it forms the conceptual basis for flood hazard assessment. From it the potential flood-hazard map is generated (Fig. III.15D). As a result of this work, suggestions were put forward for the more precise delimitation of construction sites so that they would avoid hazardous

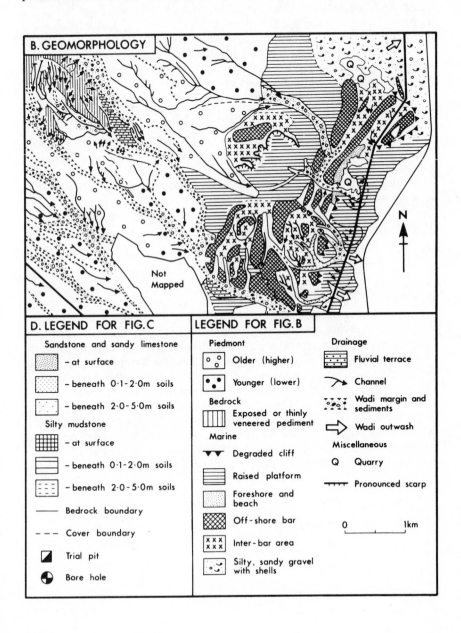

flood channels. In addition, flood-diversion barriers were proposed to protect certain development sites from alluvial-fan flooding.

The definition of mapped geomorphological features clearly has a strong bias towards the character of the soils and sediments that constitute the landforms themselves. This relationship was most useful when the bedrock and soil map of the area was compiled. In this task, the geomorphological team supplemented their field observations with borehole and trial-pit data. The boundaries between units on the bedrock geology and cover materials map are, however, very much dependent on the original geomorphological mapping; indeed, they could not have been established so quickly without it. Figure III.16B is an extract from the geomorphological map, and Fig. III.16C is an extract from the bedrock geology and cover materials map for the same area: the coincidence of boundaries between the two maps is clear.

The Suez study illustrates fully the value of geomorphological mapping as a basis for preparing derivative maps of value to the planner and developer. The whole analysis was compared with proposals set out in the master plan and recommendations were made for modification of the plan to avoid hazardous areas, and each development zone was described in terms of specific site conditions with cross-reference to the maps and to the general environmental context of each area. This allows more detailed planning to take place, taking account of the character and variability of the ground. In this way hazards can be anticipated and avoided, and natural advantages used in cost-effective planning.

(f) Summary

Land-systems survey and geomorphological mapping constitute a pair of techniques that are capable of yielding useful geomorphological information at a variety of scales relevant to urban planning in drylands. Both provide data summaries that are relatively inexpensive to prepare and relatively easy for environmental managers to comprehend. Land-systems mapping provides synoptic, generalized data summaries for discrete mapped regions. It is particularly pertinent to small-scale terrain evaluation in the early stages of the planning process, although it also facilitates hierarchical classification of landform areas and is thus also appropriate to larger-scale studies. Geomorphological mapping can provide a comprehensive geomorphological map at most practical scales, but it is especially suited to regional and town planning scales. It forms a basis for derivate maps of hazards and resources that are of direct value in the planning process.

IV

AGGREGATE RESOURCES FOR THE CONSTRUCTION INDUSTRY IN DRYLANDS

(a) Introduction

The natural materials required for the construction of new or growing urban aeas in drylands, as elsewhere, fall into four main categories; 'hard' rock, sand, gravel, and clay and monomineralic rocks (Fig. IV.1). Such minerals are generally of low intrinsic value but, because of their bulk and the large quantities required, they are expensive to transport over long distances. Thus they are normally exploited as close as possible to the place where they are to be used.

The geomorphologist can play an important role in the search for raw materials within and close to urban areas, chiefly because many of these materials occur as superficial sediments that are intimately related to landforms. The geomorphologist's role may be particularly important in

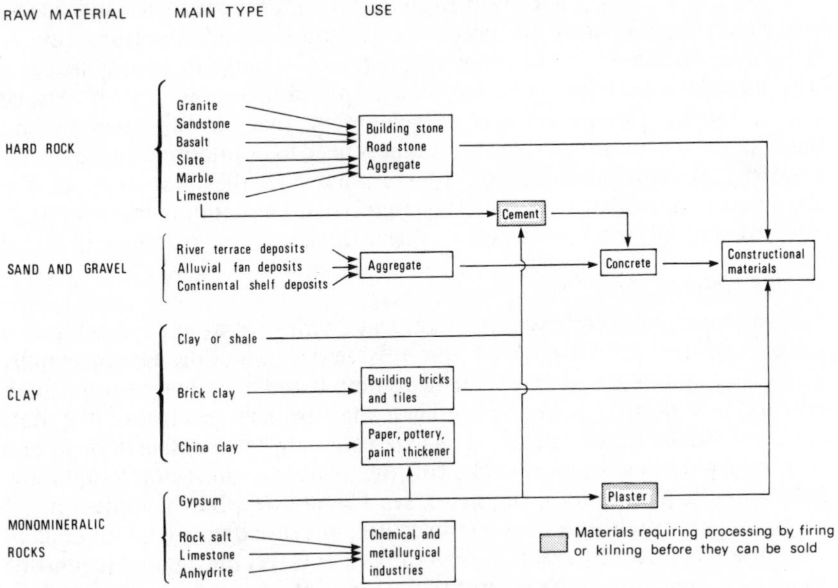

FIG. IV.1 Flow diagram to show the interrelationships between the major bulk materials used in the construction industry (after the Open University, 1974)

drylands where, to the surprise of some, suitable raw materials for construction purposes or of high quality are not always common (Allison, 1977; Fookes, 1977; Fookes and Higginbottom, 1980). It is in the search for aggregate resources that the geomorphological contribution is most valuable.

If the geomorphologist is to help in the search for raw materials it is of course essential that he is aware of the properties of the materials of importance, and of the quality specifications and standards required by developers. Thus, the first part of this chapter examines some of the properties, standards, and specifications of aggregate for the construction industry in drylands, placing emphasis on the situation in South Africa, which includes extensive drylands, and where the need for control of material quality in such environments has been recognized. This examination is followed by an appraisal of the geomorphological context of aggregate resources in drylands, and by case studies that illustrate geomorphological contributions to the search for aggregate in the Middle East.

(b) Material Properties, Standards, and Specifications

1. *Introduction*

The specific properties that render a natural aggregate resource suitable, suitable after treatment, or unsuitable, need to be determined either at the site of the potential resource or from samples taken to the laboratory for analysis. The criteria of suitability can, in the first instance, be taken as those currently being applied through published national 'standards' and 'specifications', if they exist (e.g. Libyan Arab Republic, 1975). If such standards do not exist, then it may be necessary to adopt standards from similar environments elsewhere, to make arbitrary assessments, or to establish as a priority standards relevant to the area in question. In any case, it may be shown during the progress of development that given standards are inadequate, and more rigorous controls may need to be imposed. Of particular concern to urban construction programmes are the properties of the aggregates – aggregate used for (i) concrete, (ii) other building materials (e.g. rip rap), and (iii) road construction. These three uses are examined in turn.

2. *Aggregate for concrete*

Good-quality concrete requires aggregate with several important characteristics (Harrison, 1970): (i) the compressive strength of the aggregate must be greater than that of the cement or lime-based bonding material with which it is to be mixed; (ii) the aggregate must be hard and free of materials which are likely to decompose or change volume when weathered or wetted (e.g. clay particles, pyrites, coal); (iii) the aggregate must not contain any substances (e.g. chlorides, sulphates, see Chapter V) that will affect metal reinforcing bars within a concrete structure, or cause damage by subsequent volumetric changes by reaction with the cement; (iv) the aggregate must be clean and free from organic impurities or coats of clay or dust as these prevent a proper bonding with the cement; (v) low absorption properties are required in aggregate, especially those used in reinforced concrete or exposed to liquids; (vi) angular materials make the best aggregate because they permit the establishment of a strong bond by interlocking between particles: (vii) aggregate particles should ideally not be too platey or elongated as this reduces their workability and bonding capability.

It is advantageous for the maximum size of the coarse particles to be as large as possible. The size and shape of aggregate particles may be established by the tests outlined in British Standard BS 812, section 1 or ASTM C33 in the USA. The permissible maximum size of the coarse particles is determined by the shutter (mould) size for unreinforced concrete, by the spacing of the bars in reinforced concrete (Table IV.1), and by consideration of the mix design for the concrete needed in a particular structure. Concrete properties can to some extent be controlled by using appropriate aggregate (e.g. Table IV.2). Tables IV.3 and IV.4 list the physical properties of material suitable for use in concrete as specified in South Africa. In general, a good concrete aggregate contains less than 3% 'fines' (i.e. particles < 0.06 mm diameter), and equal proportions of fine aggregate (0.06–4.0 mm) and coarse aggregate (4–40 mm) (see Table IV.5 and BS 882). Materials larger than 40 mm are usually discarded as they are unsuitable except for mass concrete work (Table IV.1). Particles of coarse aggregate that are elongated or flaky are less workable, and values for these two variables are specified in BS 812 and BS 63. Furthermore, the fine aggregate for concrete must be angular ('sharp') to promote interlocking, although fine aggregate for mortar or plaster may be 'softer'. In drylands, the presence of salts, principally as sulphates or chlorides, are an especially serious problem (see Chapter V for details).

The South African standards in Tables IV.3 and IV.4 represent a fairly comprehensive set of national guidelines. In many other dryland countries no such regulations exist and standards are at present often determined by contractors or consultants. For example, Fookes and Higginbottom (1980) identified the group of aggregate properties of importance in assessing Middle Eastern sources of aggregate for concrete, and proposed a schedule of tests and acceptable limits (Table IV.6). Each of the recommended tests is derived from British or American authorities.

Many of these requirements and desirable properties of aggregate for concrete have a direct relationship to geomorphological conditions in drylands. For example, the characteristically 'mixed' sediments of alluvial fans may provide excellent aggregate, but the distinctively rounded sand grains of aeolian deposits are often not sufficiently 'sharp' (see section (c) below).

3. *Aggregate for other building purposes*

(*i*) Heat insulation panels (e.g. as an inner leaf in a cavity wall or in a roof) are often composed of concrete made from lightweight aggregate. An aggregate is classified as lightweight if its bulk density is less than 1200 kg/m^3 in the case of fine aggregate, and less than 960 kg/m^3 for coarse aggregate (Teychenne, 1968; BS 3797; BS 877) (Table IV.7). The grading of such aggregates should conform to those shown in Tables IV.8a and IV.8b, and the density of the cured concrete must be less than 1600 kg/m^3. Tests for sulphate content, volatile matter, loss on ignition, and carbon content should

TABLE IV.1

Approximate maximum coarse aggregate size for different types of concrete

Concrete type	Maximum aggregate size
Un-reinforced concrete	$<\frac{1}{4}$ of least dimension of member
Reinforced concrete	20 mm
Densely reinforced concrete	10 mm
Mass concrete	up to 150 mm

Source: Harrison (1970).

TABLE IV.2

Properties of aggregate in relation to required properties of concrete

Property of aggregate	Property of concrete				Type of aggregate	
	weight	strength	ease of compaction	durability	sand and gravel	crushed rock
high density	x			x		x
good grading			x		x	
rounded fragments			x		x	
rough surface		x		x		x

Source: Open University (1974).

be carried out in accordance with BS 3681 and values should fall within the limits set out in BS 3797. Great care must be taken, however, over the choice of chemical testing methods: tests developed for conditions in temperate lands may be quite inappropriate for conditions in hot, dry lands. In particular, some tests are insensitive to high concentrations of salts, and hence may under-record the actual amounts present.

(*ii*) *Rip rap* is an important bulk construction material composed of large angular blocks of untrimmed natural rock or very large alluvial boulders. It is widely used as unbound stonework to dissipate the energy of breaking waves on eroding shorelines or along breakwaters, to reduce erosion along river banks or in estuaries (often in conjunction with a membrane), or as a revetment at the base of unstable slopes. Rock used for rip rap must be resistant to weathering in the environment to which it is moved, and a high resistance to abrasion is required if it is used for coastal or river-bank defences. In particular, resistance to chemical weathering (decomposition) is necessary in material used in the inter-tidal zone where alternate wetting by brine and drying by evaporation provides a particularly aggressive weathering environment (Fookes and Poole, 1978).

There are no standard specifications for rip rap materials; their suitability therefore needs to be assessed in the context of the environments in which they will be placed and the uses to which they will be put.

TABLE IV.3

Stone for concrete

1	2
Property	*Requirements*
Grading	The grading requirements given in Table IV.5 shall apply
Dust content, material passing a 75 μm sieve, % (m/m), max.	1.5
10% FACT value*, kN, min.† (a) Stone for use in concrete subject to surface abrasion (1) Dry (2) Wet	110 At least 75% of the determined dry value
(b) Stone for use in concrete not subject to surface abrasion (1) Dry (2) Wet	70 At least 75% of the determined dry value
Aggregate crushing value, (ACV), dry, %, max.†‡	29

*The term is derived from the title of the corresponding test, i.e. the 10% fines aggregate crushing test. Its use is preferred as it obviates confusion with the aggregate crushing value. (See SABS 842 or BS 812 for test technique.)

† Applicable to the minus 9.5–13.2 mm fraction of a stone only.

‡ This requirement is an optional alternative (that is applicable only to aggregates that have an ACV not exceeding 30%) that may, if acceptable replace the requirement for the 10% FACT value. (See SABS 841 or BS 812 for test technique.)

Source: South African Bureau of Standards 1083.

4. *Aggregate for road construction*

Modern surfaced roads are often designed in four layers – sub-base, base course, road base, and wearing course. Specifications, and hence aggregate requirements, vary in detail for each of these, but they have certain requirements in common. For example, all aggregate must be hard and clean, and all aggregate (especially in the sub-base and in the other unbonded layer, the base course) must be kept free of soluble salts. In hot drylands such salts may occur within the original aggregate, or may rise into the road by capillary migration (see Chapter V): in either case, they may dissolve and recrystallize under ambient atmospheric conditions, causing serious damage. For example, the dissolved salts may migrate by capillary action into the upper section of the unbonded road layers where they are capable of destroying the bond between the road base and the bonded surface layers. South African limits for soluble salt content in road aggregate are given in Table IV.9. Salts known to be troublesome include sodium chloride, sodium sulphate, and gypsum (Weinert and Clauss, 1967; Netterberg, 1970). The

TABLE IV.4

Sand for concrete

1	2	3
Property	*Class*	
	Sand derived from the natural disintegration of rock	Sand derived from the mechanical crushing or milling of rock
Grading, % (m/m) passing sieves having square apertures of nominal size, μm		
4 750	90–100	90–100
150	0–15	0–20
Dust content, material passing a 75 μm sieve, % (m/m), max.	5	10
Fineness modulus*	1.6–3.5	
Chloride content,§ expressed as Cl−, % (m/m), max.	Sand for concrete for prestressing: 0.01 Sand for all other classes of concrete: 0.03	
Organic impurities†	The colour of the liquid above the sand shall not be darker than the colour of the reference solution, except that this requirement shall not be mandatory if the sand complies with the requirement for soluble deleterious impurities	
Presence of 'sugar'‡	Free from sugar unless the sand complies with the requirement for soluble deleterious impurities	
Soluble deleterious impurities‡	The strength of the mortar made with the sand shall be not less than 85% of that of the mortar made with the same sand after it has been washed	

* When a specific fineness modulus is specified by the purchaser, the actual value shall not differ from the specified value by more than 0,2.
† Applicable only to sand derived from the natural disintegration of rock.
‡ Applicable only to sand derived from the natural disintegration of rock and mandatory only when the sand fails to comply with either or both of the requirements for organic impurities and freedom from sugar.
§ See SABS 830 or BS 812 part 4 for test technique.
Source: South African Bureau of Standards 1083.

approximate amount of soluble salt in aggregate material can be estimated in the field from a soil-paste conductivity test (Netterberg, 1970): results are semi-quantitative, and an interpretation of the conductivity readings obtained is given in Table IV.10. In Libya, by comparison, the chloride content (expressed as equivalent sodium chloride by weight) must also not exceed 0.5%, but the sulphate content (expressed as sulphate radical by weight) may be as high as 1.5% (Libyan Arab Republic, 1975).

TABLE IV.5

Grading requirements for stone

1	2	3	4	5	6	7	8	9	10
Grading category	Nominal aperture size of sieve, mm	% (m/m) passing							
		Nominal size of stone, mm							
		53.0	37.5	26.5	19.0	13.2	9.5	6.7	4.75
Category 1	75.0	100							
	53.0	85–100	100						
	37.5	0–30	85–100	100					
	26.5	0–5	0–30	85–100	100				
	19.0	–	0–5	0–30	85–100	100			
	13.2		–	0–5	0–30	85–100	100		
	9.5			–	0–5	0–30	85–100	100	
	6.7				–	0–5	0–30	85–100	100
	4.75					–	0–5	0–30	85–100
	3.35						–	0–5	0–30
	2.36							–	0–5
Category 2	75.0	100							
	53.0	85–100	100						
	37.5	0–50	85–100	100					
	26.5	0–10	0–50	85–100	100				
	19.0	–	0–10	0–50	85–100	100			
	13.2		–	0–10	0–50	85–100	100		
	9.5			–	0–10	0–50	85–100	100	
	6.7				–	0–10	0–50	85–100	100
	4.75					–	0–10	0–50	85–100
	3.35						–	0–10	0–50
	2.36							–	0–10

Source: South African Bureau of Standard, 1083.

Mineral composition of aggregates is also important in determining their suitability for any layer in a road. Montmorillonite clay is undesirable and should be very low or absent because of its high water-absorption properties and the associated volumetric expansion that takes place when dry material is wetted – a change that may have a lubricating effect (Weinert, 1976). Muscovite mica may also disrupt compaction, and it should be kept as far below 10% by volume as possible. Sulphide minerals (commonly pyrite, muscovite, and chalcopyrite) decompose readily and may yield sulphuric acid; therefore rocks with over 1% sulphide content should also be rejected.

The specific requirements for each of the four layers are as follows:

(*i*) *Sub-base* is the unbonded drainage layer of the road which derives its strength from compaction. Aggregate used in this layer must be well graded (i.e. have an even mixture of grain sizes), have a high permeability, and consist of round, smooth particles. Aggregate used in the sub-base usually ranges from 0.075–38 mm, but a most important requirement is that it is suitable for compaction.

(*ii*) *Base course* is the main load-bearing and load distribution layer, and for this reason materials need to have a high crushing strength. If the base course

TABLE IV.6

Suggested schedule of tests on Middle East aggregates for use in concrete under exposed conditions

Scope	Test	Authority	Suggested limits	Remarks
Physical properties and classification	Grading	BS 812 Part 1	BS 882	—
	Elongation index	BS 812 Part 1	Not exceeding 25%	—
	Flakiness index	BS 812 Part 1	Not exceeding 25%	—
	Specific gravity	BS 812 Part 2	—	Limits depending on rock type
	Water absorption	BS 813 Part 2	Not exceeding 20%	
	Soundness	ASTM C-88	Loss not exceeding 16%	5 cycles using magnesium sulphate solution
	Aggregate shrinkage	BRE Digest No. 35	Not exceeding 0.05%	—
	Petrographic examination	ASTM C-295	—	—
Contamination and reactivity	Silt, clay and dust	BS 812 Part 1	BS 882	—
	Clay lumps	ASTM C-142	Not exceeding 2.0%	—
	Organic impurities	ASTM C-40	—	
	Sulphate content	BS 1377 Test 9	Not exceeding 0.4% by weight	Subject to overall limits on total mix
	Chloride content	BS 812 Part 4 or C & CA Advisory Note 18	Not exceeding 0.06% by weight	
	Potential alkali reactivity	ASTM C-227 or C-586	—	—
Mechanical properties	10% fines value	BS 812 Part 3	Not less than 8 tons	
	Aggregate crushing value	BS 812 Part 3	Under 22%	Officially superseded by 10% fines test but still widely used
	Aggregate impact value	BS 812 Part 3	Not exceeding 25%	
	Los Angeles abrasion	ASTM C-131 or C-535	Not exceeding 40%	Relevant for wearing surfaces only

Key to abbreviations: BS, British Standard; ASTM, American Society for Testing Materials; BRE, Building Research Establishment (UK); C & CA, Cement and Concrete Association (UK)

Source: Fookes and Higginbottom (1980).

TABLE IV.7

Summary of requirements of British Standard 3797 for lightweight aggregate

Chemical requirements	Maximum bulk density kg/m³
Sulphate <1.0% Loss on ignition <4.0%	Exfoliated vermiculite: 130 Expanded clay: 240 Others – Coarse: 960 Fine: 1200

TABLE IV.8a

Grading requirements for fine lightweight aggregate

BS 410 test sieve		BS 877 % mass passing through	BS 3797 % mass passing through
3/16	4.76 mm	90–100	90–100
No. 7	2.40	70–100	55–95
14	1.20	45–90	35–70
25	600 μ	20–60	20–60
52	300	10–30	10–30
100	150	5–20	5–19

TABLE IV.8b

Grading requirements for coarse lightweight aggregate (BS 3797)

BS 410 test sieve	% mass passing through
10.0 mm	100
5.0 mm	90–100
2.36 mm	55–95
1.18 mm	35–70
600 μ	20–50
300	10–30
150	5–19

is unbonded, then the materials also require to have a good grading with spherical or rounded particles. The aggregate must not be porous as volume changes due to wetting and drying can lead to disintegration of the road. If the base course is bitumen bonded then the aggregate particles should ideally have irregular shapes and rough surfaces so as to provide a strong bond with the bitumen. Indeed, aggregate used in the bonded layers of a road needs a rougher surface (see BS 812 for tests) than is necessary for that used in concrete, for most materials bond less readily with bitumen than with cement (Weinert, 1976). Materials such as flint, quartzite, and fine-grained volcanic rocks are unsuitable for this reason.

TABLE IV.9

A *Limits for soluble salts in road construction (South African Bureau of Standards, 1083)*

	Stabilized bases	*Unstabilized bases*
Total soluble salts	1.0%	0.5%
Mg, Na or K		
Sulphates	0.1%	0.05%
Chlorides*	0.5%	0.5%
Gypsum†	0.25%	0.25%

* After Cole and Lewis (1960)
† After Fulton (1964)

B *Suggested provisional guide to limits of salts in salty aggregate in Middle Eastern deserts* (salts expressed as weight percent anion of dry aggregate which is extractable from crushed aggregate with dilute acids)

	Bituminous wearing course/basecourse		*Unbound base under*		*Unbound sub-base under*	
	Thick/Dense (A)	*Thin/Porous (B)*	*(A) or (B)*		*(A) or (B)*	
Soluble chloride anions in the aggregate, in possible presence of groundwater with very low dissolved salts.	0.5 (0.2)	0.4 (0.1)	1.0 (0.5)	0.4 (0.1)	1.0 (0.5)	0.4 (0.1)
Soluble anions in the aggregate where more than 10% anion is sulphate, in possible presence of groundwater with very low dissolved salts.	0.3 (0.2)	0.2 (0.1)	0.3 (0.2)	0.3 (0.1)	0.5 (0.4)	0.2 (0.1)
If local groundwater contains salts in solution.	The design must also minimise moisture migration in road section.	The design must not allow capillary moisture to reach the road surface.				
Soluble chloride anions in the aggregate in possible presence of transient moisture (rain, dew, etc.) only.	0.6 (0.3)	0.5 (0.2)	2.0 (0.8)	1.0 (0.4)	3.0 (2.0)	
Soluble anions in the aggregate, where more than 10% anion is sulphate, in possible presence of transient water (rain, dew, etc.) only.	0.5 (0.2)	0.3 (0.1)	0.5 (0.2)	0.3 (0.1)	2.0 (1.0)	

Notes:
(i) Figures in parenthesis refer to soft or friable aggregates of high hazard potential, e.g. clayey or chalky limestones (say water absorption (BS 812) > 5%; soundness loss (ASTM C88) by $MgSO_4$ > 30: ACV (BS 812) > 25.
(ii) Groundwater with very low dissolved salts is arbitrarily defined as containing less than 0.1% TDS.
(iii) The thick/dense surface is assumed to be a non-cement bound asphalt at least 40 mm thick.
(iv) Moisture movement in the road section from groundwater sources can be minimised by raising the road on embankment, use of impervious surface course, use of an impervious membrane (e.g. heavy duty polythene) or sand/bitumen layer or prime coat in the lower part of the road.
Source: Fookes and French (1977)

TABLE IV.10

Suggested interpretation of conductivity readings on base and sub-base materials

Conductivity	Total salt content	Interpretation
> 1.5 μmho	0.2%	Reject
0.6 to 1.5 μmho	0.06–0.2%	Test for So_4^{--} with $BaCl_2$. Accept if absent. If present test for SO_4^{--} and accept if < 0.05%
< 0.6 μmho	0.06	Accept

Source: Netterberg (1970).

(*iii*) *Road base*, which is the lower of the two surface layers, distributes the traffic load onto the base course and provides the shaped surface on which the wearing course is laid. It is bitumen bonded, and it too requires aggregate with irregular rough surfaces and a high crushing strength. Particle sizes should fall in the range 0.075–25 mm (Open University, 1974; BS 1984). Close tolerance on grading, flakiness (Tables IV.11 and IV.12, and BS 1984), and angularity (BS 812, section 1) are necessary for aggregates in the surface layers.

(*iv*) *Wearing course*, or the uppermost, surfacing layer not only requires aggregate with a high crushing strength and the ability to bond with bitumen but it must also have a resistance to polishing and abrasion. Quartz is a particularly good aggregate mineral for the wearing course because it is the hardest of the common minerals and when it is combined with less resistant minerals, a rough surface is maintained. The basic requirement for aggregate used in the wearing course, in addition to those listed in Table IV.11, is that they must maintain skid resistance. Polished stone values (PSV) of > 62 are desirable, for instance, on roundabouts, sharp bends, and approaches to traffic lights; PSV values of > 45 are desirable for normal road conditions (see BS 812 and BS 1984). The ability to resist abrasion is denoted by an aggregate abrasion value (AAV) which should be less than 10% (BS 812). Strength of aggregate should be high, with a value of 16 for the aggregate crushing test (BS 812) and at least 25 tons (210 KN/mn.) for the 10% fines test (Table IV.11, SABS 842, BS 812).

(c) The Geomorphological Context of Aggregate Resources

1. *Introduction*

Experience in the use of natural aggregate in the construction industry in drylands has shown that great care has to be exercised in choice and extraction, and that quality control has to be rigidly applied if expensive and sometimes even dangerous mistakes are to be avoided. The lax and carefree attitudes that prevail in some countries over aggregate quality are a cause for

TABLE IV.11

Single-sized crushed stone for roads

1	2	3
Property	Requirement	
	Grade N stone	Grade S stone
Grading	The grading requirements given for Grading Category 1 shall apply (see SABS 1083)	The grading requirements given for Grading Category 2 shall apply (see SABS 1083)
Fines content, material passing a 425 μm sieve, % (m/m), max.	Nominal sizes 6.7 mm–19.0 mm: 0.5 Nominal size 4.75: 1.0	Nominal sizes 9.5 mm–26.5 mm: 2.0 Nominal sizes 4.75 mm and 6.7 mm: 3.0
Dust content, material passing a 75 μm sieve, % (m/m), max.	–	All nominal sizes: 1,0
Flakiness index, max. (a) Stone for surface dressing and bituminous paving mixtures (b) Stone for rolled-in chips	Nominal sizes 26.5 mm and larger: 35 Nominal sizes 13.2 mm and 19.0 mm: 25 Nominal size 9.5 mm: 30 All nominal sizes: 20	Nominal sizes 37.5 mm and larger: 35 Nominal sizes 26.5 mm and smaller: 30 All nominal sizes: 20
10% FACT value*, kN, min.† (a) Stone for surface dressing 　(1) Dry 　(2) Wet	210 At least 75% of the determined dry value	
(b) Stone for bituminous paving mixtures 　(1) Dry 　(2) Wet	160 At least 75% of the determined dry value	
(c) Stone for rolled-in chips 　(1) Dry 　(2) Wet	250 At least 75% of the determined dry value	
Aggregate crushing value (ACV), dry, %, max.†‡ (a) Stone for surface dressing (b) Stone for bituminous paving mixtures (c) Stone for rolled-in chips	21 25 18	

* The term is derived from the title of the corresponding test, i.e. the 10% fine-aggregate crushing test. Its use is preferred as it obviates confusion with the aggregate crushing value. (See SABS 842 or BS 812 for test technique.)

† Applicable to the minus 13.2 mm plus 9.5 mm fraction of a stone only.

‡ This requirement is an optional alternative (that is applicable only to aggregates that have an ACV not exceeding 30%) that may, if acceptable replace the requirement for the 10% FACT value. (See SABS 841 or BS 812 for test technique.)

Source: South African Bureau of Standards, 1083.

considerable concern, a concern that is amply justified by numerous building, foundation, and road failures in the Middle East and elsewhere. The geomorphological contribution towards improving this situation lies in

TABLE IV.12

Aggregates for base courses

1	2	3
Property	Nominal size of aggregate, mm	
	37.5	26.5
Grading, % (m/m) passing sieves having square apertures of nominal size, mm		
37.5	100	
26.5	84–94	100
19.0	71–84	85–95
13.2	59–75	71–84
4.75	36–53	42–60
2.00	23–40	27–45
Fines content, material passing a 425 μm sieve, % (m/m)	11–24	13–27
Dust content, material passing a 75 μm sieve, % (m/m)	4–12	5–12
Flakiness index, max. (applicable only to the fraction $-26.5+13.2$ mm)	35	
10% FACT value*, kN, min.† (a) Dry (b) Wet	110 At least 75% of the determined dry value	
Aggregate crushing value (ACV), dry, %, max.‡ †	29	
Atterberg limits (a) Liquid limit, max. (b) Plasticity index, max. (c) Linear shrinkage, %, max.	25 6 3	

* The term is derived from the title of the corresponding test, i.e. the 10% fine-aggregate crushing test. Its use is preferred as it obviates confusion with the aggregate crushing value. (See SABS 842 or BS 812 for test technique.)
† Applicable to the minus 13.2 mm plus 9.5 mm fraction of a stone only.
‡ This requirement is an optional alternative (that is applicable only to aggregates that have an ACV not exceeding 30%) that may, if acceptable replace the requirement for the 10% FACT value. (See SABS 841 or BS 812 for test technique.)
Source: South African Bureau of Standards 1083.

improving the search for aggregate from superficial sediments through the use of geomorphological information in the light of the criteria defined in section (b) above. In addition the geomorphologists' tasks includes the study of spatial variations in resource quality, and the identification of locations which should not be used for aggregate extraction even if the materials are themselves suitable. This contribution is based essentially on the fact that there is normally a close functional relationship between superficial sediments

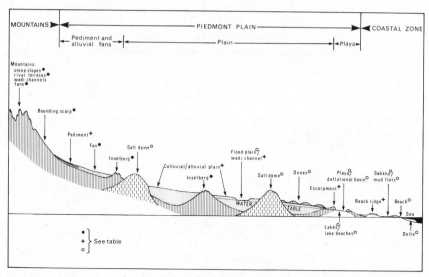

FIG. IV.2 Some landforms of drylands and their potential suitability as sources of aggregates (see Table IV.13 for explanation of symbols)

on the one hand, and the landforms, contemporary processes, and geomorphological history on the other. In general, the geomorphological contribution is specific to particular locations and aggregate requirements, so that only broad guidelines can be formulated. The following discussion, which is based on Fig. IV.2, concerns only aggregate resources. Site problems for construction are discussed in subsequent chapters. Figure IV.2 classifies dryland topography into mountains, pediments and alluvial fans, plains, playas, and coastal areas (cf. Table IV.13). Before considering each of these zones, two general comments are important. First, most of the superficial sediments of potential value as aggregates are fossil deposits of Quaternary age, and appraisal of them invariably benefits from a knowledge of Quarternary geomorphological evolution (e.g. Chapter II, section (c) 2), and especially from a study of the interrelationships of fluvial, aeolian, and marine processes. Second, there is a fundamental distinction to be drawn between areas of downward and upward leaching (Fookes and Higginbottom, 1980). Downward leaching dominates the mountain valleys, some upper piedmont slopes, and the main, contemporary wadi channels: it arises from percolation by periodic floodwaters and restricts the build-up of deleterious salts in sediments. Sediments in such locations are therefore generally preferable to those elsewhere. Beyond these areas, especially on the lower plains, where evaporation exceeds the supply of water, upward leaching predominates, duricrust development is common, and problems of salt contamination prevail.

2. Mountains

The topography of mountain areas in drylands is very varied, and sites suitable for the production of aggregate have to be selected on the basis of

TABLE IV.13

Major landforms as aggregate resources in drylands

Mountains
Including peaks, ridges, plateau surfaces, steep (excluding precipitous) slopes,* deep valleys and canyons, wadis, river terraces* and alluvial fans,* bounding scarp slopes.* Forms vary with rock type and the evolutionary history of the area.

Pediments and Alluvial Fans
Rock pediment,+ fan* and bajada,+ with occasionally inselbergs* or salt domes⁰ forming locally high ground.

Plains
Occur downslope of pediments or alluvial fans without a distinct boundary and may include a whole variety of features including: alluvial+ and colluvial plains,+ wadi channels and flood plains, dune fields,⁰ salt domes,⁰ inselbergs,* and extensive stone pavement surfaces.⁰

Playa Basins
Enclosed depressions receiving surface runoff from internal catchments or within escarpment zones.+ They frequently contain lakes (either temporary or permanent), lake beaches, evaporite deposits⁰ and may be strongly influenced by aeolian, fluvial and salt processes in their baseland zones.

Coastal Zones
These include beach ridges+ (formed at periods of higher sea-level or during exceptional storms), sabkhas,⁰ mud flats,⁰ beach⁰ and foreshore,⁰ estuaries⁰ and deltas.⁰

---

* Normally a major source of aggregate, conditional on suitable mineralogy
† May be a reasonable source, depending on specific characteristics
⁰ Normally should *not* be used for aggregates.
(Symbols refer to fig. IV.2)

local conditions. Several sites may be appropriate for aggregate extraction (Table IV.13), but not all may be easily accessible or within an economic haulage distance of the construction site. Suitability of resource depends in part on bedrock lithology (Table IV.14, Fig. IV.3). Bedrock, of course, normally requires crushing to produce suitable aggregate, whereas alluvial deposits may require only washing, screening, and sorting.

Duricrusts, which are relatively hard material formed near to the surface often in association with soil-forming processes, occur widely in drylands and may provide a suitable source of aggregate. Such duricrusts include deposits dominated by calcium carbonate (calcrete or caliche), iron (ferricrete), silica (silcrete), or gypsum (gypcrete). Of these, calcretes and gypcretes are the most common in drylands (the latter occurring generally in the drier regions) (Goudie, 1973). In some bedrock areas, limestone or dolomite bedrock may be associated with calcareous duricrusts that may be useful, for example, in road construction (e.g. in the western part of the South African plateau (Netterberg, 1971), Lebanon and eastern Arabia (Fookes, 1976a)). Calcareous duricrusts, however, commonly occur in association with soluble salts (Table IV.15) and this is one of their main disadvantages as sources for aggregate. In addition, many calcretes are physically fairly weak, may have clay inclusions, have variable and often unpredictable properties, and, as a result, they usually only fulfil the specifications for road base material (Table IV.12); only rarely have they been used for surfacing chips (Netterberg, 1967, 1971). Calcrete has a self-stabilizing tendency when it has

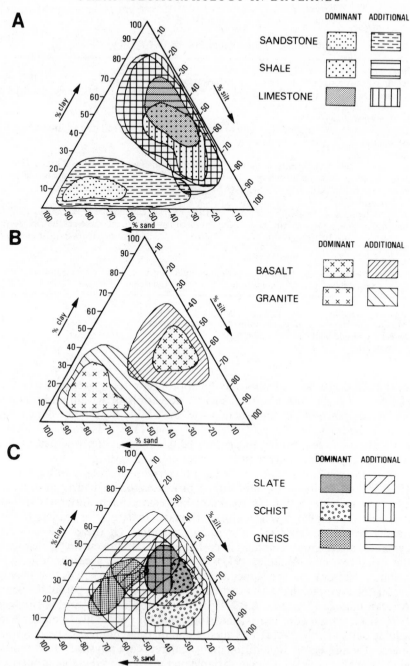

FIG. IV.3 Dominant materials formed from different bedrock types: (A) Sedimentary rocks; (B) Igneous rocks; (C) Metamorphic rocks (modified from Way, 1973)

TABLE IV.14

General characteristics of bedrock types in dry lands for aggregates and other construction purposes

Rock type	Sand resources	Gravel resources	Aggregate quality	Suitability for surfacing	Suitability as borrow	Suitability as building stone
Sandstone	Fair: especially where locally drifted by winds	Not suitable	Excellent if well cemented Fair to poor if weakly cemented	Poor but fair if some clays available as binding agent	Variable: good for coarse fill if spoilings from blasting and rock removal is used	Excellent An important source, but quality may be very variable
Shale	Not suitable	Not suitable	Not suitable	Poor	Good to poor may be used for road fill if compacted	Not suitable
Limestone	Not suitable	Not suitable	Excellent (NB. Coral poor)	Good to excellent Crushed limestone rock good for surfacing secondary roads	Fair to poor Residual soil poor for road fill because hard to work and plastic	Fair to good
Granite (unweathered)	Not suitable	Not suitable	Poor to good Prone to chemical weathering, hence poor for concrete. Good for earthfill	Good The sandy clays of residual soils provide a good wearing stable surfacing for secondary roads	Poor to good Blasted rock good for large scale-operations	Excellent
Basalt	Not suitable In general few sand-sized particles produced by weathering	Excellent to poor In arid areas stony gravels normally form part of the weathering residue	Good to poor Excellent to poor The denser varieties of basalt provide excellent aggregates	Good to poor Dependent on grade characteristics Good if wide range of sizes and sufficient binder	Good to poor Poor if high water content	Good to not suitable
Slate	Not suitable	Not suitable	Not suitable	Not suitable	Good to fair Slate bedrock good for fill, highly weathered residual soils less good	Good – as roofing materials
Schist	Not suitable	Not suitable	Not suitable	Fair to poor fair if residual soil has sufficient fines to provide binder. Mica-rich soils are poor	Good to fair unless high mica content. Strict control on moisture content required	Not suitable
Gneiss	Not suitable	Not suitable	Good to fair	Good to poor Requires both coarse and fine particles in residual soils. Mica renders poor	Good to fair Moisture limits critical, less good with mica content	Excellent to good

Source: modified from Way (1973).

TABLE IV.15

Range of Conductivity, pH and estimated total salt content of the major calcrete types

Material type	Range of -0.420 mm Conductivity millimhos/cm	Range of approximate salt content of whole sample %	Range of pH	Number of samples	SO_2*	
					%	No. of samples
Calcified sands	0.25–2.3	0.03–0.25	7.7–9.2	3	0.01	1
Powder calcretes	0.14–21†	0.02–2.4	8.1–8.5	5	0.00	1
Modular calcretes	0.17–7.4‡	0.02–0.8	7.7–9.6	34	0.02–0.07	4
Tufaceous hardpan calcretes	0.18–0.61	0.02–0.07	8.4–10.0	5	?	0
Other hardpan calcretes	0.09–0.15	0.01–0.02	8.1–9.5	7	0.00–1.02	8
Boulder calcretes	0.06–0.16	0.007–0.02	8.2–9.9	3	?	0

* By silicate analysis, not on water extract
† Only one over 0.6
‡ Only one over 2.5
Source: Netterberg (1970).

been in place for some time, and so in its case some of the physical specifications (such as those for strength and grading) may possibly be relaxed. No specific data are available on such relaxation standards. Calcretes and other duricrusts are by no means restricted to mountain areas – indeed, they are often most extensive and dominate the broad alluvial plains (see below).

More important as sources of aggregate than duricrusts in mountains are the alluvial sands and gravels in mountain valley floors. These deposits are often relatively well graded and, as indicated above, largely free from salts. They therefore often provide an excellent aggregate if they are accessible, as is the case, for example, in Wadi Aday, near Muscat.

3. Pediments and alluvial fans

The bedrock beneath pediment surfaces is only likely to be a good aggregate source if its lithology is suitable (Tables IV.16 and IV.17), otherwise the expense of extraction, which has to be by open-cast mining, may be prohibitive. In addition, the pediment bedrock is only likely to be useful if it has not been weathered. Unfortunately, the weathering of bedrock beneath pediments is a fairly common dryland phenomenon (e.g. Mabbutt, 1966; Cooke and Mason, 1973); on the other hand, pediment bedrock is worth investigating because it is often conveniently located for construction sites. It has to be added, also, that developers and others seeking bedrock are often unaware that it commonly exists near to or at the surface of piedmont slopes.

Whereas bedrock excavation may be expensive and difficult, many alluvial fans, composed as they are of fluvially deposited sediments, may be relatively easy to excavate and thus provide an attractive resource. As discussed in Chapter VI the size and shape of alluvial fans is variable (e.g. Anstey, 1965), and the incorporated deposits are equally variable, often ranging from angular boulders and cobbles near the mountain front to finer gravels, sands, and silts downslope, and often including mudflow deposits and other debris flow deposits as well as well-stratified channel sediments (Fig. IV.4). In general, the upper parts of fans, and deposits in entrenched wadis, are the most suitable for gravel extraction, being relatively salt free, but if the material is now well graded (due to deposition under torrential flood conditions) some crushing of boulders may be required to obtain fractions missing from the natural deposit. Once crushed, the resulting material will have more fractured faces and be more angular than the original deposit and hence may be used as a higher-quality aggregate. It is usually essential for fan gravels to be carefully screened and possibly also washed before use. Where the source rock includes volcanic material, such as tuff or pumice, fan gravels may have a relatively low specific gravity making them suitable for lightweight aggregate (as in parts of Idaho and California (Witkind, 1972)).

Inselbergs of bedrock in the piedmont plain may provide suitable aggregate quarries. Salt domes, such as those in parts of Iran and the Atacama Desert, are, of course, wholly unsuitable as sources of aggregate.

4. Plains and playas

Spreads of silts, sands, and gravel occur within desert plains in a variety of forms (Fig. IV.4) and with a variety of origins (e.g. alluvial outwash from fans, channel deposits etc.). Such deposits are normally the result of ephemeral-runoff deposition, and have often been reworked by successive flows and even channelled from time to time. Their grain-size character is therefore difficult to predict, except that grain size usually decreases

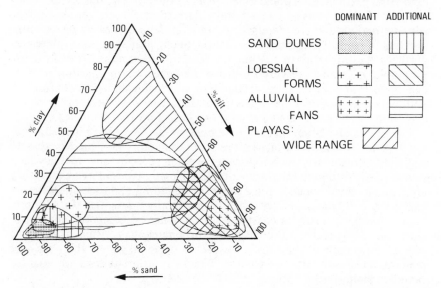

FIG. IV.4 Dominant materials associated with major landforms in drylands (after Way, 1973)

TABLE

Geomorphological features in drylands and

Geomorphological feature	Bedrock mountains	Duricrust	Alluvial fan deposits			Other Piedmont Plain alluvia
			Upper	Lower	Small	
Aggregate type	Crushed rock suitable for all types of aggregate	Road base and sub-base	All aggregate; when crushed for road base concrete	Road base and fill or not useful	Fill or not useful	Variable: locally good fill and concrete aggregate
Nature of material	Angular, clean and rough texture	Often contains deleterious salts	When crushed angular and clean. Otherwise dirty and rounded	Often high fines content	High fines content	Dirty, rounded an well graded. May contain salts
Engineering properties and problems		Self-stabilizing by re-solution	Good compaction			Good bearing capacity (dense sedi ments) and compaction. Locally poor due to clay and silt layers
Potential volume of deposits	Very extensive	Often only small deposits of good quality	Often very extensive but good quality material in small deposits			Very extensive bu good quality material in small deposits

Source: Way (1973).

downslope and stratification is normally good. Where these superficial sediments are subject to contemporary erosion and deposition, their surfaces are often unstable, so that soil development and duricrust formation are relatively unusual. From the resource point of view, usable aggregate may be derived from mixing together sediments from the stratified layers or extracting particular horizons: much depends on local circumstances.

Some of these alluvial deposits may date from earlier, perhaps wetter climatic periods and are therefore non-renewable resources whose exploitation has to be justified prior to extraction. Such is the case, for example, in northern Kuwait (Kuwait Institute of Scientific Research, 1978). In many areas alluvial plains are very extensive, and the task of charting suitable aggregate resources in a sea of alluvium is difficult and time-consuming. Without doubt, however, the task is facilitated if a model of geomorphological evolution can be formulated and used as a basis for exploration.

The plains also often include sand-dune areas, in which rounded sand-sized particles, usually of quartz, are superabundant and generally easy to exploit. Unfortunately such sediments are not of high value for they lack coarse particles that make them suitable for many types of aggregate (they are typically fine grained – in the range 50–600 mm), and their characteristically rounded character gives them poor 'binding' qualities. In addition, many dunes, especially in coastal areas, may not be composed entirely of silica, but include relatively undesirable carbonates etc. Any materials on the plains, especially if they are or were close to a water-table, may be contaminated with salts (see Chapter V) and appropriate tests are essential before exploitation.

The lower areas of enclosed drainage basins and alluvial plains often consist of playas that are composed of fine-grained, highly saline soils (see

IV.16

their potential for aggregate materials

	Old river deposits	Old lake deposits	Dunes		Salt playas and sabkhas	Coastal dunes	Storm beach	Foreshore
			Dune	Interdune				
Concrete		?	Not suitable	Fine aggregate blasting sands	Aggressive ground	Not suitable	Fine aggregate	Not suitable
Variable		?	Too fine and rounded	Coarse and angular		Too fine and rounded	Sufficiently coarse. Clean	Fine, round sand. Salt contaminated
Difficult to locate in field		?	Poor compaction and bearing capacity			Poor compaction and bearing capacity	Good sharp sand for concrete	
Often deposits are patchy and thin		?		Very localized			Extensive deposits in North Africa and Middle East may be local on rocky coasts	

Chapter VI) and are on both counts unsuitable as sources of aggregate, and in almost all circumstances their exploitation is best avoided. Salts and fine silt may be removed from the basin floors by wind action, possibly leading to the contamination of nearby potential aggregate resources. On the other hand, the salts may themselves be a useful resource.

5. *Coastal zone*

A high salt content normally characterizes the deposits of the coastal zone (see also Chapter V); this, together with the fact that coastal dryland sediments are often fine grained, and sometimes composed predominantly of carbonates, restricts their value. A notable exception is that of storm beaches that may provide a useful aggregate resource. The variability of exposed coastal sediments in terms of both type and quality is often great, and detailed geomorphological mapping is thus particularly useful in identifying aggregate sources (e.g. Fig. III.11). Duricrusts are sometimes associated with coastal sands (Fookes and Collis, 1975a; Fookes and Poole, 1978) and may make acceptable construction material. The high salinity of deposits in the coastal zone (as in playa basins) results from the close proximity of the watertable to the ground surface and the relatively high evaporation rates that promote salt crystallization. Predominantly carbonate coastal sands are used extensively in the Middle East, but they are unsatisfactory in many ways – for example, they tend to be absorptive, to have poor gradings, and to be chemically unstable (Fookes and Higginbottom, 1980), and thus special concrete mix designs may be necessary.

(d) Case Studies

1. *Introduction*

The general account given above has illustrated the close coincidence between aggregate resources and some specific desert landforms. A geomorphological investigation, initially through geomorphological mapping

TABLE IV.17

General characteristics of landform soil materials in drylands for aggregates and other construction purposes

Landform	Sand resources	Gravel resources	Aggregate quality	Suitability for surfacing	Suitability as borrow	Suitability for buildings
Sand dunes	*Excellent*: normally clean but restricted range of particle size, and particle shape not ideal for many purposes	*Not suitable*	*Not suitable*	*Not suitable*	*Good to poor* Requires addition of fines	*Not suitable*
Loessial forms	*Not suitable*	*Not suitable*	*Not suitable*	*Not suitable*	*Fair* Moisture content critical	*Not suitable*
Flood plains	*Excellent* e.g. along stream beds, within point bars or along levees	*Poor* Usually small pockets well scattered	*Not suitable*	*Excellent to fair*	*Excellent to poor* Less suitable with increasing fines	*Not suitable*
Deltas	*Good to poor* Only young arc deltas are good for well-graded sands	*Good to not suitable* Only good in young arc deltas	*Poor* Only young arc deltas contain potentially suitable sources	*Excellent to fair* Excellent when good mix of fine and coarse materials	*Good to poor* Young arc deltas good, rest tend to be poor	*Not suitable*
Alluvial fans	*Excellent to fair*	*Excellent to fair*	*Excellent to fair* Depending on material source and position on fan	*Good*	*Excellent to good* Stratified deposits may have to be mixed on extraction to provide range of particle sizes	*Not suitable*
Playas	*Poor*	*Not suitable*	*Not suitable*	*Good* source for fines to use as binder when mixed with other deposits (e.g. fans)	*Poor* Moisture content critical	*Not suitable*

Source: modified from Way (1973).

(see Chapter III), may thus be a primary basis for identifying potentially useful (or otherwise) areas of aggregate resources. The way that this has already been done may be illustrated by examples from Ras al Khaimah, Dubai, and Bahrain. The techniques described are equally applicable in other drylands.

2. *Ras al Khaimah*

Ras al Khaimah lies in the United Arab Emirates at the north-eastern end of the Oman Peninsula, between the Oman Mountains and the Arabian Gulf. Geomorphological methods were used in a study of Khor Khwair, in the northern part of the state, where the Oman Mountains approach close to the coast. The three main topographic components (see Fig. IV.5) are mountains, alluvial fans, and a coastal zone, there being inadequate space between the shore and the tectonically controlled mountain front for a broad piedmont plain to have developed. The area thus includes several environments that are likely to provide good aggregate resources (see Table IV.16). Geomorphological mapping was used to identify the nature and extent of each landform zone (Halcrow Middle East, 1977b). Soil and rock materials were examined in each of the mapped geomorphological units so that their suitability for aggregate could be assessed. This study revealed a close association between the landforms and their material composition, and so a materials map was compiled using the geomorphological mapping boundaries as surrogates for material boundaries (see Fig. IV.6). The coalescing fans at the mountain front consisted of two large fans (emerging from Wadi Ghalilah and Wadi Rahabah) and a number of smaller ones. Only the two major fans provided a sufficient volume of material to justify extraction, in addition to which the smaller fans had a high percentage of fines rendering them too dirty to make good aggregate. Indeed, even on the bigger fans the lower, distal portions had the same drawback. Even the materials from the upper parts of the big fans required screening and washing, and they proved not to be very well sorted, with the finer and medium-sized gravels being platey and elongated while the larger boulders required crushing. As a whole, the fan deposits were less suitable as aggregate than material derived from the local, high-grade crushed rock sources. It was found that the wadi-floor gravels could also be exploited but only at the risk of creating a flood hazard by the re-routing of the natural drainage channels.

Within the coastal zone only the fine-to-medium calcareous quartz sands of the storm ridge were suitable as a fine aggregate, and then only after screening to remove shells and washing to remove soluble salts. The consequence of extracting this material, however, would be to increase the potential for sea-storm floods reaching inland beyond what is now a natural protective barrier. All other parts of the coastal zone consist of deposits that are too fine or saline.

3. *Dubai*

Further south on the Oman Peninsula, the mountains and the sea become separated by an increasingly wide intervening plain (see Fig. III.10). This

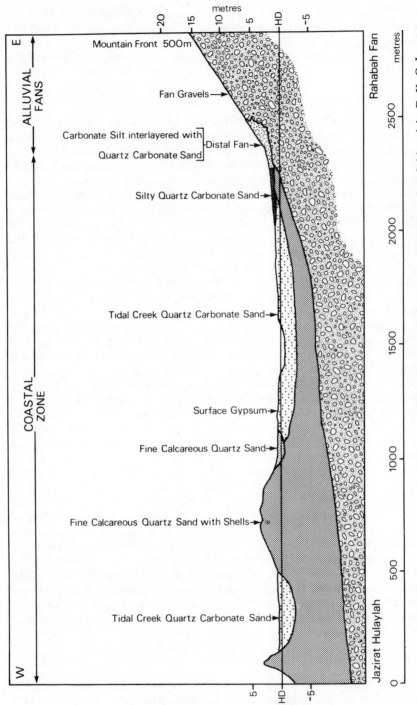

FIG. IV.5 Topographic cross-section of Ras al Khaimah (from Halcrow Middle East 1977b, based on fieldwork by D. K. C. Jones, D. Brunsden, J. C. Doornkamp, and P. R. Bush). Line of profile is located on Figure IV.6C

plain consists for the most part of old alluvial-fan deposits buried by sand dunes of varying ages, composition, and dune forms. By and large no suitable aggregate resources exist among these dunes, but in Dubai a search was made for suitable materials among the deposits of the coastal zone. In this case a substantial amount of field sampling had already been carried out before a geomorphological survey was commissioned (Brunsden *et al.*, 1976).

A difficulty had arisen during the field sampling of materials in that no information was available on the extent to which a particular soil sample was representative of materials over a wider area. The purpose of the geomorphological work was to provide a landform interpretation of the coastal zone so as to provide the prima-facie boundaries for each of the soil types already identified. The budget available for this task did not permit any geomorphological fieldwork. Instead the whole of the analysis had to be carried out from aerial photographs. The result was a geomorphological map which was used as a basis for locating soil pits and judging the lateral extent and significance of the data they revealed. In addition, the genetic interpretations of the landforms provided by the geomorphologists gave a valuable insight into the reasons why particular soil or sedimentary conditions had been found. Thus, not only could data on suitable aggregate resources be extrapolated laterally, but the investigation could now explain the material characteristics that had been found in terms of their genesis and present environments.

4. *Bahrain*

One of the main purposes of the 1 : 10 000 scale geomorphological mapping of Bahrain was to identify sources of aggregates, especially sand and gravel deposits, for the Bahrain construction industry.

Figure IV.7B shows an extract from the original mapping and Fig. IV.7C shows part of the derivative map of types of fine aggregate available, classified by their site and in terms of likely aggregate quality. The geomorphological survey enables further quality-control investigations and economic analyses to be concentrated in only those areas where there is any hope of finding a worthwhile resource. No further effort or time needs to be spent in the areas shown by the mapping to have no potential for fine-aggregate resources.

The geomorphological interpretation of Figure IV.7B relies on the identification of alluvial fans derived from suitable source areas. Fans derived from intensely gypsiferous areas are clearly chemically unsuitable. This type of assessment of superficial sediments was carried out for the whole of Bahrain, and culminated in a fine-aggregate resources map (Fig. IV.8). Further studies of the geomorphological history of Bahrain (Bahrain, Ministry of Works, Power and Water, 1976; Doornkamp *et al.*, 1979) indicated, however, that many potentially usable aggregates had been blown south-eastwards off the island by the 'Shamal' winds and now lie below water off the south-east coast.

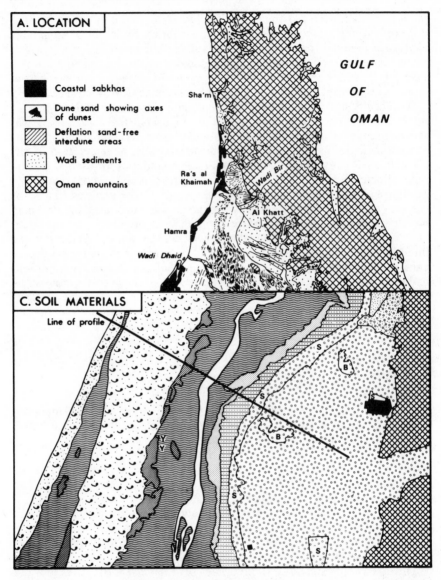

FIG. IV.6 Ras al Khaimah: sections of geomorphological and soils maps (after Halcrow Middle East 1977b, based on mapping by D. K. C. Jones, D. Brunsden, J. C. Doornkamp, and P. R. Bush). (A) is based on Glennie (1970)

(e) Summary

Urban development in drylands, as elsewhere, demands large quantities of aggregate, chiefly for concrete and road construction. Yet, perhaps

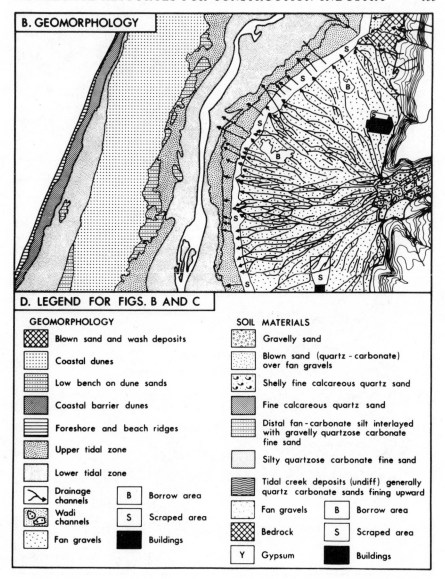

B. GEOMORPHOLOGY

D. LEGEND FOR FIGS. B AND C

GEOMORPHOLOGY

Blown sand and wash deposits

Coastal dunes

Low bench on dune sands

Coastal barrier dunes

Foreshore and beach ridges

Upper tidal zone

Lower tidal zone

Drainage channels

Wadi channels

Fan gravels

B Borrow area

S Scraped area

Buildings

SOIL MATERIALS

Gravelly sand

Blown sand (quartz - carbonate) over fan gravels

Shelly fine calcareous quartz sand

Fine calcareous quartz sand

Distal fan - carbonate silt interlayed with gravelly quartzose carbonate fine sand

Silty quartzose carbonate fine sand

Tidal creek deposits (undiff) generally quartz carbonate sands fining upward

Fan gravels

Bedrock

Gypsum

B Borrow area

S Scraped area

Buildings

surprisingly, good-quality aggregate is not widely available, and many apparently suitable deposits suffer from such serious problems as inappropriate grading, salt contamination, and easily weathered materials. The two main sources of aggregate are superficial sediments and bedrock. It is in the evaluation of the former that the geomorphologist can and does play an important role. Given an understanding of the qualities of aggregate required by clients, and a knowledge of local regulations (where they exist), and

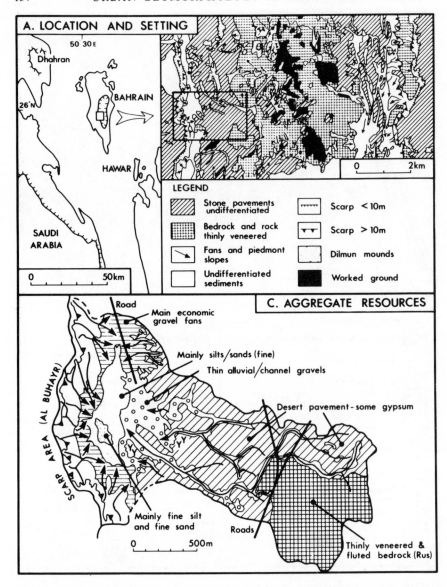

FIG. IV. 7 Geomorphological mapping for fine aggregate resources in Bahrain: (A) Location and geomorphological setting; (B) Geomorphological map; (C) Aggregate resources; (D) Mapping legend (after Bahrain, Ministry of Works, Power and Water, 1976 and Doornkamp *et al.*, 1979)

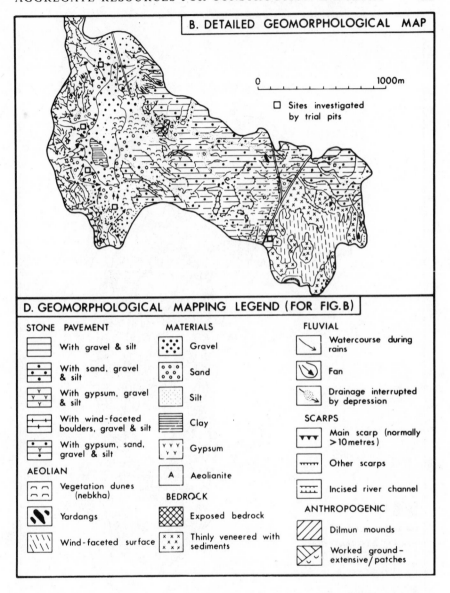

B. DETAILED GEOMORPHOLOGICAL MAP

0 _____ 1000m

☐ Sites investigated
by trial pits

D. GEOMORPHOLOGICAL MAPPING LEGEND (FOR FIG. B)

STONE PAVEMENT

With gravel & silt

With sand, gravel & silt

With gypsum, gravel & silt

With wind-faceted boulders, gravel & silt

With gypsum, sand, gravel & silt

AEOLIAN

Vegetation dunes (nebkha)

Yardangs

Wind-faceted surface

MATERIALS

Gravel

Sand

Silt

Clay

Gypsum

Aeolianite

BEDROCK

Exposed bedrock

Thinly veneered with sediments

FLUVIAL

Watercourse during rains

Fan

Drainage interrupted by depression

SCARPS

Main scarp (normally >10 metres)

Other scarps

Incised river channel

ANTHROPOGENIC

Dilmun mounds

Worked ground– extensive/patches

standard tests suitable for determining those qualities, the geomorphologist can locate and evaluate superficial sediments, normally in the context of their fluvial, aeolian, or marine environments of deposition, in a way that provides a sound basis for the sensible sampling of deposits and the detailed appraisal of sites selected from within the areas of potential resources identified by the geomorphologist.

Urban areas

Escarpments

AGGREGATE MATERIAL
EXTRACTION AREAS

Rock

Fine aggregates

Unconsolidated sediments

Ra's Al Aqr formation
(Pleistocene)

Jabal cap formation
Eocene carbonates

0   Km   5

FIG. IV.8 Aggregate resources of Bahrain (after Brunsden *et al.*, 1979)

V

SALINITY, GROUND-WATER, AND SALT WEATHERING IN DRYLANDS

(a) Introduction

The breakdown of rock by salt attack has been recognized as an important geomorphological process for many years (e.g. Evans, 1970). Similarly, the deleterious effects of salts on natural building stones have also been appreciated for a long time (e.g. Schaeffer, 1932), particularly with respect to historical monuments. However, the full significance of salt-weathering attack as an engineering hazard has only become apparent recently during the course of ambitious construction programmes underway in many drylands, such as those of the Middle East (e.g. Fookes and Collis, 1975a). Here the use of concrete and natural stone for buildings, structures, and roads in the presence of salts has resulted in numerous problems ranging from unsightly blemishes to serious failures which, in some cases, have required costly remedial works. Consequently, considerable research has been undertaken into the nature of salt weathering and the assessment of the salt-weathering hazard is increasingly coming to be seen as an important element in planning and engineering decisions in drylands. Before discussing the ways in which salts affect urban structures and the methods by which such attack can be minimized, it is first useful to examine the occurrence of salts and the characteristics of salt weathering as a geomorphological phenomenon.

(b) Salt Weathering in Drylands

1. *The nature and occurrence of salts*

The climatic conditions experienced in most hot drylands usually include high day-time air and surface temperatures, low and erratic rainfall, and very high potential evaporation and evapo-transpiration. As a result, capillary rise of water in surface sediments is commonly very pronounced and leads to the concentration of salt crystals in the upper parts of soil profiles (Fig. V.1) and, often, on the ground surface as salt efflorescence. The most commonly occurring salts include calcium carbonate ($CaCO_3$), calcium sulphate either in the hydrous form (known as gypsum, $CaSO_4.2H_2O$) or its anhydrous forms (hemihydrate, $CaSO_4.\frac{1}{2}H_2O$ or anhydrite, $CaSO_4$), and sodium chloride (NaCl).

Many other salts may also be present both as crystals and in solution in groundwater. Some indication of the range of salts likely to be encountered

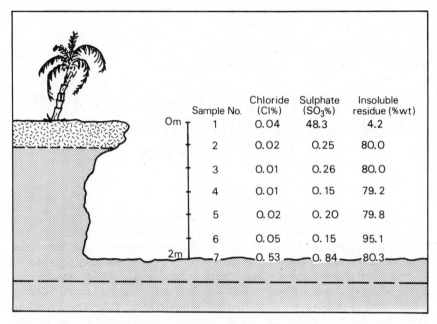

Sample No.	Chloride (Cl%)	Sulphate (SO₃%)	Insoluble residue (%wt)
1	0.04	48.3	4.2
2	0.02	0.25	80.0
3	0.01	0.26	80.0
4	0.01	0.15	79.2
5	0.02	0.20	79.8
6	0.05	0.15	95.1
7	0.53	0.84	80.3

FIG. V.1 Chloride and sulphate profile from a typical Middle Eastern sand borrow pit showing increased salt concentrations near the surface and immediately above the water-table (dashed line) (after Fookes and Collis, 1975a)

in drylands is shown in Table V.1 which was compiled by Netterberg (1970) based on the analysis of 48 inland groundwater samples undertaken by Joffe (1949). Several important points emerge from Netterberg's study. First, the occurrence of salts is extremely variable from place to place – some only occur rarely, whereas others, like NaCl, are apparently ubiquitous. Second, the mix of salts in solution is also highly variable, as is clearly shown by the great range of values recorded for NaCl (0.2–90.0%) and the fact that six out of 14 salts (43%) locally achieved sufficient concentrations to exceed 40% of the measured total salt content. Third, the method of analysis emphasized the importance of highly soluble salts and, therefore, significantly underestimated $CaSO_4$, large quantities of which are stored as relatively insoluble anhydrite and gypsum. Finally, sea-water intrusion beneath coastal areas significantly alters the groundwater chemistry so that precipitates might there include Fe_2O_3, $CaCO_3$, $CaSO_4$, NaCl, $MgSO_4$, and $MgCl_2$ in that order, with NaCl and $MgCl_2$ predominating if evaporation is completed (Netterberg, 1970).

The concentrations and relative significance of these salts varies with geographical, geological, climatic, and topographical factors. Salt concentrations are highest in areas characterized by surface evaporation of surface or sub-surface saline waters. Thus coastal regions, enclosed depressions, and other low-lying areas that receive the drainage of internal catchments, all tend to have high salt concentrations, whereas uplands and free-draining

TABLE V.1

The abundance of soluble salts in samples of saline soils from drylands

	Salt	% Occurrence in samples (48)	Range of proportion of total salt content (%)
Common	NaCl	100	0.2–90
	Na_2SO_4	83	0–93
	Na_2CO_3	73	0–99
Significant	K_2SO_4	38	0–21
	$MgSO_4$	33	0–44
	KCl	13	0–13
Local	Na_2HPO_4	10	0–8
	$MgCl_2$	6	0–6
	$CaCl_2$	4	0–58
	Na_3PO_4	2	0–2
	K_2CO_3	2	0–7
	$NaNO_3$	2	0–87
	$(NH_4)_2 CO_3$	2	0–2
	$Mg (PO_4)_2$	2	0–2

Source: Netterberg (1970).

deep sands are usually low in salt. Surface salts may also be moved by the wind, resulting in surface contamination downwind of a source area and the occasional development of localized drifts and dunes formed of salt crystals. The movement of salts by both fluvial and aeolian processes has resulted in the identification of a 'salt cycle' (Fig. V.2). In this, salts dissolved from rock and soil in the upper parts of a drainage catchment are transported by stages across intervening areas to depositional zones near to local base-levels. Here they are often concentrated in temporary lakes and precipitated out as water is lost through evaporation. Thereby, they become exposed to wind action which may subsequently result in removal by deflation, either to the upper parts of the same catchment or into an adjacent basin. This sequence of salt movement may be repeated many times.

An additional important airborne increment is that of oceanic salt (Eriksson, 1958), mainly NaCl. The scale of this increment varies with the salinity of the sea or ocean water adjacent to the coast and the local climatic conditions. In general this source of contamination results in coastal soils tending to be more saline than those of interior regions. Thus, in Australia, for example, annual increments of 125–750 kg/ha/yr are estimated to be deposited near the coast of Western Australia (Jackson, 1957), whereas inland the increments decrease to 12.5 kg/ha/yr (Jennings, 1955). Salt fall-off curves have also been plotted for Israel (Yaalon and Lomas, 1970) where the annual increment of NaCl for the country as a whole is estimated as 101 800 tonnes (Yaalon, 1963). The study by Yaalon (1963) revealed two interesting patterns: (i) while increments of salt are greater in moist areas, accumulation is greatest in dry regions and (ii) although NaCl deposition predominates

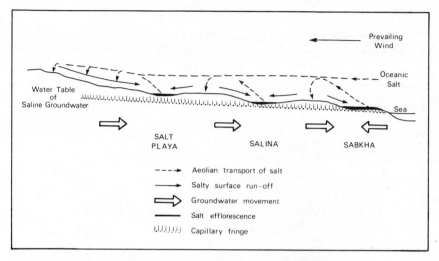

FIG. V.2 Cross-section showing a 'salt cycle' and the usual relationships of the three main types of surface salt concentrations (salt playa, salina and sabkha). In such a situation, surface soils will become increasingly saline towards the shore

over the whole country, sulphates increase in importance with increasing aridity.

In addition to the contemporary patterns of salt movement and accumulation outlined above, there exists a number of relict salt-rich deposits produced under more pluvial conditions during the Pleistocene Period and preserved at or near the present surface. These deposits range from small elevated erosion residuals of former lake floors (commonly comprising gypsum) to extensive tracts of slightly elevated duricrust (see also Chapter IV (c) 2). Of the duricrusts, the gypcretes are particularly important because they can provide increments of $CaSO_4$ to the contemporary salt cycle.

2. *Processes of salt weathering*

Despite the fact that salt weathering has long been recognized as an important process in coastal environments under a variety of climatic conditions, the significance of surface salts as a cause of rock breakdown in drylands has only recently become fully appreciated (e.g. Wellman and Wilson, 1965; Birot, 1968; Cooke and Smalley, 1968; Evans, 1970). Laboratory studies have clearly revealed the potential efficacy of this form of weathering in desert environments (e.g. Goudie *et al.*, 1970; Kwaad, 1970; Goudie, 1974; Cooke, 1979), with the result that many geomorphological features of drylands such as rock splitting, granular disintegration, and the creation of basal overhangs (Fig. V.3), are often being re-interpreted as the product of salt attack rather than of the traditionally advocated processes of 'wetting and drying' weathering, insolation weathering ('heating and cooling' processes), and aeolian processes (Cooke and Warren, 1973).

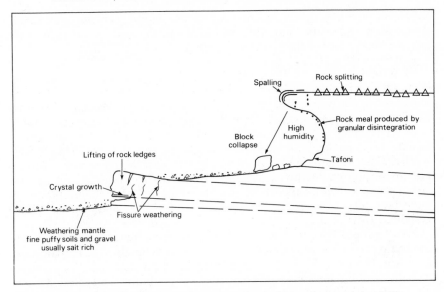

FIG. V.3 Effects of salt weathering noted in Bahrain (Doornkamp *et al.*, 1980)

Rock disintegration due to salt-weathering attack results mainly from stresses developed by the expansion or growth of salts in confined spaces, such as pores, joints, and fissures, or by chemical changes between salt and rock. Three main pressure-inducing processes of crystal growth have been recognized (Cooke and Smalley, 1968; Evans, 1970; Kwaad, 1970): (i) *the growth of salt crystals* through time due to the evaporation and/or cooling of saline solutions; (ii) *the thermal expansion of salt crystals* as a consequence of pronounced day-time heating; and (iii) *the hydration* of certain salt crystals. In addition, several chemical processes of salt weathering may result in volume changes that can cause disintegration, especially in association with concrete (Fookes and Collis, 1976). For example, chemical exchange between sulphate radicals in solution and calcium aluminium hydrate in cement may produce reactive products of larger volume. Reactions between cement and gypsum may produce expansive portlandite (calcium hydroxide). Accelerated corrosion of ungalvanized steel embedded in concrete and in the presence of chlorides may disrupt the concrete, and chloride-aluminate reactions may also create expansive forces. In addition, as French and Poole (1976) have described, reactions between silica-reactive aggregate and cement paste may produce a gel that can absorb water and swell to many times its volume to initiate cracking, and this process may be exacerbated by salts contaminating the aggregate.

It is important to recognize that considerable uncertainty still exists about the nature, spatial variability, and relative importance of the different salt-weathering mechanisms. It is expected that further field and laboratory investigations, including the use of 'climatic cabinets' which facilitate the reproduction of 24-hr or accelerated sequences of ground-surface climatic conditions for any desert location by means of independent control of temperature and relative humidity, will yield valuable data on the nature and relative importance of the processes involved (Cooke, 1979).

(*i*) *Salt-crystal growth*. Salt-crystallization pressures (*Salzsprengung*) have been accepted as an effective weathering process, and they are notably illustrated by the widespread occurrence of ground heave associated with the development of salt polygons. Crystal growth pressures have been summarized by Buckley (1951). Laboratory studies (Evans, 1970; Goudie, 1974; Cooke, 1979) have confirmed field observations (e.g. Jutson, 1918; Blackwelder, 1940; Wright and Urzua, 1963; Beaumont, 1968) that rock splitting and disintegration can be caused by salt-crystal growth in confined spaces and there is growing support for the view that this is probably the most important salt-weathering process (Evans, 1970; Cooke and Warren, 1973; Goudie, 1974).

Investigations have revealed that crystals can continue to grow against a confining pressure as long as a film of solution is maintained at the salt-rock interface (Evans, 1970). This condition is dependent on the relative magnitude of the tensions at the salt–rock, salt–solution, and solution–rock interfaces. If the sum of the latter two tensions is less than the former then the solution can penetrate between the salt crystal and the surrounding rock, thereby facilitating further crystal growth and the generation of pressures which vary depending on the degree of supersaturation of the solution.

Laboratory studies into the efficacy of salt-crystallization weathering have attempted to simulate the process in desert conditions (Goudie *et al.*, 1970; Goudie, 1974; Cooke, 1979). In one of these experiments (Goudie *et al.*, 1970) 3-cm rock cubes were immersed in a range of saturated salt solutions ($NaCl$, Na_2CO_3, Na_2SO_4, $NaNO_3$, $CaCl_2$, $MgSO_4$) at 17–20 °C for one hour. They were then removed and dried in an oven at 60 °C for six hours (a temperature approximating to day-time soil temperatures in warm deserts) and 30 °C for the remainder of the 24-hour cycle (a temperature approximating to night-time soil temperatures in warm deserts). The cycle was then repeated up to 40 times and the weight loss from specimens was recorded after each cycle. A comparison of the effectiveness of different salts in inducing disintegration clearly showed that the most rapid breakdown of the 3-cm sandstone cubes used in the experiment was achieved by Na_2SO_4, followed closely by $MgSO_4$, while the other salts, including $NaCl$, caused little or no change. The experiment was then repeated to determine the breakdown rates of a number of rock types when treated with Na_2SO_4. The results indicated that limestones and sandstones are the most susceptible to this form of attack, while shales and igneous rocks show little or no response. The reasons for such variability are not certain but they appear chiefly to be the result of a combination of rock properties including water-absorption capacity (Cole, 1959; Minty, 1965), porosity and microporosity, surface texture, and surface area/volume ratio (Minty and Monk, 1966).

In addition to the above rock properties, spatial variability in intensity of salt-crystallization weathering also reflects such factors as climate, location, and the concentration and combination of salts present. For example, the outstanding efficacy of sodium sulphate weathering in drylands is probably related, *inter alia*, to the fact that the wide diurnal temperature range in deserts encourages crystallization of the salt whose distinctive solubility curve shows a sharp increase of solubility with increasing temperature (e.g. Kwaad, 1970). Similarly, it seems possible that a combination of NaCl and $CaSO_4$ may be more effective than either salt alone (Cooke and Doornkamp, 1974). Again, evaporation and diurnal temperature range – probably two of the major climatic controls on the process – are both high (but spatially variable) in drylands.

(*ii*) *Thermal expansion of salt crystals*. The expansion of salt crystals when heated depends on the thermal properties of the salts and the temperature range to which they are subjected. Diurnal air-temperature ranges in desert areas are often in excess of 25 °C and individual instances as high as 54 °C have been reported (Keller, 1946). Ground-surface temperatures are usually significantly higher than air temperatures during the day and may reach 80 °C (e.g. Cloudsley-Thompson and Chadwick, 1964; Peel, 1974). At night the reverse is often the case due to radiation cooling. As a result, diurnal ground-surface temperature ranges may exceed 50 °C and even approach 80 °C at times. The magnitude of the diurnal range rapidly diminishes with depth within both rock and soil profiles so that little fluctuation can be recognized at –0.5 m (e.g. Hörner, 1936) (for examples, see Cooke and Warren, 1973).

Although the significance of salt weathering as a result of thermal expansion has yet to be evaluated, the potential for disruptive stress generation appears considerable, particularly in the case of salt crystals lodged in near-surface pores and crevices. One reason for this is that the coefficients of thermal expansion of common desert salts are greater than those of most rocks (Cooke and Smalley, 1968) and in the case of NaCl the coefficient has been shown to be similar to that of ice (Hunt and Washburn, 1966). Thus in the high surface day-time temperatures in drylands, a salt, such as NaCl, could expand by up to 1% during a diurnal cycle and produce stresses which, together with those created through the differential thermal expansion of the rock itself, might conceivably induce further cracking and splitting. As yet, however, laboratory experiments carried out on this mechanism (Goudie, 1974) have proved disappointingly inconclusive.

(*iii*) *The hydration of salts*. Disruptive stresses are thought to be generated by the rehydration of anhydrous salts which have developed through the dehydration of hydrous salts under desert day-time temperatures and humidities (Kwaad, 1970; Winkler and Wilhelm, 1970). Such a change is of particular significance in the case of salts that occur widely in drylands and may convert from one form to another several times in a season and possibly even during a single day. The hydration of Na_2SO_4 is an example. The mechanism of hydration and dehydration is complex and depends principally on atmospheric relative humidity and temperature conditions and the dissociation vapour pressures of the salts concerned. It is possible, however, that capillary-rise salt solutions may yield crystals under high day-time temperatures which are low in water of crystallization but which subsequently absorb water vapour from the air during the cooler night-time conditions when the pressure of atmospheric aqueous vapour pressure exceeds the dissociation pressure of the hydrate involved. Under such circumstances anhydrous salts may be converted to higher hydrates.

Calculations of hydration pressures using a prediction equation show that certain salts under ideal conditions may generate stresses that approach those associated with frost weathering (Winkler and Wilhelm, 1970). Kwaad (1970) pointed out that at least three conditions must be satisfied for rock disruption by hydration: (i) the hydration process should be accomplished in at least 12 hours, or else it cannot be completed by diurnal temperature changes; (ii) the hydration salt must not be able to escape from the pore it is in; and (iii) hydration pressure must exceed the tensile strength of the rock.

(*iv*) *Spatial implications.* Although the relative significance of the physical and chemical salt-weathering mechanisms has yet to be evaluated by field observations, it seems likely that all operate in drylands. The preceding discussion makes it clear that salt weathering is spatially variable. The most intense activity is likely to occur in the following four situations: in *humid deserts of western littorals* where the supply of salts from fogs and chemical weathering is accompanied by frequent wetting and drying; in *coastal areas adjacent to virtually land-locked seas,* such as the Red Sea and Arabian Gulf, where high humidities result in frequent dews and fogs, and salinities are high; *around the margins of playas* (*playas* are areas of fine clastic sediment in the floors of enclosed drainage basins, see Chapter VI) fed by salty runoff (see Fig. VI.7) and *along fluvial channels* where salts tend to be concentrated and wetting and drying is most frequent; in *low-lying areas where the ground surface is affected by capillary rise* from shallow saline groundwater which in certain cases may have salinities up to eight times that of sea water. Such a situation is commonest in coastal regions, although small areas may occur inland due to deflational lowering of the ground surface down to the moister materials affected by the capillary fringe (see Fig. VI.7). The greatest concentrations of salts occur in the low, flat, damp salt-encrusted plains that are normally called *sabkhas* in coastal areas, and *salinas, internal sabkhas, chotts,* or 'moist playas' (Motts, 1965) in inland areas (Fig. V.2). Such areas are relatively restricted in extent, covering, for example, only about 5% of the Middle East deserts (Fookes and Collis, 1975a).

(c) Salt Weathering in Urban Areas

1. *General*

The presence in drylands of saline solutions and mobile salts often results in damage to buildings, roads, and engineering structures. The usual signs of salt-weathering attack are cracking, spalling, and granular disintegration and they have been reported from such widely differing geographical locations as Australia, south-western USA, North Africa, and the Middle East. The review in this section is based heavily on the experience of Fookes and Collis (1975a), as well as on studies made by the authors, in the Middle East.

The most serious causes of damage due to the activity of salts in urban areas include the following:

(i) Alteration of the chemical composition of the materials used in construction, which can lead to reduction in their strength and durability, or can promote volume changes that exercise a disruptive effect. For example, concrete can be chemically altered in the presence of salts. One illustration of this is the reaction of free tricalcium aluminate (C_3A) in concrete with sulphates and chlorides dissolved in water to form such new minerals as *etringite* (calcium aluminium sulphate), a reaction that can result in the disintegration of the concrete. Similarly, the corrosion of steel reinforcing bars due to the presence of salts can lead to an increase of three to seven times in the volume of the bars and thereby exert very great pressure on the surrounding concrete and result in its disintegration. Of all serious causes of salt weathering these are probably by far the most significant to engineers.

(ii) Salts and saline solutions in buildings and structures may cause weakening and disintegration due to volumetric changes caused by the processes of salt weathering outlined in the previous section (i.e. thermal expansion, hydration, and crystal growth).

(iii) Changes in the volume of salts contained in foundation fills can result in damage to structures due to ground movements. For example, leaching of salts (e.g. gypsum) may cause an increase in void space and subsequent surface subsidence, while hydration can lead to volumetric expansion and result in ground heave.

Potentially destructive salts may be introduced into buildings, structures, and roads in one or more of the following three ways (which were originally identified by Schaeffer (1932) and are elaborated here in the light of recent construction experience):

(i) Salts present in materials prior to their use in construction (e.g. salts emplaced during the deposition of sedimentary rocks and superficial deposits used as aggregates; salts deposited in rocks and superficial deposits from sea spray, rainfall, or evaporating groundwater; salt contamination of construction materials (including steel bars) resulting from storage on saline ground or from windblown salts; and salts formed during the manufacture of tiles).

(ii) Salts formed in structures (post-construction) as a result of chemical changes to the original materials.

(iii) Salts emplaced in buildings, structures, and roads from external sources (post-construction), the most important additions being from the atmosphere, capillary rise of saline groundwater, and the washing or inundation of buildings with weakly saline water from piped water supplies.

The magnitude of the actual or potential salt-weathering damage is dependent on three main groups of variables. First, there are the *environmental conditions* such as climate, geology, and groundwater characteristics, that determine the spatial variability of salt type, concentration, and mobility. Second, there is the nature of the *building materials* used, for differing materials have differing degrees of susceptibility to salt weathering. The third main variable is *workmanship* (Fookes and Collis, 1975a) which includes the experience of the manual labour involved in construction, as well as the adequacy of supervision and the level of design and construction practices employed. Bad workmanship or faulty design can significantly increase the susceptibility of structures to salt attack by, for instance, providing cracks, crevices, and hollows within which salts can be concentrated.

Both the *environmental* and *material* variables are examined in greater detail in the following sections, together with an indication of the nature of the damage sustained by man-made structures and suggestions as to the counter-measures that can be employed to minimize the impact of this engineering hazard. Such a discussion is best approached by subdividing the subject into three main sections: buildings composed of natural stone and brickwork; buildings constructed of concrete; and roads. It is important to appreciate at the outset that although the methods employed to reduce salt-weathering hazard losses may vary in detail depending on geographical, economic, material, and type-of-construction considerations, they generally fall into four main groups of measures:

(*i*) *Planning solutions,* which involve the modification of development plans and the re-siting of structures so that construction does not take place on hazardous ground.

(*ii*) *Engineering solutions,* where buildings, structures, and roads are specifically designed to withstand salt-weathering attack.

(*iii*) *Quality control,* where the materials used in construction are carefully screened so as to minimize the quantities of deleterious salts introduced into structures.

(*iv*) *Maintenance,* where the upkeep of buildings, structures, and roads is maintained at an adequate level so as to minimize the avenues available for salt attack. Under this heading are included the washing of buildings with salt-free water and the eradication of surface cracks by repointing.

2. *Building composed of natural stone and brick*

Sandstone and limestone blocks bound by cement or mortar are traditional indigenous building media in many drylands. Both the blocks and the cementing materials may be subjected to salt attack through volumetric changes (see section (b) 2 above) by salts emplaced before, during, or after construction (Schaeffer, 1932). This is clearly illustrated by the rapid disintegration of exposed brickwork at Mohenjo-Daro in the Indus Valley (Pakistan), which has required extensive conservation measures (Van Lohuizen de Leeuw, 1973; Goudie, 1977). The most serious damage occurs in low-lying areas underlain by shallow saline groundwater (playas and sabkhas) where capillary rise into buildings up to a height of about 2 m above ground-level may result in damp staining, salt efflorescence, and weathering attack. A survey of the old town of Suez (Halcrow and Partners, 1975) revealed numerous instances of building deterioration through salt attack (both mechanical and chemical), the most common forms of damage being:

(i) Groundheave of tiles and floor blocks, both inside and outside buildings attributed to crystal growth from underlying gypsum-rich desert fill; (ii) crumbling and disintegration of the *surface* of granite, sandstone, and limestone blocks up to 1.5 m above ground level; (iii) destruction of cement and mortar by chemical sulphate attack, thereby facilitating salt weathering within walls in contrast to the normal situation where attack is limited to the outside surface; (iv) salt attack of brick foundations emplaced in the zone of capillary rise, where brickwork is sometimes reduced to powder, with serious consequent settlement damage to buildings.

While the effects of salt weathering on stone buildings are unsightly, little work has so far been undertaken to assess the seriousness of the problem in terms of safety considerations or the rapidity of attack. A recent survey of part of the Suez old town (Egypt, Ministry of Housing and Reconstruction, 1978), described as an 'old, salt-infested, mature district', has gone some way towards such an assessment by gathering information on buildings concerning the type of construction, height of capillary rise in walls, the state of construction, and the degree of weathering (Table V.2). A random sample of 347 buildings was used in the survey and yielded data on the relationships between degree of weathering and type and age of construction, and degree of weathering and the state of construction (Table V.3). The main conclusions of the survey, which could well be characteristic for locations underlain by moderately saline groundwater at shallow depth in drylands, were as follows:

(i) While salt weathering causes unsightly discoloration and decoration change, it does not appear to result in the structural destruction of buildings; (ii) salt weathering does not appear to display a preference for any particular one of the types of construction encountered; (iii) the degree of salt weathering damage appears closely related to 'state of construction' (ranging from 'uninhabitable' to 'high quality') – the poorer the state of construction, the greater the degree of damage; (iv) rate of weathering appears to diminish with age in this area underlain, as it is, by moderately saline groundwater (i.e. about 10‰, or roughly one-third the average salinity of sea-water, and characterized by capillary rise in walls to between 1.2 and 1.4 m above ground level), although the reason for this is not clear.

Most of the salt-weathering damage to stone buildings is due to the action of salts derived from capillary rise, although additional inputs may result from washing and leakages. Significant reductions in damage can therefore result from avoidance of low-lying, damp, saline areas for construction purposes, or from the construction of damp courses. In addition, varying degrees of protection can be achieved by improving standards of workmanship, washing with salt-free water (if available), and the application of surface treatments that seek to seal the outer surface of buildings, thereby preventing or reducing the access of saline solutions. Unfortunately, many of the surface treatments have proved both expensive and disappointingly ineffective, mainly because in preventing access of solutions, they also tend to prevent the escape of solutions already contained in the stone, so that sub-surface disintegration may continue after an application (e.g. Cooke and Doornkamp, 1974).

TABLE V.2

Survey of building damage by weathering, Suez, Egypt

Category	I. *'Type of construction' classification* Description
A	Load-bearing stone walls throughout sometimes rendered, often to 75 cm above ground
B	Load-bearing stone under brick walls often rendered throughout, sometimes with concrete upper floors through walls. Stone extends to 3 m above ground
C	Concrete frame, brick infill, usually on stone base 0 to 0.5 m above ground
D	Suezi type, commonly mixed stone and brick load-bearing walls with a varying proportion of walling as clay bonded rubble within light timber cages
E	Mixed construction other than Suezi

Measure	II. *'Degree of weathering' classification* Description
1	Barely perceptible efflorescence
2	Distinct discoloration but no visible surface damage
3–4	Heavy discoloration combined with inception of decay by spalling of rendering along lower edges, slight retraction of mortar in joints and the like
5–6	Widespread loss of rendering, 25% missing and some loosened. Substantial mortar loss in joints or spalling away of base stone and brickwork
7–8	Rendering largely gone, mortar lost from lower joints, about 1 cm of base stone and brickwork spalled away
9–10	Structural breakdown of wall leading towards settlement of living areas sufficient to curtail their utility

Measure	III. *'State of construction' classification* Description
0	Uninhabitable for structural reasons
1–2	Long neglected crumbling structure, probably best demolished
3–4	Neglected and generally weakened structure, probably capable of restoration
5–6	Stable and generally serviceable structure in average state of repair and decoration
7–8	High quality specification solid and well finished structure with apparent indefinite life
9–10	High quality specification design specifically for Qalzam conditions. (A fully effective damp proof course would be obligatory. None was seen and hence no buildings came into this category.)

Source: Egypt, Ministry of Housing and Reconstruction (1978).

Reduction of salt-weathering damage can also be achieved by quality control of building materials. This includes the use of sulphate-resistant cement, employment of building stones with low initial salt content, and use of building materials that show a high level of resistance to salt-weathering attack. Laboratory studies have already shown that rocks display varying resistances to salt weathering (see section (b) 2 above). It is possible,

TABLE V.3

I. Relation of 'degree of weathering' to 'age and type of construction' in survey of building damage by weathering, Suez, Egypt

Age and type of construction	Low −	1	2	3	4	5	6	7	High 8	9	10	Totals
Over 15 years												
A Stone	−	1	6	4	4	5	2	3	3	1	−	29
B1 Brick, Stone Base	−	5	13	13	7	13	7	4	1	−	−	63
B2 Brick, No Stone Base	−	−	3	4	2	4	2	−	−	−	−	15
C Concrete Frame	1	3	3	−	3	5	−	1	1	−	−	17
D Suezi	1	6	5	7	2	17	5	1	2	1	−	47
E Mixed	−	−	2	3	−	2	2	−	1	−	−	10
	2	15	32	31	18	46	18	9	8	2	−	181
3 to 15 years												
A Stone	1	1	1	−	−	2	−	−	−	−	−	5
B1 Brick, Stone Base	4	8	14	16	16	12	11	1	1	−	−	83
B2 Brick, No Stone Base	3	4	3	4	1	4	−	−	2	−	−	21
C Concrete Frame	2	4	6	10	7	3	4	−	−	−	−	36
E Mixed	−	−	1	−	−	−	−	−	−	−	−	1
	10	17	25	30	24	21	15	1	3	−	−	146
Less than 3 years												
B1 Brick, Stone Base	−	1	−	−	−	−	−	−	−	−	−	1
C Concrete Frame	5	7	6	1	−	−	−	−	−	−	−	19
	5	8	6	1	−	−	−	−	−	−	−	20
Totals	17	40	63	62	42	67	33	10	11	2	−	347

II. Relation of 'degree of weathering' to 'state of construction' in the survey of Suez, Egypt

State of construction		Low −	1	2	3	4	5	6	7	High 8	9	10	Totals
Derelict	−	−	4	4	5	−	3	2	−	1	−	−	19
	1	1	2	2	8	−	14	6	2	1	−	−	36
	2	−	3	3	6	3	6	6	4	4	−	−	35
	3	5	5	13	15	16	18	4	2	3	−	−	81
	4	−	3	9	5	7	10	10	−	−	−	−	44
	5	8	13	22	17	15	14	4	2	1	2	−	98
	6	2	8	5	4	−	1	1	−	1	−	−	22
	7	−	2	3	1	−	1	−	−	−	−	−	7
	8	1	−	2	1	1	−	−	−	−	−	−	5
	9	−	−	−	−	−	−	−	−	−	−	−	−
Sound in all respects	10	−	−	−	−	−	−	−	−	−	−	−	−
Summary	<3	1	9	9	19	3	23	14	6	6	−	−	90
	3 to5	13	21	44	37	38	42	18	4	4	2	−	223
	>5	3	10	10	6	1	2	1	−	1	−	−	34
Totals		17	40	63	62	42	67	33	10	11	2	−	347

Source: Egypt, Ministry of Housing and Reconstruction (1978).

therefore, to select 'resistant' building stones by subjecting samples of locally available, potentially useful materials to laboratory-simulated salt-weathering tests.

Such an approach was adopted as part of the Bahrain Surface Materials Resources Survey (Bahrain, Ministry of Works, Power and Water, 1976; Brunsden et al., 1979). Although specifically directed towards the search for aggregate materials, the methodology is also applicable to the selection of building stones. In this survey, 15 different rock samples were subjected to examination. Rock samples were chosen for a variety of reasons: some were already used as aggregates or in 'fill'; some were known to be non-resistant (e.g. the Yellow Siltstone, see Table V.4) and thus provided a contrast with the more durable rocks examined, and some appeared relatively durable because they produced higher ground and could, therefore, be potentially useful as building materials.

The salt-weathering test employed in this study was the same as that carried out by Goudie et al. (1970), and the climatic conditions simulated were considered similar to those found in Bahrain. In this test, three 3-cm cubes from each rock type were subjected to the test described above (section (b) 2(i)). The salts used were Na_2SO_4 and $CaSO_4$. Na_2SO_4 was used because previous experiments (Goudie et al., 1970) revealed that it works relatively quickly; calcium sulphate was selected because it is one of the most important salts present in arid areas such as Bahrain; and distilled water was used to provide a control on the experiments. In addition to the standard test, the

TABLE V.4

Experimental salt weathering test results for rock samples from Bahrain

Sample no.	Material	Water absorption %*	Rebound no.†	% Remaining‡	Cycles elapsed	Rank
SW2	Yellow Siltstone	42.74 (1)	27.5 (11)	0.00	7	14=
SW3	Khobar Bed 3	9.79 (6)	31.8 (9)	0.00	9	13
SW4	Numulitic Limestone	11.06 (5)	22.1 (12)	0.00	7	14=
SW5	Crust on Bed 5 Bedded Limestone	4.15 (11)	34.3 (7)	75.20	50	6
SW6	Khobar Bed 4	11.71 (4)	28.3 (10)	0.00	15	12
SW7	Blue Askar	1.94 (13)	49.1 (3)	90.38	50	5
SW8	Flinty Khobar	6.78 (7)	18.9 (16)	0.00	34	9
SW9	Miocene Limestone	6.13 (8)	34.0 (8)	0.00	36	8
SW10	Blue Askar	2.37 (12)	38.3 (6)	103.18	50	1
SW11	Alat Dolomite	5.17 (9)	43.1 (5)	101.65	50	2
SW12	Alat below cap rock	5.04 (10)	44.0 (4)	100.52	50	3
SW13	White Askar (Govt. Quarry)	1.55 (14)	57.50 (1)	0.00	19	11
SW14	Sharkstooth 9	21.07 (3)	17.10 (14)	0.00	23	10
SW15	Sharkstooth 24	23.07 (2)	12.80 (15)	2.78	50	7
SW16	Blue Askar	1.17 (15)	50.30 (2)	99.84	50	4

Notes
*Determined by British Standard Method (BS 1377). Figures in brackets are ranks.
†Determined by Schmidt Hammer (as described in text).
‡After treatment with sodium sulphate (as described in text).
The material names used are based on those of Willis (1967).
Source: Bahrain, Ministry of Works, Power and Water (1976).

samples were subjected to two additional tests; one for water absorption (BS 1377), the other for compressive strength, known as the Schmidt Hammer Test. The Schmidt Test hammer is intended for non-destructive testing of concrete and natural materials (e.g. Kazi and Al-Mansour, 1980). It measures the Rebound Number (R) which is related to hardness. R-values have been found to correlate with compressive strength. The hammer allows strength testing to be carried out rapidly, on a large number of samples, cheaply, and in the field. Clean, non-flaky, non-jointed, approximately flat surfaces were selected for field testing of rock types subsequently used in salt-weathering tests. Ten R-values were recorded at each location, and the mean value ($R\bar{x}$) was derived. Corrections to the R-values need to be made according to whether one is dealing with vertical or horizontal surfaces. These correction values are shown on the nomogram attached to the instrument.

The results of this investigation are shown in Table V.4 which summarizes data relating to types of material, water absorption capacity, rebound numbers, and the destruction of samples during Na_2SO_4 tests. The response of rock samples to Na_2SO_4 tests over time is shown in Fig. V.4.

The samples tested were found to have water-absorption capacities which ranged from as little as 1.2% to as much as 42.7%. Similarly, rebound numbers determined with the Schmidt Hammer gave a range of from 57.50 (equal to a cube compressive strength in excess of about 11 000 p.s.i; $7.5845 \times 10^4 KN/m^2$) at the strongest end to about 12.80 (equal to a cube compressive strength of no more than about 1 000 p.s.i.; $0.6895 \times 10^4 KN/m^2$) at the weakest end.

This variability in material character was reflected in the highly variable response to Na_2SO_4 crystallization tests (Fig. V.4). Samples SW10, SW11, SW12, SW16, SW7, and SW5 (Table V.4) showed a relatively high degree of resistance to this process and might thus stand up well to crystallization of salts on exposed surfaces within the zone of salt-weathering hazard. On the other hand, other materials responded badly, including the highly porous and soft Yellow Siltstone (sample SW2).

All samples tested, with the exception of the Yellow Siltstone, responded well to the effects of simple wetting and drying, none of them experiencing a weight loss above 2% over 40 cycles. The effect of gypsum crystallization, probably because of the low solubility of calcium sulphate in water, was minimal and only three samples (SW2, SW4, SW15) showed any measurable response.

Neither the water-absorption capacity nor the hardness of samples appear to offer good predictors of resistance to Na_2SO_4 crystallization. This clearly shows the desirability of subjecting materials to laboratory weathering tests rather than simply relying upon surrogate properties if the materials are to be used in areas subject to salt-crystallization hazards.

3. Buildings and structures made of concrete

The impact of salt weathering has been most apparent in concrete buildings and structures and has resulted in widespread reports of surface discoloration, disfigurement, and damage on a scale sufficient to require expensive renovation or to cause a reduction in the life of buildings (Fookes and Collis, 1975a, 1976). Such damage is the result of salt accumulation within buildings

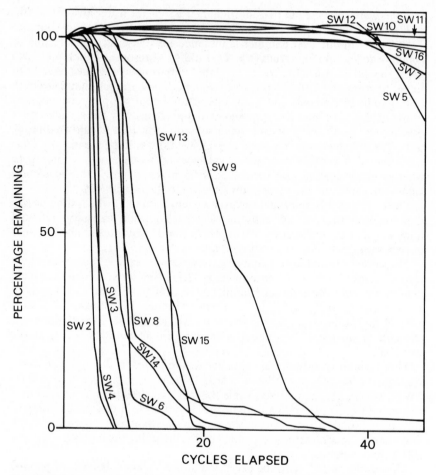

FIG. V.4 The response of rock samples from Bahrain to sodium sulphate weathering tests (see also Table V.7) (after Doornkamp *et al.*, 1980)

and structures, either by contamination of building materials (aggregates, mixing water, steel bars) or by post-construction introduction from saline groundwater (foundations) or saline capillary water. Damage is most apparent in those areas underlain by shallow saline groundwater.

A wide variety of cracking phenomena have been reported for concrete structures in drylands (Table V.5 and Fig. V.5) and there is general agreement that cracking density is much greater under these conditions than for similar concretes in temperate latitudes (Fookes, 1976c). The causes of cracking are highly variable (Table V.5) and include both physical and chemical processes many of which are encountered by the construction industry in other environments and are not restricted to hot drylands. However, Fookes and Collis (1975a) indicated that concrete appears to deteriorate more rapidly under dryland climatic conditions (Table V.6). The greatly increased

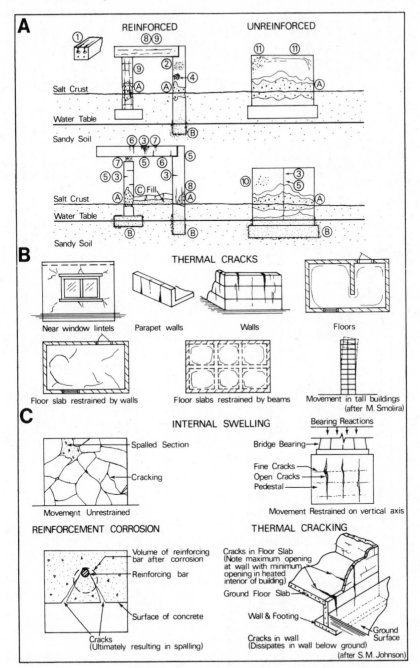

FIG. V.5 A highly idealized illustration of the various crack types outlined in Table V.5 (A) Crack type numbers refer to Table V.5; (B) Examples of thermal cracks; (C) Some details of various crack types (after Fookes, 1976c)

TABLE V.5

A preliminary classification of various crack types in Middle East concretes

Type of cracking and Fig. V.5	Age of onset	Principal cause	Principal location and frequency	Principal confirmatory tests	Potential risks	Remedial measures if required	Avoidance
(1) Plastic settlement	First few hours before setting	Vertical movement, bleeding, loss of moisture	Rectilinear pattern – coincides with top (upper side) of reinforcement. Fairly common	Coring and inspection. Timing of crack development	Loss of bond strength to top steel. Exposure of reinforcing steel	Epoxy grouting, or similar to seal but restoring bond strength problematical	Improve mix design. Reduce retardation. Air entrainment. Improve curing, especially use of windbreaks
(2) Plastic shrinkage cracks	First few hours	Plastic movement (not in sense of settlement). Loss of moisture	Any surfaces. Any concretes. Fairly common	Ditto	Exposure of reinforcing steel	Epoxy grouting	Improve curing, especially windbreaks
(3) Drying shrinkage cracks	Days to months	Shrinkage from loss of moisture	Central position walls, columns, beams and floors. Common	Ditto	Ditto	Ditto	Improve curing. Improve mix – lower w/c ratio. Crack inducers and grout
(4) Crazing	Ditto	Minor surface shrinkage of laitance	Any smooth cast surfaces. Common	Inspection and timing of development	Unsightly	Surface cosmetics	Improve curing. Do not overwork
(5) Thermal cracks	Days to weeks for initial movement. Months to years for diurnal and seasonal movement	Short- and long-term expansion and contraction movements	Thin members in particular. Fairly common	As (1) plus microscopic examination	As (1)	Epoxy grouting	Reduce peak temperature and temperature differentials, plus allowance in structural design for long-term movements
(6) Tensile cracks	Ditto	Tensile stress movements	Structural members. Not too common	As (1), but coring may be risk to integrity of structure	As (1) – may be critical depending on situation	Ditto, or replace affected member or strap up or box	Improve structural design
(7) Shear cracks	Ditto	Shear stress movements	Ditto	Ditto	Ditto	Ditto, or pin with bolts	Ditto

(8) Reinforcement corrosion cracks	Many months to many years	Expansion of reinforcement by rusting, especially in porous concretes and in marine environments	Reflects pattern of reinforcement. Fairly common	Coring and/or hand samples for chemical tests. Observation of condition of steel	Ditto	Replace affected members	Good cover to steel 50–75 mm minimum Dense concrete
(9) Reinforcement corrosion related to high chloride content of concrete	Weeks to years	Ditto	Ditto. Very common	Ditto	Ditto	Replace affected members *or* completely clean down to and all around whole length of the reinforcement and reinstate with clean concrete	Ditto, plus chemically clean aggregates. Sweet water for mixing
(10) Popouts and later possibly map cracking	Few months to many years	Expansion of aggregates (on wetting)	Any concrete surface Not too common	Coring and/or hand sampling. Microscope examination	As (2)	As (8)	Use non-expansive aggregate
(11) Ditto, plus extruded gel and major cracking ('alkali aggressive')	Few to many years	Expansion of aggregates on chemical reaction	Any concrete Rare	Ditto, plus chemical tests	Ditto, but can lead to complete loss of structure	As (8). This condition will continue unless affected members are replaced	Ditto, consider using low alkali cement or pozzalanic content
(A) Physical 'salt weathering'	Months to several years	Salts in ground and atmosphere	From just below ground surface up to about 2 m above in porous materials. Very common	Observation, plus hand samples and chemical tests (these not usually necessary)	Unsightly. May cause collapse of thin and/or porous materials (e.g. block wall)	Make good and use avoidance methods	DPC plus tanking by membrane (not paint or emulsion)
(B) Chemical 'sulphate attack'	Years to many years	Sulphate salts. Damp or wet ground (inc. garden watering)	From just above ground water-table to well below. Not too common	Observation plus coring and/or hand samples and chemical tests	May cause eventual collapse of foundations	Replace or make good and use avoidance methods	Use dense concrete SRC and/or impeccable tanking by membrane
(C) Floor heave	Months to many years	Sulphate and other salts under floor in ground or fill. Dampness	Tiled or concrete floors. Locally common	Observation plus hand samples and chemical tests	Irregular surface	Replace or make good and use avoidance methods	Tank floor on ground or use chemically clean fill

Source: Fookes (1976c).

TABLE V.6

Some aspects of concrete deterioration related to certain environmental factors and time

Feature	Possible effects	Time before effects may be observed in hot, arid maritime climates. Temperate climates shown — — — for comparison					
		hours	days	weeks	months	years	decades
Plastic shrinkage (including plastic settlement)	Cracking, localized loss of bond and/or cover to reinforcement	■ —					
Initial drying shrinkage	Crazing and/or cracking, reduced effective/actual cover to reinforcement		■	■	—		
Initial thermal contraction	Cracking in depth, localized loss of cover to reinforcement			■	■ —		
Salt hydration	Cracking in depth, loss of cover to reinforcement			┄	■		
Drying shrinkage	Cracking in depth, localized loss of cover to reinforcement				■	■ —	
Sulphate attack external	Expansion and cracking, surface disintegration, general loss of effective and actual cover to reinforcement					■	—
Unsound aggregates	Surface disintegration and general loss of effective and actual cover to reinforcement					■	—
Salt crystallization	Surface disintegration and localized loss of effective and actual cover to reinforcement					■	┄
Reactive siliceous aggregates	Pop-outs, expansion and cracking, loss of effective and actual cover to reinforcement					■	┄
Carbonation	Reduction in alkalinity, loss of effective cover to reinforcement						■ —
Sulphate attack internal	Expansion and cracking, loss of effective and actual cover to reinforcement						■ —
Reactive carbonate aggregates	Expansion and cracking, loss of effective and actual cover to reinforcement						■ ┄
Loss of effective or actual cover to reinforcement	Reinforcement corrosion cracking and spalling					■	—
Excessive chloride concentrations	Accelerated reinforcement corrosion cracking and spalling				■	■	—

Source: Fookes and Collis (1975a).

significance of crack development in tropical and sub-tropical drylands appears to be a consequence of several factors: (i) *Climate*, particularly with respect to the prevalence of high temperatures and low relative humidities, which results in rapid surface drying and the production of 'early shrinking movements' in concrete. (ii) *Poor workmanship and design.* (iii) *Use of unsuitable aggregates*, both in terms of size grading and content (Fookes and Collis, 1975a). The latter is of particular significance as aggregates containing hydratable materials (expansive aggregates) can result in 'popouts'. The use of alkali-reactive rocks (including silica-reactive and carbonate-reactive types) can lead to even more serious damage through chemically generated volume changes (French and Poole, 1974, 1976; Fookes and Collis, 1975a). (iv) *Physical and chemical processes* associated with the presence of significant concentrations of a wide variety of salts.

Because of the numerous reasons for concrete cracking it is often difficult to identify the significance of salinity-induced mechanisms. Further complexity results from the fact that cracking due to non-salt processes provides suitable avenues for subsequent salt-weathering attack. Similarly, cracking caused by salt weathering facilitates the ingress of the atmosphere into the concrete mass and can result in the initiation of other breakdown mechanisms such as the rusting of reinforcing bars. Thus all forms of cracking must be seen as elements of the salt-weathering problem, and every effort should be made to minimize their development.

Of the many different salts present in drylands the most widely occurring aggressive ones affecting concrete are halite and gypsum, with Na_2SO_4 and $MgSO_4$ also of significance. Their potential impact is best considered in the context of *internal* attack (i.e. introduction in building materials), and *external* attack (i.e. post-construction introduction by saline groundwater) (Fookes and Collis, 1975a, 1976).

(i) *Internal problems.* Most of the problems associated with salts *in* concrete are due to the presence of chlorides and sulphates that have been introduced by contaminated aggregates and mix water (Harrison, 1971; Newman, 1971; Fookes and Collis, 1976).

Chloride salts. The presence of chlorides (usually of sodium) can result in several problems (Table V.7) of which the corrosion of steel reinforcing bars is generally considered to be the most widely developed and serious (Fookes and Collis, 1976). The reason for chloride attack is that as the inhibitive action of lime and alkali in solution is significantly reduced in the presence of chlorides – the higher the chloride concentration the more rapidly rusting appears to occur. These views have mainly emerged as a result of the analysis of concrete samples obtained from various parts of the Middle East. One such study (Fookes and Collis, 1975a, 1976) revealed that over 50% of the cracked concrete sampled adjacent to corroded reinforcements had chloride contents (as NaCl) in excess of 4% by weight of cement (maximum 13%) and over 95% of samples were in excess of 1%. However, by no means all concretes with over 1% NaCl had suffered cracking, which clearly indicates that further research is required to ascertain the reasons for this differential response. The three most likely explanations appear to be variations in

TABLE V.7

Some possible effects of 'internal' chloride salts

Accelerated corrosion of unprotected steel due to chemical and electrochemical reactions caused by internal and external chlorides on steel.

Reduced sulphate resistance due to complex chloride reactions with certain hydrates which in turn are more susceptible to sulphate attack than the original hydrates.

Disintegration due to crystallization of internal and external chloride salts on and in exposed concrete surfaces and within aggregate particles.

Increased salt content near exposed surfaces due to the migration of internal chloride salts towards exposed surfaces, where evaporation of water from within is taking place.

Efflorescence due to the deposition of internal chloride salts as water evaporates through exposed concrete surfaces.

Increase in normal bleeding tendency due to reduced surface tension and other possible effects of internal dissolved salts.

Undue expansion due to possible expansive reaction between internal and external chloride salts and the aluminates present as normal products of cement hydration.

Increased alkali metal content due to the chloride salts being principally sodium chloride, with secondary amounts of potassium chloride, and of relevance where aggregates are indicated to be potentially reactive with alkalies.

Others: speculative or as yet unidentified reactions, influences or catalytic effects, where concrete also includes excesses of other salts originating from the indigenous materials and/or the local environment.

Source: Fookes and Collis (1976).

thickness of concrete cover, variations in the porosity of the concrete cover, and chemical reactions (as yet not completely identified). Work on this problem is currently being carried out by government bodies, engineering firms and materials consultants (e.g. Messrs Sandbergs of London, and Messrs Kenchington, Little and Partners (1974–5)) in the hope of identifying the relative importance of the various possible causative factors.

As a result of these and similar investigations it has been suggested (Fookes and Collis, 1976) that total acid-soluble chloride levels (as NaCl) for general purpose reinforced concrete should not exceed 0.50% by weight of cement and preferably be less where other factors may accentuate corrosion and cracking. Fookes and Collis (1976) also recommended a limit of 0.05% by weight for coarse aggregate, and 0.10% by weight for fine aggregate. Local standards and guidelines may be even more severe, as in Qatar where the Engineering Services Department in 1976 laid down maximum permitted levels of 0.06%, 0.10%, and 0.4% respectively for fine aggregates, coarse aggregates, and concrete (see also section IV (a) 2, and Tables IV.4 and IV.6). Even more stringent levels are recommended by many engineering firms who suggest that maximum chloride contents in concrete should be as low as 0.20%. However, as the majority of construction activity takes place in coastal regions, most easily available aggregate materials (both superficial deposits and rocks) show high levels of contamination due to inputs of airborne oceanic salt (see section (b) 1) as well as from saline groundwaters.

The production of chemically suitable aggregates therefore requires the use of extremely careful and selective extraction and quarrying practices, careful processing (with tight control of water quality), and sensible stockpiling on uncontaminated surfaces away from, or protected from, coastal exposure (Fookes and Collis, 1975a).

If these precautions are taken then it is suggested (Fookes and Collis, 1976) that the realistic minimum sodium chloride level in concrete likely to be attainable using aggregates from very salty areas is about 0.3% by weight of cement if demineralized water is used (NaCl not greater than 0.05% by weight). Values of less than 0.3% are only likely to be achieved if chloride-free (less than 0.01% NaCl) aggregate is used.

Sulphate salts. The presence of sulphates (usually of calcium) can also result in several problems (Table V.8). However, the damage produced by internal sulphate attack appears to be less than that resulting from the presence of chlorides, although they may be of significance if anhydrite is present in the aggregate materials. This uncertainty about the importance of sulphate attack is due to the identification of fairly high sulphate contents in undamaged concretes. For example, the Qatar samples mentioned above revealed sulphate levels of 3.50–7.70% (the sulphate content is the total sulphate soluble in dilute hydrochloric acid and expressed as sulphur trioxide (SO_3) by weight) and yet no correlation between sulphate content and damage could be identified.

It is clear that considerably more research into this form of weathering is urgently required. In the meantime precautionary measures should be taken to limit the sulphate content of concretes. Fookes and Collis (1976) suggested that the sulphates present in aggregates should not be more than double that in the cement (Cement and Concrete Association, 1970) and they suggested that the total sulphate content of reinforced and unreinforced concrete should not exceed 4.0% by weight of cement, including the sulphate present in the cement, thereby indicating that the maximum levels for both coarse and fine aggregates should be 0.4% by weight of aggregates. Similar limits have been

TABLE V.8

Possible effects of 'internal' sulphate salts

Undue expansion and disintegration due to chemical reactions between external and internal inclusions of sulphates and certain aluminates present as normal products of cement hydration.

Undue expansion due to post compaction hydration of inclusions of calcium sulphate in a dehyrated condition.

Disintegration due to the crystallization of soluble sulphates, principally magnesium and sodium sulphates, on and in exposed concrete surfaces, and within aggregate particles.

Increased alkali metal content due to the presence of some dissolved alkali metal sulphates in ground waters.

Other: speculative or as yet unidentified reactions, influences or catalytic effects where concrete also includes excess of other salts originating from the indigenous material and local environment.

Source: Fookes and Collis (1976).

established by governments and engineering firms. As most easily won aggregate, especially superficial deposits, has sulphate levels in excess of 0.4%, the enforcement of such standards requires the careful choice of extraction sites, adequate supervision of extraction methods (Fookes and Collis, 1975b), and frequent checks on aggregate chemistry.

(ii) *External problems*. External attack of concrete can result from two distinct sets of processes, '*physical*' *salt weathering* and *sulphate attack*, which may operate together or separately (Fookes and Collis, 1976).

'*Physical*' *salt weathering*. This form of salt weathering is produced by the growth of salt crystals in surface cracks which cause damage by the exertion of pressures due to crystal growth, hydration, and thermal changes (section (b) 2). It usually occurs in a relatively narrow vertical zone extending from just below ground level to up to 2 m above the ground surface and results from the emplacement and accumulation of salts by capillary rise of saline solutions and aeolian processes.

The existence of cracks in the concrete, due to either poor workmanship or the mechanisms outlined in Table V.5, is an important factor for salt weathering which can result in crack enlargement and surface disintegration. This, in turn, can facilitate the penetration of air into the body of the concrete and may result in the initiation of further damage-inducing processes such as the rusting of reinforcements. Prevention of salt weathering can be achieved by good workmanship and design practices, quality control of materials, employment of 'tanking' membranes, and careful maintenance.

Sulphate attack. Sulphate attack has been widely reported from drylands where there is capillary rise into buildings and other structures from highly saline groundwater (Fookes and Collis, 1975a, 1976). Although encountered in temperate latitudes, both the scale and rate of attack are greatly increased under the climatic and groundwater conditions encountered in drylands, particularly dry coastal areas (Table V.6). Several damage-inducing chemical exchanges have been identified. The most significant and widespread is the reaction of the calcium aluminium hydrate in cement with the sulphate radicals in groundwaters which results in a volume increase due to the formation of etringite. Similarly, if magnesium salts are present, then faster volume changes occur due to reaction with the calcium hydroxide of the cement paste (Fookes and Collis, 1975a, 1976).

The commonest result of such reactions is the upwards bulging (heave) and disintegration of floors composed of bricks, stone blocks, paving stones, tiles, and concrete (Table V.5 and Fig. V.5) and this is particularly apparent in buildings constructed on damp sulphate-laden 'desert fill' (Fookes and Collis, 1976). Expansion of concrete, or the mortar bedding under tiles, may even cause walls to be thrust sideways if conditions are particularly aggressive. Conversely, buildings constructed on gypsum-rich 'desert fill' in areas with low water-tables can also suffer cracking due to foundation settlements resulting from downward leaching of the gypsum by water from pipeline leaks, waste pipe leaks, soakaways, etc.

The spatial variation in intensity of sulphate weathering is largely explicable in terms of the relationship of the water-table to the topographic surface (i.e. 'height' or 'depth' of water-table), groundwater chemistry, and the character of foundation materials. However, its widespread occurrence and the magnitude of the resultant damage has proved unexpected, for although sulphates are plentiful in desert environments, particularly in coastal situations, most of them are locked up as gypsum which has a rather low solubility in water. The undoubted significance of sulphate attack suggests that gypsum is more readily dissolved by migrating waters containing other salts, one suggestion being that the presence of bicarbonates (locally common in the Middle East) facilitates the removal and redistribution of sulphates (Fookes and French, 1977).

Preventative measures with respect to this form of attack in saline arid environments are similar to those used in temperate environments (e.g. use of dense concretes, sulphate-resisting cement, and tanking), except that more stringent precautions should be taken owing to the range and concentration of salts, the higher ground and air temperatures, and the generally greater permeability of the soils (see Fookes and Collis, 1976). A summary of the various preventative measures that may be used in aggressive conditions is shown in Table V.5 and Fig. V.6. Sulphate damage may be further reduced through the careful selection of foundation and aggregate materials (see Chapter IV) and good workmanship. The ultimate solution is, of course, to plan development in such a way as to avoid the most hazardous sites. This approach is examined in detail in a later section.

4. *Roads*

Surfaced roads in drylands often show signs of damage in terms of cracking, potholing, scabbing, stripping, crumbling, and disintegration (Fookes, 1976b). These phenomena are particularly apparent in areas where saline groundwater is at or near the surface and where abundant soluble salts occur on the ground surface and in the soil. Nevertheless, it remains uncertain how far such damage is directly attributable to salt-weathering activity, for there are other possible causes of deterioration such as overstress in the paving layers resulting from rapidly increasing traffic flows and abnormally high axle loads, the use of poor coarse or fine aggregate, lack of quality control, and poor workmanship (Fookes and French, 1977). Break-up may also be due to thermal cracking (Tomlinson, 1978). However, the presence of salt efflorescence on road surfaces, together with the fact that salt enrichment is often achieved on small rises of ground (such as roads), suggests that salt weathering is probably of major significance and operates in conjunction with the other factors outlined above.

The potential for salt-weathering damage to bituminous paved roads has been known for some time and examples of deterioration through soluble salt attack have been noted in many drylands including Australia (Cole and Lewis, 1960), India (Mehra *et al.*, 1955), South Africa (Weinert and Clauss, 1967; Netterberg, 1970; Netterberg *et al.*, 1974; Blight *et al.*, 1974), and the Middle East (Bahrain, Ministry of Works, Power and Water, 1976; Fookes, 1976a; Fookes and French, 1977).

The presence of salts at, or near, the ground surface is not necessarily an engineering hazard if the salts exist in a virtually dry condition. Thus salty

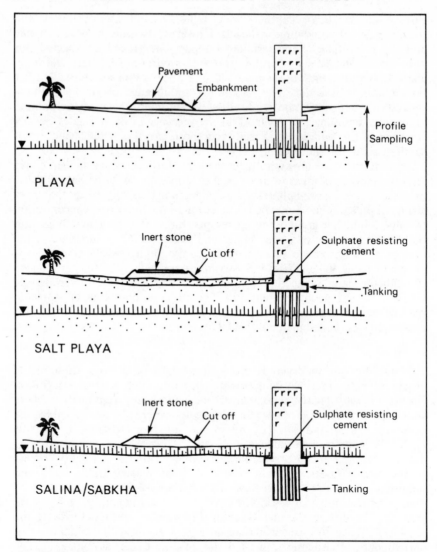

FIG. V.6 Different methods of foundation protection in different types of desert terrain (after Fookes and Collis, 1976)

soils (generally sabkha material) have been used successfully as binders for unsurfaced roads and as bases of surfaced roads in very dry areas (Baudet *et al.*, 1959; Ellis, 1973; Ellis and Russell, 1973; Russell, 1974; Tomlinson, 1978). In damp conditions, however, physical damage to the surface layer results from the upward transport and concentration of salts, especially calcium, magnesium, and sulphate, by capillary movements induced by evaporation through the surfacing (Netterberg *et al.*, 1974; Blight, 1976). Such movements may be increased by the very existence of the road, for the soil temperatures immediately beneath paving layers are often in excess of those under adjacent desert surfaces (Hatherly and Wood, 1957; Tomlinson, 1978) and may result in a concentration effect with some saline water being attracted to the road from the adjacent areas. The salts involved may be derived from saline groundwater, salts in the ground materials immediately underlying the road, or salts contained in the aggregate materials used for the road base and sub-base – with groundwaters (including capillary water) being of the greatest significance. The concentration effect is clearly shown on Table V.9, particularly with respect to calcium and sulphate and, to a lesser extent, magnesium.

The physical and chemical consequences of attack on roads are complex and largely depend on the road design, the nature of the aggregate materials used, and the physical and chemical characteristics of what can be termed the '*water–salt system*' (Fookes and French, 1977). These are summarized in Tables V.10, V.11, and V.12. Three main causes of damage have been identified:

(*i*) *The chemical attack by soluble sulphates on Portland cement stabilized road bases* (Uppal and Kapur, 1957; Sherwood, 1962; Smith, 1962), a mechanism similar to that described under 'sulphate attack' above.

(*ii*) *The physical and chemical attack of aggregate used as base or sub-base materials.* There appear to be a variety of mechanisms which could operate in these circumstances, none of which is well understood (see section (b) 2). In addition, the relative importance of such processes appears to vary over time and space in response to changes in environmental conditions. Such attack can result in the crumbling and settling of road bases and sub-bases, which, in turn, can cause reflective cracking in the wearing surface (Fookes and French, 1977). Once surface cracks appear, road traffic may generate further damage, in the form of deformation, cracks, and pot-holes, thereby

TABLE V.9

Examples of soluble salt contents in groundwater and precipitated on ground surface and in roads. (Samples from eastern Saudi Arabia; weights in %)

ion \ w%	1	2	3	4
Na	22.5	15.4	6.0	3.3
K	1.4	0.7	0.4	not determined
Ca	5.6	16.2	27.7	25.7
Mg	3.0	1.6	3.2	3.7
$CO_3^=$	Not found	8.2	2.8	19.6
$SO_4^=$	39.5	36.2	52.3	37.6
Cl^-	28.0	21.7	13.5	10.1

1. Salt content of a groundwater; 2. Salts in surface crust of a sabkha with groundwater of composition 1; 3. Salts in thin road surface associated with groundwater of composition 1. Road surface in zone of capillary moisture movement; 4. Extractable salt component of a sabkha (from Russell, 1974). All analyses represent salts extractable in dilute nitric acid.
Source: Fookes and French (1977).

TABLE V.10

Outline of potential physical and chemical problems encountered in road construction

| | Bituminous Wearing Course/Basecourse | | Unbound base e.g. wet or dry bound macadam, wet mix or all-in granular material | Unbound sub-base e.g. gravel, crushed stone |
	Thick/dense	Thin/porous		
Potential migration of water	*Very low*	*Moderate*	*High* (varies with aggregate etc.)	*High* (varies with aggregate etc.)
Potential migration of salt	*Very low*	*Moderate to high*	*Moderate* (varies with aggregate etc.)	*High* (varies with aggregate etc.)
Physical changes in presence of groundwater – permanent, intermittent or capillary	*Unlikely* in short or medium term	*Probable*	*Probable*	*Probable*
Distress	Possible long term aggregate disintegration, surface erosion & stripping	Short term aggregate disintegration, surface erosion scabbing, blisters, potholes	Short to medium term disintegration, stripping, settlement	Medium term disintegration, settlement
Physical changes in presence of transient water (rain, dew, etc.) depends on aggregate and salt content	*Unlikely* unless salt in aggregates high	*Unlikely* unless high salt in aggregates	*Unlikely* unless salt in aggregates	*Unlikely* unless salt in aggregate
Distress	Very slow aggregate disintegration	Slow aggregate disintegration	May be slight disintegration, settlement	May be slight disintegration, settlement
Chemical changes in presence of groundwater – movement intermittent, capillary. Depending on salt levels and type	*Unlikely*	*Possible* bitumen and aggregate may decompose and disintegrate if salt high	*Possible* if salt content high	*Possible* if salt content high

Distress	Unlikely	Potholes, scabbing, stripping	Volume changes and loss of strength	Volume changes and loss of strength
Chemical changes in presence of transient water (rain, dew etc.) – depends on aggregate and salts present	Unlikely	Unlikely bitumen may decompose if salt high	Possible with some aggregate and some salts	Possible with some aggregate and some salts
Distress	Unlikely	Unlikely	Small volume change or loss of strength	Small volume change or loss of strength

Source: Fookes and French (1977).

TABLE V.11

Factors affecting salt aggression on roads constructed in highly saline areas

Aggregate properties	Chemical stability	Limestone, chalks, clays etc. may be reactively replaced by salt crystals with possible loss of strength and volume change	
	Permeability	High permeability is not necessarily harmful unless coupled with low strength, high porosity or reactivity etc.	
	Porosity	High porosity coupled with small pore size will promote transfer and ppt. of salts. The capillary pressure of aggregates should be measured	
	Soundness	Measurement of soundness will give an indication of potential for physical disruption by salts	
	Strength	Rocks of high mechanical strength may resist decomposition even if salt concentrations are high	
Water–salt system	Water availability	Permanent saturation by groundwater (moisture zones A & B)	May lead to reactive decay of aggregate and bitumen
		Intermittent saturation, capillary fringe to groundwater (moisture zones C & D)	Reactive replacement and physical disintegration may occur. Salt precipitation likely. High salt concentration may develop, even if initial concentration is low, by evaporation
		Transient water by rain, snow, dew, humidity, services, etc. (moisture zone E)	Not likely to be harmful unless aggregate salt concentration is high
	Salt availability	Aggregates	Aggregate may contain salts-gypsum, anhydrite, pyrite, iron sulphate, sodium chloride etc. These may cause chemical or physical changes if wetted or heated
		Groundwater	Sulphate and chloride ions may be abundant but even low concentrations can be enhanced to deleterious levels within road structure by evaporation
	Soluble-ion types	Cations$^+$	Sodium, potassium, magnesium, and iron tend to enter groundwater or surface water preferentially. These ions may exchange with aggregate minerals and change aggregate properties
		Anions$^-$	Chloride and sulphate are common – environment is generally oxidizing. Anions may ppt. from solution or by reaction with aggregate. Sulphate may be a greater potential hazard than chloride

Note: Zones A, B, C, D, and E are defined in Fig. V.8.
Source: Fookes and French (1977).

providing avenues for additional and more rapid salt attack. Thus salt weathering of base-course materials may be slow at first but tends to increase both in extent and intensity with time. The same is true of the resultant damage, which clearly indicates the importance of prompt and adequate maintenance if road standards are to be kept at a high level.

TABLE V.12

Summary of the influence of moisture zones A to E on behaviour of roads

Zone	Moisture conditions	Salt conditions	Possible damage
Moisture zone E	Transient water from rain, dew etc	Salts may be removed in solution and may accumulate by subsequent evaporation or by vapour transfer, etc	Damage not serious unless aggregate is rich in salt or road is of thin construction with unsound aggregates, in long term
Moisture zone D	Water present by capillary moisture movement	Salts may be precipitated at all levels of road construction and in large quantities	Aggregate and bitumen may decompose, blisters may develop, small holes and cracks likely. Serious damage only in thin construction
Moisture zone C	Water present by capillary movement or ground may be saturated at times of high water-table	Salts precipitated and may be re-dissolved	Large pot-holes develop, aggregates and bitumen decompose rapidly. Irregular surface develops. Maximum damage in thin construction
Moisture zone B	Permanently saturated zone below capillary fringe	Soil and rock properties may be changed in long term	Damage by long term deformation possible
Moisture zone A	Saturated zone below water	May create sabkha conditions in reclaimed ground or embankments	Damage as moisture zones E, D, and C depending on elevation of construction

Note: for explanation of moisture zones see Figs. V.8–V.10.
Source: Fookes and French (1977).

(*iii*) *The upward leaching of salts by capillary water and the resultant evaporation through the road surface frequently leads to the concentration of crystallized salts at the base of the wearing course.* Crystal growth pressures result in the formation of salt blisters that continue to grow because the concentration of salt crystals appears to attract further salt so that the accumulation rate increases with time. Eventually the bituminous surface is heaved up into a dome and causes the development of surface cracks which allow the base-course to be attacked (Netterberg *et al.*, 1974; Blight, 1976; Fookes and French, 1977; see also Chapter IV). The occurrence of salt blistering appears to be a function of the character of the 'water–salt system', the thickness of the bituminous surface, and the traffic density. The scale of damage appears to decrease with increasing thickness of the wearing course and may be prevented by the employment of a surfacing layer with a ratio of water-permeability to thickness of $30(ms)^{-1}$ or less, which can be achieved most easily by using at least 30 mm of dense asphalt concrete (Blight, 1976). Where blistering has occurred, the scale of damage appears to be inversely proportional to traffic volume, which suggests that surface rolling is an important factor in fighting salt heave (Fookes and French, 1977).

Reduction of salt-weathering damage potential for roads can be achieved in three main ways: (i) *relocation* to avoid the most saline ground (see next section); (ii) careful choice of *aggregate* materials (see Chapter IV); (iii) good road design (cf. Fig. V.9) and a high standard of maintenance.

(d) The Spatial Variability of Salt Weathering

Of the various types of salt-weathering damage discussed in previous sections, that resulting from the external attack of buildings, structures, and roads by sulphates is the most important. Such attack occurs largely as a consequence of the post-construction accumulation of salts due to capillary rise of saline groundwater. The damp environments where this hazard is particularly prevalent have been variously described as those with 'aggressive soils', 'aggressive ground', and 'aggressive salty ground' and are best developed in sabkhas and salinas. The potential impact of these aggressive conditions may be minimized by means of *planning solutions* (e.g. relocation) and/or *engineering solutions* (e.g. membranes). But in both instances the essential prerequisites are first, the recognition that a salt-weathering hazard exists and second, an appreciation of the spatial variation of the potential hazard on the ground. The first has been very largely achieved as a consequence of the ever-mounting costs and inconvenience imposed by construction delays, the undertaking of necessary remedial action, and the decay and shortened life of many buildings, while an additional factor has been the escalation in construction costs resulting from the employment of the protective measures required in such environments. Unfortunately, less has been accomplished regarding the second vital element of the problem, for relatively few maps have been produced that attempt to show the variation in aggressive ground hazard for construction purposes, and virtually no work has been undertaken on the relationships between the various types and intensity of salt-weathering damage and the physical and chemical characteristics of the 'water–salt system'. Only when the latter is achieved will it be possible to produce accurate ground hazard maps of value to various forms of construction activity.

The main recent development with regard to the study of aggressive ground conditions has been the mapping of the spatial characteristics of the 'water–salt system'. Such studies are dependent on information concerning the distribution, composition, and temporal variability of groundwaters. The most important feature associated with water in the ground is the *water-table* which is here defined as 'the level to which water rises in observation wells in free communication with the voids of the soil *in situ*'. Above this occurs the *capillary fringe*, which is 'the zone of saturated and partially saturated material above the water-table', the term *groundwater* being applied to both 'water in the zone of saturation beneath the water-table and in the capillary fringe above'.

Aggressive soil conditions as an engineering hazard are produced when saline groundwaters are drawn upward through the soil to produce a capillary fringe that either reaches the ground surface or approaches sufficiently close to it to affect foundations. This upward movement of capillary water through surface and near-surface materials is sometimes known as 'evaporative pumping' and is essentially the product of high surface temperatures. The *height of capillary rise* (i.e. the thickness of the capillary fringe), varies with the soil temperature gradient and the nature of the soil materials, normally being less than a metre in clean gravel but, according to laboratory studies

(Lane and Washburn, 1946; Scott and Schoustra, 1968), extends up to 10 m in clay. Terzaghi and Peck (1948) suggested that the height of capillary rise (h_c) may be computed by using the formula:

$$H_c = \frac{C}{e\,D_{10}} \qquad\qquad (V.1)$$

where C = an empirical constant that depends on grain shape and surface impurities, and ranges between 0.1 and 0.5,
   e = void ratio,
   D_{10} = 'effective size', in which 10% of particles in the grain-size analysis are finer and 90% are coarser than this value.

However, the variability of surface materials often makes such calculations cumbersome, except in the case of detailed studies of small areas. Field investigations by the authors in drylands (Bahrain, Dubai, Egypt) suggest that capillary rise under desert conditions is normally no more than 2–3 m and rarely, if ever, exceeds 4 m. The upper surface of the capillary fringe, here termed the *limit of capillary fringe*, can vary in position from deep within the ground to the situation where it is potentially well above ground level (Fig. V.7). The latter results in a truncated zone of capillary rise and salt efflorescence (salt crusts) on the ground surface. However, the presence of buildings and structures in such situations can allow the full height of capillary rise to be achieved within the building materials, thus leading to the damp-stained walls that are a common feature of coastal towns in the Middle East and elsewhere.

The evaporation of water within the capillary fringe results in the deposition of salts. The relationship of the *limit of capillary fringe* to the ground surface is thus of considerable planning and engineering significance, for it determines the extent to which foundations and superstructures may be subjected to wetting by saline solutions. Of particular importance is the line where the limit of capillary fringe intersects the ground surface, for it marks the boundary between the zone where only foundations are at risk and that

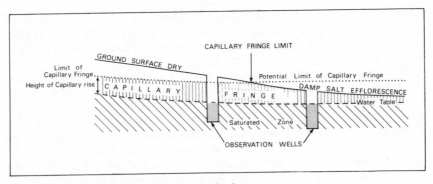

FIG. V.7 Definitions of groundwater terminology

where both foundations and superstructures may be attacked. This important planning boundary is here termed the *capillary fringe limit* (Fig. V.7).

The aggressiveness of an area is determined by a combination of factors including the chemical composition of groundwaters, depth of water-table, soil characteristics, soil-temperature characteristics and regime, and the interrelationship of the limit of capillary fringe and the ground surface. Given reasonably uniform conditions, it is usually possible to identify a sequence of hazard zones based on the 'wetness' of the ground. Three different, but closely related, schemes have been proposed for the purposes of mapping and analysing such variations in ground aggressiveness. That proposed by Fookes and French (1977) involves the recognition of five moisture zones (Fig. V.8; A–E), of which zones C, D, and E can normally be recognized on the ground surface and mapped. Such an approach is of particular relevance in the case of roads and other forms of construction involving only shallow foundations (Fig. V.9) and can be used to show the areal variation in potential hazardousness and, therefore, identify those areas where additional design precautions need to be undertaken (Fig. V.10). The second approach is that of Fookes (in Fookes and Collis, 1976), which involves the subdivision of coastal desert terrains into four zones (Fig. V.11; A–D), two of which are essentially 'dry', i.e. above the limit of capillary fringe, one 'damp', and one exceedingly 'wet' (Zone A which is inter-tidal) and for which general engineering implications are suggested *irrespective* of the actual geochemistry of the 'water–salt system'. The third approach, used in the case of the Suez City sub-surface investigation (Egypt, Ministry of Housing and Reconstruction, 1978; see case studies in next section), is an attempt at a finer subdivision of the hazard zone in those areas where the 'water–salt system' has a direct bearing on engineering works, particularly where foundations are involved. Once again, four zones are identified (Fig. V.12). These are, in generally increasing order of hazardousness:

Zone I – No hazard from groundwaters as the limit of capillary fringe is so deep within the ground as to be below the base of foundations.

Zone II – The limit of capillary fringe is below the ground surface but sufficiently close to it to affect foundations.

Zone III – The limit of capillary fringe is potentially above ground level so that both foundations and superstructures are at risk. Such areas usually show up as a dark tone on air photographs and have 'puffy' surfaces on which footprints and tyre-tracks are clearly visible.

Zone IV – Water-table within half a metre of the ground surface for most of the year so that foundations are emplaced in water and there is potential for capillary rise to well above ground level (sometimes in excess of 2 m). Such areas are very low in elevation and may be periodically inundated as a consequence of rainfall or sea surges. These areas appear as a dark tone on air photographs, often with patches of standing water. On the ground, the surface is usually either 'puffy' or crusted. Patches of salt efflorescence are often developed and, in certain instances, salt polygons may also be visible. Standing water areas are invariably rimmed with salt crystals.

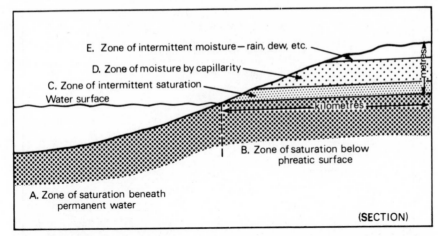

FIG. V.8 Schematic diagram to show the inter-relationships of soil moisture zones (after Fookes and French, 1977)

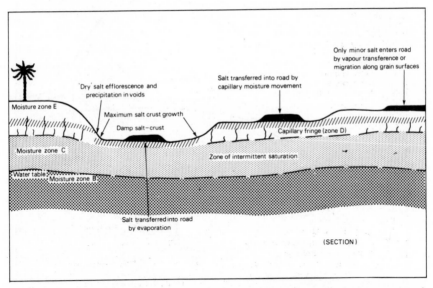

FIG. V.9 Schematic section to show the relevance of soil moisture zones to road construction (after Fookes and French, 1977)

The relationships between the zonal subdivisions proposed in the three schemes are shown in Table V.13. It is important to recognize that such subdivisions merely provide a framework for further investigation and research, because they only show the physical characteristics of the water system. Aggressiveness is a function of the 'water–salt system' and it is therefore important that data are also collected on the variety and concentration of salts within the various zones. Such studies allow the further

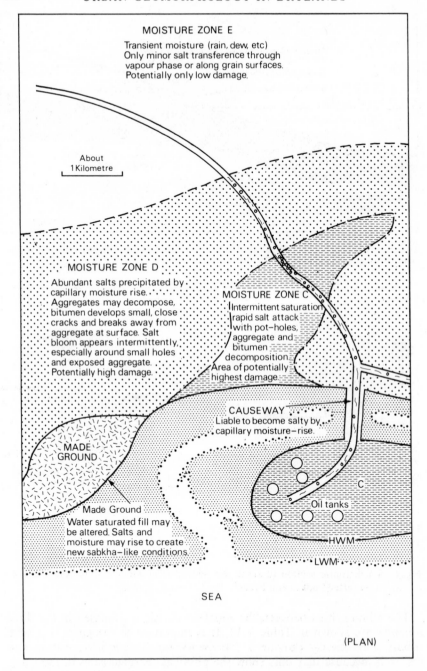

MOISTURE ZONE E

Transient moisture (rain, dew, etc)
Only minor salt transference through
vapour phase or along grain surfaces.
Potentially only low damage.

About
1 Kilometre

MOISTURE ZONE D

Abundant salts precipitated by
capillary moisture rise.
Aggregates may decompose,
bitumen develops small, close
cracks and breaks away from
aggregate at surface. Salt
bloom appears intermittently,
especially around small holes
and exposed aggregate.
Potentially high damage.

MOISTURE ZONE C
Intermittent saturation
rapid salt attack
with pot-holes,
aggregate and
bitumen
decomposition.
Area of potentially
highest damage.

CAUSEWAY
Liable to become salty by
capillary moisture-rise.

**MADE
GROUND**

Made Ground
Water saturated fill may
be altered. Salts and
moisture may rise to create
new sabkha-like conditions.

C

Oil tanks

HWM

LWM

SEA

(PLAN)

FIG. V.10 Hypothetical map to show how identification of soil moisture zones can
assist in road construction and the determination of maintenance priority areas (after
Fookes and French, 1977)

ZONE	NATURAL CHARACTERISTICS	ENGINEERING IMPLICATIONS
A	Intertidal. Near surface Chemistry in equilibrium with sea water.	Design and construct foundations as for marine conditions.
B	Sabkha or Salina. Ground surface within groundwater capillary range. Salts concentrate at surface by evaporation.	Tanking (by membrane) probably required and dense, good quality concrete for foundations.

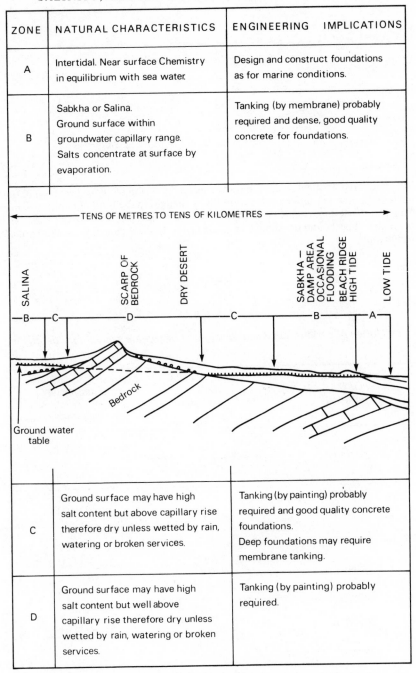

| C | Ground surface may have high salt content but above capillary rise therefore dry unless wetted by rain, watering or broken services. | Tanking (by painting) probably required and good quality concrete foundations. Deep foundations may require membrane tanking. |
| D | Ground surface may have high salt content but well above capillary rise therefore dry unless wetted by rain, watering or broken services. | Tanking (by painting) probably required. |

FIG. V.11 Idealized cross-section of coastal desert terrain showing the relationship of salty soils to the local groundwater table (after Fookes and Collis, 1976)

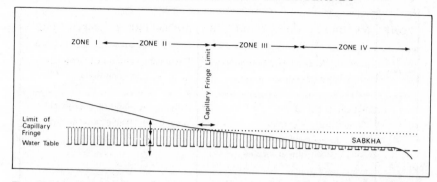

FIG. V.12 Hypothetical section to show zonal subdivision of aggressive ground (n.b. fluctuation in water table levels can result in changes in the limit of capillary fringe. This can cause movement of the capillary fringe limit and thus of the zone II/III boundary. The boundary should be fixed at the highest elevation of the capillary fringe limit)

TABLE V.13

Correlation of various zonal sub-divisions of aggregate ground (for details see text)

Fookes and French (1977) (for roads)	Fookes and Collis (1976) (for concretes)	Suez Study (Authors 1978) (for general urban infrastructure)
ZONE E	ZONE D	ZONE I
	ZONE C	ZONE II
ZONE D	ZONE B	ZONE III
		ZONE IV
ZONE C	ZONE A	
ZONE B		
ZONE A		

subdivision of the hazard zones into those of, for instance, high, moderate, and low salinity. While it seems likely that damage increases with salinity, the significance of salinity zones is unclear because there is, as yet, no empirical information on damage-inducing threshold values for total salts or individual ions. Nevertheless, such information should be gathered prior to embarking on major engineering works, for general indications as to the relative hazardousness of different sites can be deduced so that damage may be minimized through relocation. Furthermore, such studies allow comparisons to be made between areas of existing and new developments and thus facilitate the recognition of the measure of salt-weathering damage likely to be encountered and thus the level of protection that should be employed.

In this context it is important to appreciate that damage surveys of buildings may yield inconclusive and misleading data on the intensity of salt weathering because: (i) it is generally difficult, and sometimes impossible, to determine the age of affected buildings, and thus the period during which salt weathering has been active is often unknown, and (ii) the effect of salt weathering is strongly influenced by the materials used in construction and the design of the structures, both of which may vary greatly from place to place and over time.

The evidence obtained from such surveys must, therefore, be used with great care.

For a full and detailed investigation of aggressive ground conditions it is imperative that adequate time be allowed for the identification of both spatial and temporal variations. Such a study should, if it is to be of maximum value, be based on the collection and analysis of the following information:

(*i*) *Ground elevation variation*, i.e. detailed topographic maps or survey information;

(*ii*) *The nature and distribution of surface materials* (which facilitate computation of height of capillary rise);

(*iii*) *The elevational configuration of the water-table*. (Variables (*i*)–(*iii*) will allow the establishment of hazard zones);

(*iv*) *Variation in water-table configuration(s) over time* in response to precipitation, urban development, and tidal conditions;

(*v*) *Spatial variation in groundwater salinity*.

(*vi*) *Spatial variations in the concentration of ions* in the groundwater (Na^+, Ca^{++}, and Mg^{++} by atomic absorption techniques; Cl^- by Volhard titration; and SO_4^{--} by gravimetric or turbidometric methods);

(*vii*) *Variations in groundwater salinity and ionic concentrations over time* in response to the factors of change outlined under (*iv*) above.

Such information would provide a clear indication of the spatial characteristics and variable intensity of the groundwater-salt hazard. Unfortunately, long-term monitoring programmes do not exist, and investigations into the salt-weathering hazard have usually been undertaken rapidly as part of on-going site-investigation programmes, with the inevitable result that they are invariably based on an inadequate network of data points, and usually only have information on the spatial variation of salinity (often produced by portable conductivity meters which can be wildly inaccurate) rather than the pattern of ionic concentrations based on geochemical analyses, and so are only valid for the time of survey (there being no information on temporal variability and thus no indication as to the significance of the identified pattern). A fully satisfactory determination of ground aggressiveness thus depends on the establishment of long-term investigations.

(e) Field Studies

1. *Introduction*

The following case studies, based on field work in the Middle East, exemplify what can be achieved during general investigations of the salt hazard at a regional scale (Bahrain), site scale (Dubai), and a city-wide scale (Suez).

2. *Aggressive ground conditions – northern Bahrain*

This study was carried out as part of the Bahrain Surface Materials Resources Survey (Bahrain, Ministry of Works, Power and Water, 1976; Brunsden *et al.*, 1979; Doornkamp *et al.*, 1980) and was concerned with the northern part of the island where groundwater is both shallow and saline, and where urban development is concentrated.

As a first step the extent of areas affected by capillary rise were mapped on the basis of three criteria – damp surface materials, 'puffy' ground, and salt efflorescence. This facilitated the determination of the capillary fringe limit at the time of the survey (April, 1975) (Fig. V.13). Second, estimates and observations were made of the height of capillary rise in various materials *inland* of the capillary fringe limit and from these data it was possible to predict that for all planning and engineering purposes the inland limit of the hazard could be taken as the 10-m contour (Fig. V.13). Third, an attempt was made to determine the spatial variability of the hazard within the hazard zone. The hypothesis was established that the aggressiveness of the ground is related directly and equally to shallowness of groundwater and groundwater salinity. To resolve this hypothesis in spatial terms, data were collected at 147 sampling points on the depth of water-table below the surface, groundwater salinity, and ionic concentrations in groundwater of Cl^-, SO_4^{--}, Na^+, K^+, Ca^{++}, and Mg^{++}. All the data were put onto punched cards and used to produce distribution maps on the computer with the aid of a SYMAP program.

The groundwater-level map (Fig. V.13) refers to water-table depth below ground level (not below a horizontal datum), and the isolines are interpolated between sample points to produce a generalized, most probable pattern based on values at the sample points only. In general, water-table depth increases inland, although at the capillary fringe limit and the 10-m contour, its predicted depth varies considerably, mainly as a reflection of varying bedrock and superficial materials. The pattern is further complicated by geological inliers and dune fields. Estimates of the height of capillary rise indicate that is is normally less than 3.0 m above the water-table. Few areas below the 10-m contour are unaffected by capillary rise to near the surface.

Salinity of groundwater was measured by means of a conductivity meter,

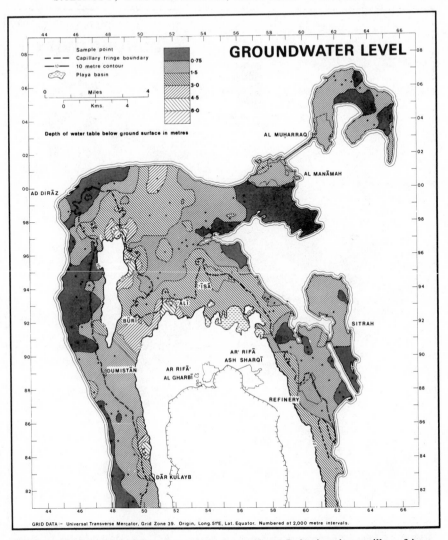

FIG. V.13 Aggressive ground conditions in northern Bahrain: the capillary fringe limit, and depth to groundwater beneath the surface (after Bahrain, Ministry of Work, Power and Water, 1976)

which indicates specific conductance (i.e. the ability of a conductor to convey an electric current). In the case of water, specific conductance is related to the concentration of ions present and to the temperature at which measurement is made, and it is measured in μmho/cm. Electrical conductivity is related directly (but in a non-linear way) to the concentration of solutions, but it gives no indication of the nature of the substances in solution. Figure V.14 shows the pattern of electrical conductivity in northern Bahrain, and clearly

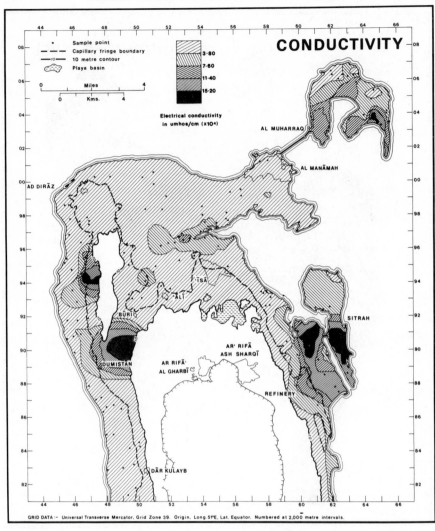

FIG. V.14 Aggressive ground conditions in northern Bahrain: electrical conductivity of groundwaters (after Bahrain, Ministry of Work, Power and Water, 1976)

reveals not only a relatively low general electrical conductivity in the hazard zone (normally less than 38 000 μmho/cm) but also five distinct areas of relatively high conductivity that are likely to be areas of potentially relatively high salt-weathering hazard. In addition to electrical conductivity, the spatial pattern of ionic concentrations was also investigated and it was shown that, not surprisingly, the highest ionic concentrations coincide with the areas of highest electrical conductivity.

On the assumption that hazard intensity is directly and equally proportional to shallowness of groundwater and the electrical conductivity of groundwater, the distributions shown on Figs. V.13 and V.14 were combined to produce a map that is a first approximation of hazard intensity (Fig. V.15). This hazard map is a valuable basis for preliminary planning and engineering decisions, but it does have several obvious limitations: it is based on an, as yet, untested hypothesis, it is based on data for a single period of time (water-table levels are changing in both the long and the short term, for instance), and the computer analysis involved a 'smoothing' of information.

3. *Aggressive ground conditions: Mina Jebel Ali port site (Dubai)*

A field investigation was carried out in the area of a major port and associated industrial development in southern Dubai (Halcrow Middle East, 1977a) to determine the prevailing characteristics of the 'water–salt system' in a hitherto undeveloped zone of coastal sabkha. The results are of interest because they show how the system can vary within a small and apparently fairly uniform geomorphological unit.

Data on the sedimentary deposits, salinity, and water-table configuration were obtained by digging 70 pits, each of which extended down to the water-table. In each, the sedimentary sequence, the depth to water-table, and the groundwater salinity were recorded. The sampling positions were selected on the basis of establishing a well-distributed set of points within each 'unit' recognized initially on the aerial photographs. Areas where the water-table was already affected by engineering works were avoided.

Eight distinctive sedimentary zones were recognized (Fig. V.16A):

Zone I – Heavily gypsified, very weathered, alluvial-fan material at shallow depth, overlain by loose, coarse crystalline gypsum. An intervening layer (up to 0.5 m) of fine quartz sand, containing some carbonate material, occurs in the lower-lying areas.

Zone II – Loose to well-cemented gypsum sand with some fine quartz sand. Gypsum forms 90–95% of the sediments and occurs as fine to coarse sand-sized particles. Cementation tends to increase with depth. The loose layer of gypsum may, in part, be an aeolian deposit.

Zone III – Surface layer of loose gypsum sand with some quartz sand, often overlying a cemented layer (gypsum), below which occur sands composed of varying proportions of gypsum and quartz with subordinate amounts of carbonate material.

Zone IV – Surface sediments contain roughly equal quantities of medium-coarse gypsum sand and fine-grained quartz sand, with subordinate amounts of carbonate material. The general stratigraphic sequence is a thick surface layer of loose, medium to coarse, gypsum sand overlying calcareous fine quartz sand, within which occurs a moderately- to well-cemented (gypsum) layer.

Zone V – A thin surface layer (usually less than 0.1 m) of loose gypsum sand with small amounts of fine quartz sand, overlying calcareous fine quartz sand within which occur cemented horizons. This zone can be subdivided into:
(i) Zone VA, where the underlying sands are rich in carbonate, cross-

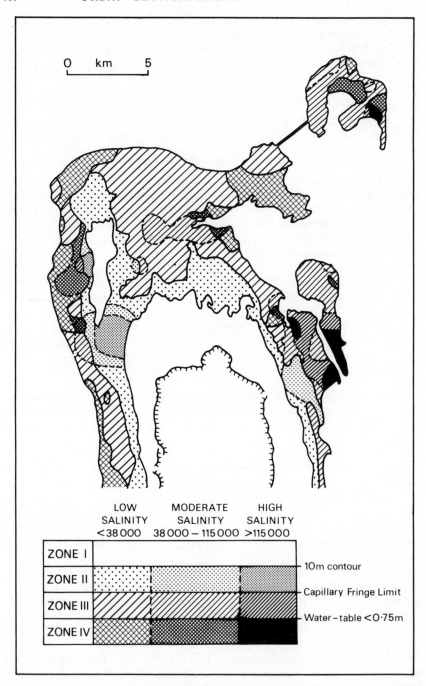

	LOW SALINITY <38 000	MODERATE SALINITY 38 000 – 115 000	HIGH SALINITY >115 000
ZONE I			
ZONE II			
ZONE III			
ZONE IV			

— 10m contour

— Capillary Fringe Limit

— Water-table <0·75m

FIG. V.15 Aggressive ground conditions in northern Bahrain: predicted hazard intensity (after Bahrain, Ministry of Work, Power and Water, 1976; Jones, 1980)

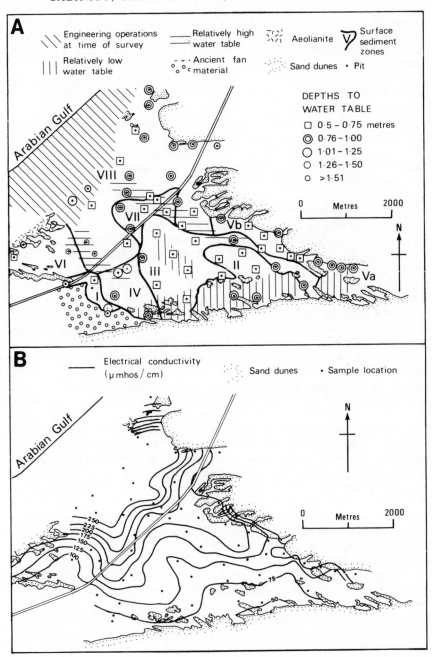

A

Engineering operations at time of survey — Relatively high water table — Aeolianite — Surface sediment zones

Relatively low water table — Ancient fan material — Sand dunes · Pit

DEPTHS TO WATER TABLE
☐ 0·5 – 0·75 metres
◎ 0·76 – 1·00
◯ 1·01 – 1·25
○ 1·26 – 1·50
○ >1·51

Arabian Gulf

VIII VII Vb VI iii II IV I Va

0 Metres 2000

N

B

Electrical conductivity (μ mhos / cm) ·· Sand dunes · Sample location

Arabian Gulf

N

0 Metres 2000

250 225 200 175 150 125 100 75 50

FIG. V.16 Aggressive ground conditions in Dubai: (A) Water-table and sediment data; (B) Electrical conductivity (after Halcrow Middle East, 1977a), figures × 1000

laminated, and contain moderately- to well-cemented horizons, and (ii) Zone VB, where the surface gypsum layer is underlain by a thin, weakly to moderately cemented, slightly calcareous fine quartz sand which becomes loose in depth.

Zone VI – Loose, slightly calcareous quartz sand with a thin discontinuous surface layer of medium to coarse gypsum crystals. Gypsum cemented layers ('desert rose' type) occur in certain areas.

Zone VII – Alternating layers of medium to coarse gypsum sand and calcareous quartz sand, passing down into cemented calcareous fine quartz sand with shell fragments and whole shells (bivalves and gastropods).

Zone VIII – The remainder of the study area within which further subdivision is impracticable because of lack of data.

The sabkha surface, which is composed of these varied sedimentary types, declines towards the coast (from 5.7 to 2.8 m) as does the water-table (from 4.4 to 1.5 m) (Fig. V.16A). Depth to water-table below ground surface thus shows relatively little variation. Study of the height of capillary rise revealed values up to 3 m, indicating that the whole area is within the zone of capillary rise. The close correspondence between ground surface and water-table configuration suggests that predominant groundwater movement is from south-east to north-west with a subordinate movement from higher ground on the north-east, and that surface elevation is determined by aeolian deflation which is, in turn, controlled by water-table configuration.

In contrast, groundwater electrical conductivity is highly variable within the sabkha (Fig. V.16B). It (and thus salinity) are generally high, ranging from 55 000 μmho/cm to 340 000 μmho/cm (for comparison, Arabian Gulf seawater values are normally c. 55 000–60 000 μmho/cm, and a 7% solution of NaCl has a conductivity of 100 000 μmho/cm). All samples were taken from the uppermost layer of groundwater at the time of survey. Conductivity values are lowest on the southern margin of the sabkha, and generally increase north-west, supporting the view that the main source of groundwater is from the south-east of the sabkha and that evaporative losses cause an increase in salt concentration as the water moves northwards. The idea of an evaporative effect is further supported by the fact that minor topographic and water-table 'lows' tend to be associated with higher conductivity values. Reduction of conductivity in the northernmost part of the study area is probably produced by the intrusion of relatively less saline seawater into the sabkha sediments.

Chemical composition of the groundwaters is also highly variable within the sabkha. For instance, surface gypsum tends to decrease as groundwater conductivity increases, so that to the west of the road gypsum is rare. This, as geochemical analyses revealed, is probably due to deficiencies of either sulphate or calcium in highly saline areas that preclude gypsum formation.

The whole sabkha is classified as very aggressive (Zones I and II, Fig. V.12), with a limit of the capillary fringe at 1.25–2.5 m above ground surface. As a result of this survey it was proposed that all possible protective measures should be used during construction, that dock walls could act as barriers to water movement and thus affect water-table level and groundwater salinity, that fill materials should be used extensively so as to raise the surface above the limit of capillary rise, and that waste water could cause gypsum solution and lead to foundation movements. In conclusion, it was suggested that as much building as possible should be located on higher ground to the north where the hazard is minimal, and that the harbour quay walls should be permeable so as to allow the free movement and interchange between groundwater and seawater.

4. *Aggressive ground conditions: Suez City development (Egypt)*

A similar reconnaissance survey to those previously described was undertaken as part of the investigation of ground conditions in the area designated for the new city of Suez (Egypt, Ministry of Housing and Reconstruction, 1978). The resulting patterns are of particular interest because they reveal how the 'salt–water system' can be modified by urbanization processes.

Groundwater observations were somewhat biased towards the coastal zone (Fig. V.17A), where access to the water-table was easier, and it proved difficult to interpolate groundwater contours because of topographic and geological irregularities. Nevertheless, it proved possible to divide the region into three generalized groundwater zones (Fig. V.17B):

(i) An area adjacent to the present shoreline where the water-table is within 2.5 m of the surface. Over most of this area depth to water-table is about one metre and in the lower-lying parts the water-table may occasionally intersect the ground surface to form pools of standing water due to tidal changes in the Gulf of Suez, so foundations are likely to be placed in water with capillary rise to well above ground level (Zones I and II, Fig. V.12).

(ii) A rather narrow and irregular belt where the water-table is between 2.5 and 5 m down. Both shallow and deep foundations are likely to encounter the capillary fringe and deep foundations may well be emplaced below the water-table. Capillary rise is unlikely to extend above ground level (Zone III, Fig. V.12).

(iii) An extensive zone where the water-table is more than 5 m below the ground surface and where the distance between water-table and ground level increases towards the north-west. Here, only very deep foundations are likely to encounter either the water-table or capillary fringe (Zone IV, Fig. V.12).

An attempt was made at a preliminary assessment of the form of the water-table, assuming it to be a single feature (rather unlikely in reality). Despite inadequate data, it was possible to identify groundwater 'mounds' in the vicinity of the fertilizer factory and the Feisal City–Suez Town area (Fig. V.17B). These 'mounds' are probably due to influent seepage from water pipes, drains, effluent channels, and surface irrigation, whereas the regional sources of groundwater recharge are the Gulf of Suez and rainfall in the nearby mountains. Anthropogenic inputs, which appear to be becoming increasingly important, are thus significant to the development of the region, because they will not only tend to raise the water-table but will also tend to move the capillary fringe limit inland, thus extending the hazard zone.

Measurements of groundwater electrical conductivity, which were based

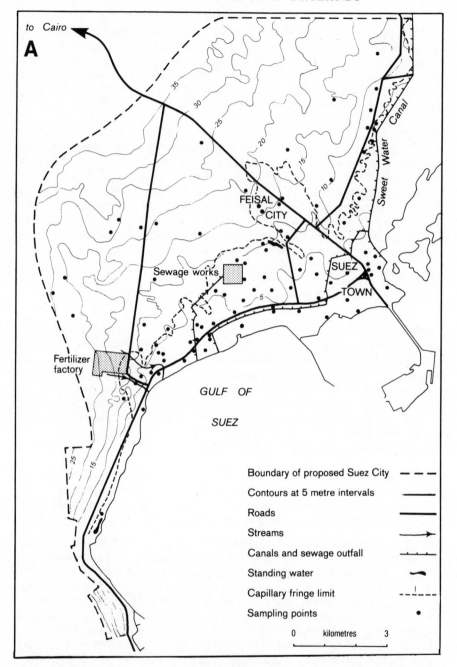

FIG. V.17 Aggressive ground conditions in Suez, Egypt; (A) groundwater sampling points and capillary fringe limit; (B) Groundwater depth (after Sir William Halcrow & Partners acting on behalf of Egypt, Ministry of Housing and Reconstruction, 1978)

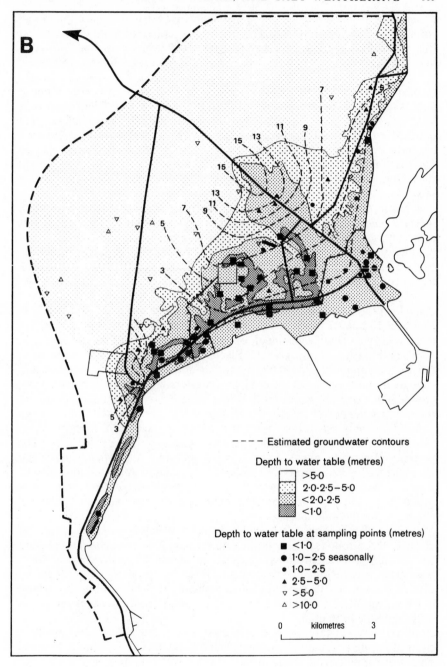

B

---- Estimated groundwater contours

Depth to water table (metres)
>5·0
2·0-2·5-5·0
<2·0-2·5
<1·0

Depth to water table at sampling points (metres)
■ <1·0
● 1·0-2·5 seasonally
• 1·0-2·5
▲ 2·5-5·0
▽ >5·0
△ >10·0

0   kilometres   3

on samples from boreholes and auger holes, revealed a remarkable range from 4 600–388 000 µmho/cm (Fig. V.18). It is apparent from the map that the generally saline groundwaters (>100 000 µmho/cm) that probably underlie the whole coastal area are overlain in places by lighter, less saline water. These places include the fertilizer factory, Suez Town, the sewage farm and gardens, and to a lesser extent the developing housing area of Feisal City (Fig. V.17A), and probably result from the fact that influent seepage is relatively 'fresh'. Thus, the probable changes in groundwater configuration outlined in the previous paragraph are likely to be accompanied by decreasing groundwater salinity, so that increasing extent of the hazard zone may go hand-in-hand with decreasing intensity. Such changes will, however, herald the introduction of a new problem – subsidence due to the solutional removal of salts in the ground.

A preliminary general assessment of the salt hazard, shown in Fig. V.19, is based on combining the information in Figs. V.17 and V.18. Although it is based on limited data, and is thus liable to continuing and major modification, it does provide a considerable advance on previous information. Four main groundwater hazard zones were recognized (Fig. V.12) and each was accompanied by recommendations as follows:

Zone I – Water-table at least 5 m below the ground surface and probably in excess of 10 m over much of the area. Height of capillary fringe is expected to be relatively small over much of the area because the water-table is at depth and mainly within gravels. Electrical conductivities are expected to fall in the range 30 000–70 000 µmho/cm although this assessment could change if, and when, more data become available. Capillary rise from groundwater was not expected to prove a hazard in this area, especially in those parts where thick layers of gravel overlie bedrock. Since deep foundations may reach the capillary fringe near the coastal margin of this zone there exists some hazard potential along the edge of this area. The presence of gypsum, halite, and possibly magnesium sulphate in surface and near-surface layers suggests that seepage may result in minor solution effects, capillary movement of saline water, and crystallization elsewhere. Thus urban and industrial development may generate salt-weathering hazard conditions in the inland areas.

Zone II – Water-table between 2–2.5 and 5 m below generalized ground surface. Deep foundations are likely to encounter water-table, and shallow foundations will reach the capillary fringe, especially in lower-lying areas. This zone was further subdivided into three on the basis of groundwater electrical conductivity, the three sub-zones being <50 000 µmho/cm; 50 000–100 000 µmho/cm; >100 000 µmho/cm. Foundations in this area are at risk and should be protected.

Zone III – Water-table within 2 to 2.5 m of ground surface and potential limit of capillary fringe above the ground surface (the maximum recorded height of capillary rise above ground surface was 1.42 m with a water-table at −0.5 m, but with signs of surface inundation). This area was also subdivided on the basis of groundwater conductivity, the three sub-zones being <50 000 µmho/cm; 50 000–100 000 µmho/cm; >100 000 µmho/cm.

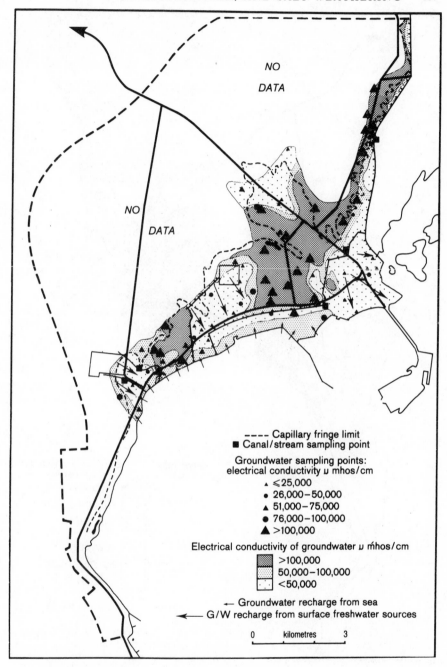

NO
DATA

NO
DATA

---- Capillary fringe limit
■ Canal/stream sampling point
Groundwater sampling points:
electrical conductivity ʋ mhos/cm
▴ ≤25,000
● 26,000–50,000
▲ 51,000–75,000
⬤ 76,000–100,000
▲ >100,000

Electrical conductivity of groundwater ʋ mhos/cm
▨ >100,000
▨ 50,000–100,000
⋅⋅ <50,000

← Groundwater recharge from sea
← G/W recharge from surface freshwater sources

0   kilometres   3

FIG. V.18 Aggressive ground conditions in Suez, Egypt: electrical conductivity (after Sir William Halcrow & Partners acting on behalf of Egypt, Ministry of Housing and Reconstruction, 1978)

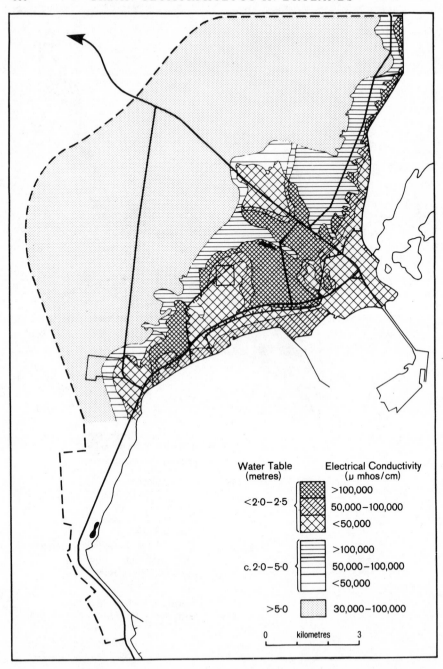

FIG. V.19 Aggressive ground conditions in Suez, Egypt: groundwater hazard intensity (after Sir William Halcrow & Partners on behalf of Egypt, Ministry of Housing and Reconstruction, 1978)

Foundations are liable to be emplaced in groundwater unless buildings are placed on fill. Highly aggressive soils occur over most of the area. An extensive and thick fill blanket could reduce aggressiveness although water-table would probably rise after development. It was recommended that either buildings be protected both above and below ground level and/or buildings be placed on up to 3 m of gravel fill. An alternative answer would be to re-zone the land and avoid residential or industrial use altogether.

Within Zone III there occur areas where the water-table is always within 1 m of the surface and may well be above the surface during high tides or after rainfall events (Fig. V.17B). Electrical conductivities are very high and the areas must be designated as Zone IV and thus very hazardous. Extensive protection, thick gravel fills, or avoidance is required.

Two main conclusions were derived from this analysis. First, salt weathering due to capillary rise of saline groundwater posed a potential threat to development over 30–40% of the city site, requiring either reconsideration of the established plans or the adoption of control measures. Second, the form of the water-table and the salinity of groundwater suggest that anthropogenic factors are of increasing significance in the growing urban area, and future changes in these variables require monitoring.

(f) Summary

Salts present a major hazard to urban development in drylands. The nature of the hazard is extremely complex, depending on the relations between environmental conditions, the types of salt present, the nature of the materials with which they come into contact, and the design and workmanship of the structures placed in the hazardous areas. The deleterious effects of salts relate chiefly to chemical reactions between salts and building materials, and the disruptive forces of various types of crystal growth. Of particular importance, especially to concrete structures, are various types of sulphate and chloride attack. Geomorphological contributions to the appraisal of this hazard include (i) surveys of damage caused by salts, (ii) laboratory studies of rock durability under controlled conditions in simulated desert environments, and (iii) the field determination of hazard zones and the variability of hazard intensity within them. Geomorphological studies of the hazard in the Middle East have provided the basis for decisions to avoid or alleviate the problems by planners and engineers through, for example, quality control of materials, foundation protection, and relocation.

VI

WATER AND SEDIMENT PROBLEMS IN DRYLANDS

(a) Introduction

Runoff events in drylands are relatively infrequent and, as a result, are often poorly perceived as environmental problems. Undoubtedly the major problems of water and sediment movement in drylands, as they affect urban areas, arise from this fact. In part, the poor perception is because useful data are scarce, and it is therefore often extremely difficult to make reliable evaluation of the frequency and magnitude of flow events. Even today, when

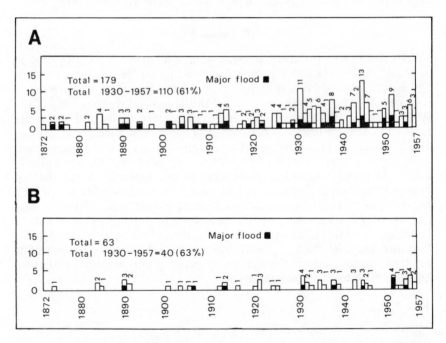

FIG. VI.1 (A) Distribution of number of floods reported per year during the period 1872–1957 in the area bounded by lines joining Mono Lake, California; Tonopah, Nevada; Baker, California; and Mojave, California (including the White Mountains); (B) The dramatic increase in the number of reported floods in the last 30 years in part of the western USA, due largely to the extension of human settlement in the region (from Kesseli and Beaty, 1959)

knowledge has improved, the possibilities are considerable for conflicting predictions or serious misjudgements to be made. Thus, damage due to runoff events in urban areas of drylands may be extensive; and the consequences of urbanization in drylands on runoff and sediment regimes may create further, often unanticipated, problems (e.g. Schick, 1979).

As a result of urban expansion in drylands, human contact with runoff events is increasing and, although there is little evidence to suggest that extreme rainfall events or flood frequency are themselves increasing, it is undoubtedly true that more damage is being caused, and that the events are increasingly being remembered and recorded (Fig. VI.1A and VI.1B); Woolley, 1946; Kesseli and Beaty, 1959). Nevertheless, scientific knowledge of dryland runoff events remains poor. With the significant exception of certain areas of the USA and Israel, there are few comprehensive studies of flood generation, and even precipitation data are usually inadequate. There are three main aspects to this problem: (i) to evaluate the nature of water and sediment movement as a constraint on planning options; (ii) to assess the nature and scale of flood risk and expectable damage from storms of given return periods; and (iii) to determine the nature of adjustments and controls required to minimize risk.

It appears essential that, as development proceeds at an increasing rate, the nature of these problems is clearly understood and that the background material be summarized so that efficient research programmes can be devised. The purpose of this chapter is to contribute towards this aim by demonstrating how geomorphological approaches and information can provide sensible statements concerning the movement of surface water and sediment in drylands and the problems they pose to urban development.

(b) Water and Sediment Systems in Drylands

1. *Introduction*

Even the driest of drylands receive some precipitation, and almost all areas – with the possible exception of extensive, active sand-dune areas – experience surface runoff and possess drainage systems that may be studied in terms of their catchments, valley-side slopes, channels, and depositional areas. Because rainfall and runoff tend to be localized and short-lived, the drainage systems differ in several respects from their more humid counterparts. But they do commonly present a hazard to urban development, and city planning in drylands ignores the problem at its peril.

Dryland drainage systems possess some or all of the following characteristics:

(i) Due to low rainfall, high evaporation, and commonly high infiltration (transmission losses) into superficial sediments, much runoff fails to reach the sea, or indeed, any permanent base level. Equally important, runoff events rarely activate whole drainage systems, and the extent of activation varies greatly according to magnitude, frequency, and duration of precipitation events.

(ii) Rivers with sources outside the drylands may flow through dry lands and introduce unusual hydrological conditions. Such allogenic rivers, which include the Nile, are invariably important foci for settlement and urban development.

(iii) Desert drainage channel networks are commonly extremely complex, often including zones of surface-flow dispersion, and apparently discontinuous patterns (especially in areas of alluvium). Not only are most of the segments of the network normally dry, but only parts of it may be used during any flow event. There is also a marked tendency on alluvial surfaces for channels to shift during flow events, and for successive flow events to follow different courses.

(iv) Many of today's drylands were formerly wetter and cooler, and it is common for drainage systems that were produced in such pluvial periods to remain largely fossilized in the present landscape (e.g. Slatyer and Mabbutt, 1964).

The fluvial landscapes of drylands are, of course, as varied as in any other climatic region, but three distinctive landscape types occur widely and together provide useful generalizations. In areas of fairly recent tectonic instability, such as the south-western United States, the dry littoral of the Andes in South America, parts of Iran, the Oman Peninsula, Pakistan, and the high-altitude deserts of Asia, the landscape is commonly composed of dissected mountains, steep (often faulted) mountain fronts, alluvial fans sometimes with entrenched and un-entrenched channels and, if the basin is enclosed, bajadas (alluvial aprons) and playas – this is the typical basin-and-range topography (Fig. VI.2A).

Where tectonic activity is less marked or absent, and the landscape may have experienced subaerial denudation for very long periods of time without serious interruption, a characteristic landscape is dominated by extensive plains, comprising bedrock pediments (bedrock surfaces thinly veneered with alluvium) or thick alluvial plains punctuated by residual isolated rock masses or duricrust on interfluves (Fig. VI.2B). Such landscapes occur widely in dryland areas of Africa, Australia, and parts of India. In such areas, runoff is commonly dispersed in a myriad of shallow channels, sheetflow is common, and large ephemeral streams migrate across the alluvial aprons.

A third common fluvial landscape occurs in areas of horizontally bedded or gently dipping sedimentary rocks, and is characterized by extensive structural plains, cuestas, escarpments, mesas, buttes, duricrusts, abrupt breaks of slope, and extensive pediments. Rivers commonly flow in deep canyons, and extensive gullying and erosion dominate areas of softer sediments with high relative relief, especially on escarpments and their fringing pediments (Fig. VI.2C). The Colorado Plateau is the type area for such landscapes, but they occur widely in semi-arid and arid areas of North Africa, Australia and elsewhere.

The world-wide distribution of landform assemblages in drylands is summarized in a preliminary way in the terrain survey by Perrin and Mitchell (1969, 1971) (an example of their classification, from Australia, is shown in Fig. III.8). An indication of the relative importance of landform types in selected desert areas is shown in Table VI.1.

Although the landscapes shown in Fig. VI.2 are visually very different,

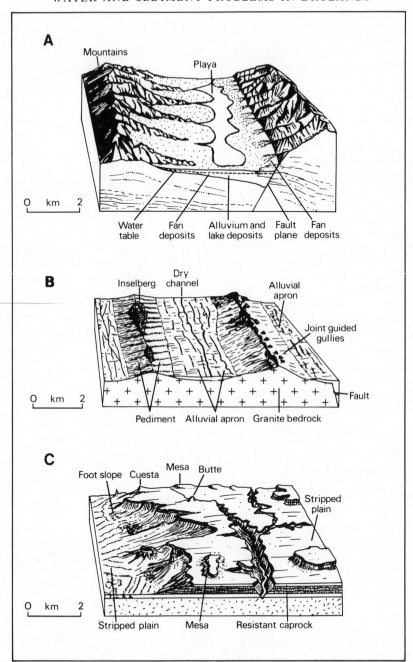

Fig. VI.2 Characteristic fluvial dryland landscapes: (A) Basin-and-range topography with alluvial fans and playas; (B) Inselberg and pediment landscape; (C) Canyon and escarpment country (after Goudie and Wilkinson, 1978)

TABLE VI.1

Comparison of desert surface types by plan area

Geographical zone	Likely occurrence commonest in engineering zone (see Fig. VI.4 for zone notation)	Sahara	Libyan desert	Arabia	South-western US
Desert mountains	I	43%	39%	47%	38.1%
Volcanic cones and fields	I	3	1	2	0.2
Badlands and subdued badlands	I/II	2	8	1	2.6
Wadis	I/II/III	1	1	1	3.6
Fans	II	1	1	4	31.4
Bedrock pavements	II/III	10	6	1	0.7
Regions bordering through flowering rivers	II/III/IV	1	3	1	1.2
Desert flats	III/IV	10	18	16	20.5
Playas and salinas	IV	1	1	1	1.1
Sand dunes	IV	28	22	26	0.6
		100.0	100.0	100.0	100.0

Source: Fookes (1976a).

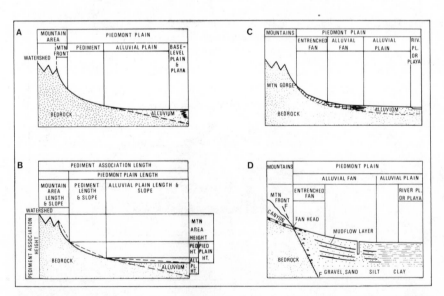

FIG. VI.3 The components and terminology of the mountain-plain model of dryland landforms: (A) Terminology; (B) Morphometry; (C) Variation 1: alluvial fans and buried pediment; (D) Variation 2: with faulted mountain front

water and sediment processes within them are in reality very similar. The visual differences arise in large measure from different combinations and geometry of a few basic forms, and on lithological and structural variations that strongly differentiate dryland topography. As a result, a simple mountain-piedmont plain model (e.g. Cooke and Warren, 1973; Figure VI.3A) can be used to represent a basic dryland fluvial system, and the dryland drainage system can be regarded as normally comprising (i) the mountain catchment with its hillslopes and channels, and (ii) the piedmont plain, with its pediments, alluvial plains, and base-level plains (especially playas). This simple model is of direct value to engineering geologists, who have used it as a basis for spatial classification of geotechnical features, as illustrated in Figure VI.4 (Fookes and Knill, 1969; Fookes, 1976b).

2. *Mountain catchments, channels, and hillslopes*

Mountain catchments of drylands have rarely been subjected to detailed analysis (Lustig, 1969). From the point of view of water and sediment production, their most important features are: (i) mountain catchment area, (ii) the nature of the channel system, (iii) slope characteristics, (iv) sediment production, (v) tectonic processes, and (vi) regional and local climate.

(*i*) *Mountain catchment area* is an important parameter since it often strongly influences the volume, instantaneous flood peak and duration of runoff (see section (d)), and because larger basins tend to yield less sediment per unit area than smaller ones. This is mainly because in larger catchments sediment storage is greater on the lower slopes and in the larger valley floors. In alluvial fan systems (Fig. VI.5), there is a strong relationship between mountain catchment area and the area of fans produced at the mountain front (e.g. Bull, 1964; Hooke, 1968). In this relationship

$$A_f = cA_d^n \qquad (VI.1)$$

where A_f = fan area,
and A_d = drainage basin area.

The coefficient, c, is the area of fan with a drainage supply area of 1 km², and varies from location to location because of local controls such as lithology, climate, slope, space available for deposition, and an exponent, n, that also varies spatially, but has a mean value of about 0.9. Very broadly, and with caution, A_f may be used as a proxy for sediment yield (Bull, 1977). The relationship is also useful because, in attempts to predict discharge of water, depositional area may also be used as a surrogate for catchment area (see section (d)).

(*ii*) *The channel system* exercises a strong control on the movement of water and sediment. Channels in dryland mountain catchments show a similar hydraulic geometry to those in other regions (Leopold and Maddock, 1953; Leopold and Miller, 1956), and the relations between channel properties can be summarized in general equations as follows:

$$w = aQ^b \qquad \text{(VI.2)}$$

$$d = cQ^f \qquad \text{(VI.3)}$$

$$v = kQ^m \qquad \text{(VI.4)}$$

$$L = pQ^j \qquad \text{(VI.5)}$$

where w = mean bankfull width, d = mean channel depth, v = mean flow velocity, Q = bankfull discharge, L = suspended sediment load, a, c, k, p are constants, and $b, f, m,$ and j are exponents, values for which are given for ephemeral and perennial streams in New Mexico and Nebraska in Table VI.2 Thornes (1976) has shown that mountain-catchment channels appear to conform to a pattern of regular downstream variation in geometrical properties ('channel auto-geometry'), responding to inputs of water and sediment received from tributary channels with pulsed changes to width and depth in a downstream direction. These changes persist downstream: care must be taken therefore in using values of channel-geometry variables for predictive purposes since they will probably be more complex than the simple linear models suggested by earlier work.

Channels in mountain catchments may consist of rock swept clean by every flood; of rock thinly veneered with weathered debris, aeolian deposits, or lagged flood deposits; or of alluvium with braided channel systems, flood terraces that are often of limited duration (Schick, 1970b, 1976b), and debris-flow deposits. The proportion of each type largely determines the nature of flood flow since the deposits provide important controls on transmission losses (see section (d)), recharge, base flow, and sediment transport, or debris-flow initiation. These features should therefore form a part of the landform inventory of any dryland area to be investigated for water and sediment problems.

It is useful to know the long- and short-term tendencies of channel change. The long-term metamorphosis of channels due to climatic change is well known in many drylands, especially semi-arid regions (e.g. the Great Plains of the USA and the Riverine Plains of Australia (Schumm, 1960, 1961a, b, 1963, 1968, 1969; Butler, 1960; Langford-Smith, 1962)). The theme of long-term metamorphosis is more important on piedmont plains than in mountains, but it should be noted that ancient river channels in the sediments of both mountain and plain regions can affect sub-surface foundations, water flow, recharge and salinity conditions, and this emphasizes the need for their study in the course of assessing foundation capability designs, or of water losses from irrigation schemes. Subsidence under load may be an associated problem. In the short term, changes in channels can be related to such features as scour following a large flood event (Thornes, 1976); the effect of overgrazing, agriculture, or construction in catchment headwaters (e.g. Duce, 1918; Bahrain, Ministry of Works, Power and Water, 1976); short-term climatic changes and desertification (Bryan, 1925; Leopold, 1951;

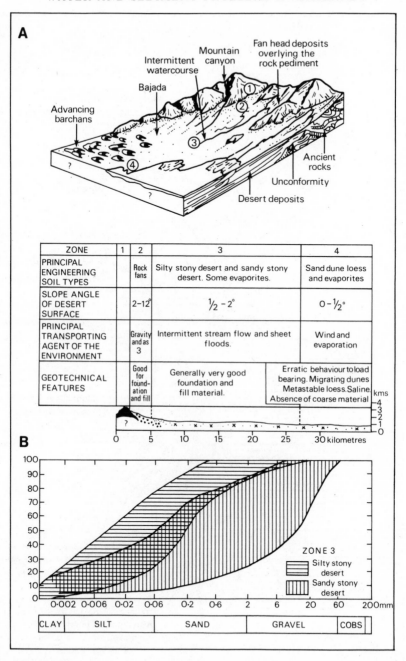

ZONE	1	2	3	4
PRINCIPAL ENGINEERING SOIL TYPES		Rock fans	Silty stony desert and sandy stony desert. Some evaporites.	Sand dune loess and evaporites
SLOPE ANGLE OF DESERT SURFACE		2-12°	½ - 2°	0 - ½°
PRINCIPAL TRANSPORTING AGENT OF THE ENVIRONMENT		Gravity and as 3	Intermittent stream flow and sheet floods.	Wind and evaporation
GEOTECHNICAL FEATURES		Good for foundation and fill	Generally very good foundation and fill material.	Erratic behaviour to load bearing. Migrating dunes. Metastable loess. Saline. Absence of coarse material

FIG. VI.4 (A) Block diagram of the mountain-plain model of dryland topography, and associated features of engineering significance; (B) Summary of particle-size analysis of surface samples from Iran, for different materials shown in zone III of A (after Fookes and Knill, 1969)

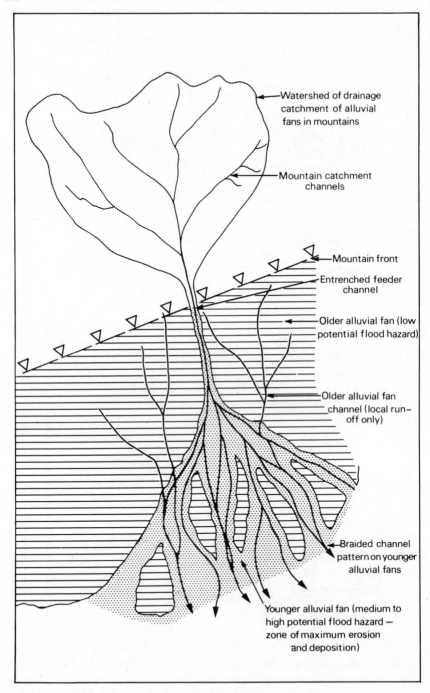

Watershed of drainage catchment of alluvial fans in mountains

Mountain catchment channels

Mountain front

Entrenched feeder channel

Older alluvial fan (low potential flood hazard)

Older alluvial fan channel (local run-off only)

Braided channel pattern on younger alluvial fans

Younger alluvial fan (medium to high potential flood hazard — zone of maximum erosion and deposition)

FIG. VI.5 Some components of an alluvial fan system

TABLE VI.2

Values for exponents b, f, m and j (after Cooke and Warren, 1973)

| | Average downstream relations | | | | Average at-a-station relations | | | |
| | Ephemeral streams | | Perennial streams | | Ephemeral streams | | Perennial streams | |
	New Mexico[1]	Nebraska[2]	Average[3]	Nebraska[2]	New Mexico[1]	Nebraska[2]	Average[3]	Nebraska[2]
b	0.5	0.03	0.5	0.69	0.26[4]	0.35	0.26	0.24
f	0.3	0.48	0.4	0.12	0.33[4]	0.43	0.40	0.56
m	0.2	0.45	0.1	0.19	0.32[4]	0.22	0.34	0.20
j	1.3		0.8		1.30[4]		1.5–2.0	

[1] Source: Leopold and Miller (1956). (The authors express reservations about the small number and accuracy of their observations, but the data probably give a reasonable indication of reality.)
[2] Source: Brice (1966). (Data from the Medicine Creek basin. At-a-station data may be considered reliable; downstream data only indicate the correct order of magnitude.)
[3] Source: Leopold and Maddock (1953).
[4] Unadjusted median values.

Antevs, 1952); and flow concentration due to road or irrigation works (Cooke and Reeves, 1976).

(*iii*) *The slope characteristics of mountain catchments* are fundamental to understanding water, flood, and sediment problems. Since runoff is usually extremely rapid in dry areas, attempts have been made by many authors to explain the characteristics of the flood hydrograph in terms of the steepness and length of the contributory slopes (see section (d)). Although recent work suggests that, for example, flood peak is best explained in terms of catchment area, and that slope data add little additional explanation, some slope variables are a strong component of many predictive methods and it is essential that they are included in geomorphological mapping projects of terrain concerned with runoff and sediment problems (see section (f)).

Sediment production is strongly affected by the resistance of rocks to weathering and erosion, the nature and rate of weathering, and the nature of slope processes. From a practical point of view, three main types of slope can usefully be recorded during geomorphological surveys:

(*a*) *Gravity-controlled slopes* are usually bare-rock features of relatively high inclination and relief, with a form dominated by the influence of rock lithology and structure (the escarpments in Figure VI.2C provide an example). The main processes are rockfall and rock slides, controlled in part by weathering, unloading (in coarse-grained igneous rocks), and basal erosion (Twidale, 1964; Schumm and Chorley, 1964, 1966). Active landsliding is not as common in arid areas as it is in more humid regions, but there was more slope failure during pluvial periods of the recent geological past, and fossil landslides may present serious problems if disturbed by constructional activity. For example, research in the central Sahara (e.g. Grunert and Hagedorn, 1976; Grunert, 1978; Hagedorn *et al.*, 1978) has shown that the western and south-western rim of the interior of the Murzik Basin is marked by massive slides for nearly 300 km. There are two principal types – rotational slides and multiple slips of mudflow form in an echelon arrangement. Some of the rotational slides are as large as 1 km long, 100 m wide, and 100–350 m high. Similar forms have been recorded in Iran (Haars, *et al.*, 1974; and Harrison and Falcon, 1937a and 1938b). Such processes can pose serious problems, especially to roads and road construction.

(*b*) *Debris slopes* usually occur downslope of gravity slopes wherever accumulation from higher slopes exceeds local disintegration and removal. Slope angles normally range from 12 to 38°, with 38° considered as a commonly occurring limiting angle (Melton, 1965; Carson, 1971), the modal class value usually lies between 25 and 28°, and unstable slopes generally occurring above 33°. The modal class value is close to the stability angle for coarse regoliths when in a saturated state – higher angles may occur because saturation may not be reached in some drylands, or because the angle of repose of dry angular material (i.e. at full frictional strength) is higher. The range of slope angles is often related chiefly to the grain size of the original material and its subsequent disintegration or decomposition. On these slopes, mass movement, especially debris slides, rolling, and sliding are the dominant processes. Fluvial processes only come to dominate on lower slopes, on finer materials, or where the slope is cemented by duricrusts. Open-textured deposits should be recorded because they have high infiltration capacities and, therefore, have a restrictive effect on flood peak. Where they include fine-grained materials, and are subject to saturation by rainfall or snowmelt, they may also be an important contributor to debris flows, a particularly hazardous flow feature of many drylands (Pack, 1923; Blackwelder, 1928; Alter, 1930; Bailey *et al.*, 1934; Sharpe, 1938; Eckel, 1958).

(c) *Wash slopes* are dominant on low-angle slopes and where there are fine-grained weathered mantles and soils, including stone pavements, and where lack of vegetation results in rapid runoff. Runoff processes range from sheet flow to turbulent flow in rills and gullies, and are generally considered to be efficient erosive agents, especially for fine sediment. The extent, angle, length, and hydrological features of such slopes should all be recorded during hydrological surveys because they strongly influence infiltration, storage, and runoff character on these slopes. Such slopes cover a higher proportion of drylands than any other type.

(*iv*) *Tectonic processes* in seismically active regions may be important controls on channel behaviour and on relief-slope-area relationships. Uplift will, for example, probably lead to channel entrenchment and move the loci of deposition away from the mountain front (e.g. Bull, 1977). Tectonic activity may also strongly affect the thickness of alluvial deposits beyond the mountain front – a feature of great relevance to water resources, transmission loss, flooding, aggregate potential, and foundations. The evidence of tectonic activity should be recorded for these reasons and also because earthquake hazards are of direct concern to foundation and structural engineers, and to architects and planners.

(*vi*) *Regional and local climates* are further controls on the water-sediment system, and are considered in section (c).

3. *Alluvial fans*

One of the most important depositional landforms in the mountain-piedmont landscape model is the alluvial fan (see Fig. VI.5). Fans occur widely throughout drylands, especially in areas of basin-and-range topography, and because of their low-angle slopes, aggregate potential, water resources, and suitability for construction are highly favoured areas for development. The literature on fans is large; extensive summaries are provided by Bull (1968, 1977), Cooke and Warren (1973), and Mabbutt (1977); engineering and other applied problems are considered in Kesseli and Beaty (1959), Schick (1971, 1974a), and by Cooke and Doornkamp (1974). This account will concentrate only on those features of alluvial fans of importance to the understanding and control of the water and sediment system.

Alluvial fans are cone-shaped depositional landforms. As a mountain stream leaves its catchment, discharge tends to decrease due to transmission loss, there is an accompanying increase in sediment concentration, a release of bedload, a decrease in channel depth and flow velocity, and a tendency for flow to be dispersed into unstable, shifting channels. The main conditions for alluvial fan formation are (i) lack of vegetation, so that sediment production is high and channels remain unfixed; (ii) relatively rapid debris accumulation and weathering within the catchment; (iii) intense precipitation or snowmelt with rapid runoff of sufficient magnitude to remove a large sediment load; and (iv) the focusing of mountain catchments on to a limited number of outlets at the mountain–plain boundary.

Alluvial fans normally comprise several distinctive features (Fig. VI.5), most of which should be recorded during the course of a geomorphological survey.

(i) *Abandoned, old or elevated fan surfaces* which usually possess varnished or cemented stone pavement surfaces and local drainage channels. These areas only generate water and sediment from their own restricted catchments, and their water availability and flood risk is usually low.

(ii) *Channels* include entrenched feeder channels, distributory networks, and even dendritic networks that originate on the fan itself. They have relatively high flood risk, and are subject to scour, fill, lateral migration, and braiding. They are also unstable forms which can be abandoned quickly if an easier course or slope presents itself (e.g. if one channel becomes blocked with coarse debris). The shifting of drainage lines in the complex channel network is a major hazard to development on fan surfaces. In mapping these features it is important to note features indicating recent activity (e.g. scour pools) or lack of it (e.g. dune accumulations).

(iii) *Contemporary depositional areas* occur at the terminal ends of active channels either within an old fan surface or, if the loci of deposition have moved as a result of climatic, tectonic, or other causes, as segmented (telescopic) fans (e.g. Denny, 1967). Fan surfaces of current activity can be distinguished from older and more stable areas by the use of morphological criteria, stone pavement form, soil-profile characteristics, vegetation pattern, weathering phenomena, height relationships, and degree of dissection. These criteria can form the basis of both qualitative and quantitative assessments of flood hazard on alluvial fans.

Although many fans are produced by sediment-rich stream flows that yield typical water-laid deposits, many fans in semi-arid lands are built by debris flows and mudflow deposits. Water-laid sediments are generally well sorted (Fig. VI.6), with occasional lamination and cross-bedding; they include scour and fill channel deposits that are characteristically poorly sorted and coarse-grained; clasts are generally rounded and separated by a fine matrix.

Debris flows, of which mudflows are a fine-grained version, are more common in dryland areas than is generally appreciated, especially in the south-western USA and many semi-arid mountainous areas of Asia and South America. They are a very serious hazard to urban development since their frequency, magnitude, velocity, and spatial location are at present largely unpredictable.

The phenomena range from turbulent and destructive mud floods (hyperconcentrated flows of >50% sediment concentration) to true, viscous

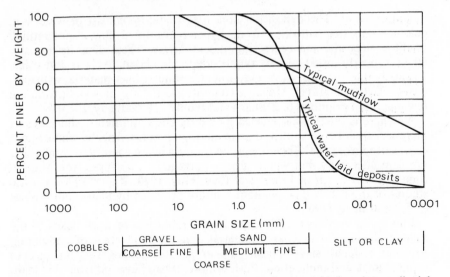

FIG. VI.6 Typical grain-size curves for mudflow and water-laid deposits on alluvial fans

debris flows of >80% concentration. They usually originate in small (10–100 km²) mountain catchments that have a dense, branching network of gullies cut in loose debris and may be fed by hill slope flows (Sharp, 1942; Curry, 1966). These latter are usually much smaller and less dramatic. The movement usually originates in a single or multiple head, follows a long narrowing track, and spreads out on the lower slopes of the valley. The flows generally occur on hillside slopes of greater than 30° and characteristically build levées alongside the track. They rarely erode deeply into the slope except in loose scree slopes. The deposits are normally composed of 1–5% clay, 40–60% sand or finer, and carry any coarse clasts which occur in the slope deposit. The reported moisture contents are often less than 10% by weight. Intense rain and snowmelt are the main causes of movement and the subsequent descent of material occurs as single or pulsating surges at velocities as high as 10–16 ms⁻¹.

The material supplied by the gullies, hillside debris flows, and other mass movements accumulate in the main valley track. This may normally operate as a stream channel but when there is sufficient material it may be fully occupied and reorganized by fill and cut processes as a true debris flow takes place. The valley flows follow long, sinuous courses along the major channels, also build levées where they overflow and drain, are capable of overriding obstacles, carry huge boulders, and form lobate accumulations on the fans. There is a rapid change of slope angle down valley from the >30° hillslopes to 15–20° in the supply channels and 6–9° in the runout areas. They can move at high speeds (<16 m/sec⁻¹) but 0.5–4.0 m/sec⁻¹ is more common. The runout can be as long as 30 km, even on the gentle slopes of a fan but if the material of the fan is very permeable then rapid drainage leads to the diagnostic levées or short lobate sieve deposits which remain close to the mountain front. The deposition area may therefore consist of water-lain channel deposits, silt-skins from hyperconcentrated flows, unstratified mudflow deposits, and coarse hummocky layers. These deposits are sub-angular, poorly sorted, and weakly bedded with abundant clasts and voids. Typically they include <15% clay and 60% sand or finer material with 40% coarse clasts and gravel. Moisture contents are 10–30% by weight from the few measured examples (e.g. Sharp and Nobles, 1953; Beaty, 1963).

Nowhere is the damage caused by debris flows and debris slides events more serious or better studied than in metropolitan Los Angeles where major engineering works have been required for over half a century to combat the problem effectively, but where serious damage is still caused from time to time (e.g. US Corps of Engineers, 1969; Guy, 1970; Rantz, 1970). Classic studies of the debris flow phenomenon are provided by Sharp and Nobles (1953) and Beaty (1963).

Finally, in very permeable deposits the rapid loss of water leads to the sudden sieving out of coarse material and the production of undulating irregular, residual, 'sieve-deposits' topography. Such topography is subject to rapid change and inundation. Furthermore, it can have unfavourable load-bearing characteristics for engineering structures built on it. For both of these reasons it should be recorded and, if possible, avoided.

4. *Pediments*

Pediments, gently sloping bedrock surfaces that are usually cut across geological structures and may be thinly veneered with alluvium, occur in all drylands. They usually occur as modest-sized footslopes near to mountain fronts or escarpments, as extensive planation surfaces of almost continental proportions (e.g. King, 1953, 1962), or as terraces along major drainage lines. The literature on pediments is more extensive than it is illuminating, and recent reviews include those by Cooke and Warren (1973) and Mabbutt (1977). Here attention is drawn only to water and sediment processes, although the occurrence of bedrock near to the surface is itself an important consideration in planning urban development, as discussed briefly in section II.2.

Pediments are relatively smooth landforms, but they also commonly possess a low relief formed by a ramifying channel system. On the upper part the channels may be in distributary form, and straddle (or run parallel to) the mountain–piedmont junction. Downslope, channels deepen and then either become shallower and fade out, or deepen still further in response to some lowered baselevel. The drainage net is normally developed partly on bedrock and partly on alluvium, and it is usually complex, with frequently changing patterns and a multitude of shallow rills. Sheetwash is common on inter-channel surfaces.

The most important fluvial processes are raindrop impact, splash, sheetwash, rilling and, where discharge of water and sediment are adequate, gullying. Severe erosion can develop, as in the Badlands National Monument, Dakota, and in burnt or grazed areas of South Australia. Perhaps the most important processes are channelled flow and sheetwash, both of which transport debris from the mountain areas and remove debris on the pediment surface itself, causing it to be regraded. Such debris transfer is an important component of runoff quality and sediment yield from dryland catchments, and severely restricts agricultural development on pediment areas.

There are few measurements of fluvial processes on pediments, and monitoring has only been attempted on a limited scale, for example by using erosion pins and sediment traps in a very few areas. Precise data are very scarce. Rahn (1967) for instance, demonstrated that runoff from storms across sediments in part of southern Arizona, was in streamflow form near the mountains, in sheetflood form further downslope, and was supercritical, at discharges in the range 0.2–7.28 cumec. But as yet, discharge on pediments cannot be adequately predicted, despite the fact that it is clear from the few observations that pediment channels are occasionally active, liable to flood, subject to cut and fill, and tend to change position – evidence which points towards the existence of a potential hazard to urban development. In the absence of 'process' data, geomorphological surveys should record stream nets carefully and record form and deposits on pediment areas with a view to attempting to define development risk.

5. *Playas*

The lowest areas of enclosed mountain–piedmont drainage systems in drylands normally consist of a flat plain that is largely free of vegetation, composed of fine-grained sediments, and may be subject to seasonal flooding; such features are known as playas in the USA and by many other local names elsewhere (Cooke and Warren, 1973; see also Chapter V). They occur extensively in drylands, especially in basin-and-range-deserts, in areas such as the western Americas, western Australia, Iran, the Middle East, and Tunisia.

The hydrological regime of playas is closely related to the relative proportions of surface and near-surface water derived from precipitation, runoff from within the surrounding catchments and groundwater discharge. The position of the water-table reflects the balance between inflow through infiltration and loss by capillary rise, evaporation or human extraction: its relationship to the surface is critical in determining surface-crust characteristics, desiccation phenomena, and flood characteristics. Flooding features may range from temporary inundation to the formation of perennial lakes (e.g. Lake Eyre Committee, 1950). Where evaporation dominates the surface, soils may become saline, and saline ground conditions pose special problems to urban development (see Chapter V).

There are relatively few data readily available on the engineering characteristics of playa surfaces (and their coastal equivalents, sabkhas), but several studies reveal the kind of field measurements that may be of value (Motts, 1965; Langer and Kerr, 1966; Neal, 1969; Krinsley, 1970; Ellis, 1973; Fookes, 1976a). Tables VI.3 and VI.4, and Fig. VI.7 summarize some useful data on playas.

Playas are often of economic potential since they may contain lacustrine evaporite deposits such as chlorides, sulphates, carbonates, nitrates, and borates (e.g. Neal, 1975), some of which may be of commercial value, as in northern Chile (e.g. Mueller, 1960). In addition, playa surfaces tend to be flat

TABLE VI.3a

Hydrological classification of playas

1 Increase of surface-water discharge	2	3 increase of groundwater discharge	4	5 All groundwater discharge
All surfacewater discharge	Relatively small amount of ground-water discharge and large amount of sur-face-water discharge	Approximately equal proportions of groundwater and surface-water discharge	Relatively large amount of ground-water discharge and small amount of sur-face-water discharge	
1A *1B* Small Large amount of amount of surface- surface-water water discharge discharge				

increase of surface-water discharge

Source: Motts (1965).

TABLE VI.3b

Classification of playa surfaces

(i) Total surface-water discharging playas where the watertable is so deep that no groundwater discharge occurs at the playa surface
(ii) Playa surfaces where discharge occurs at the surface by capillary movement
(iii)Playa surfaces where discharge occurs directly at the watertable
(iv) Playa surfaces where discharge occurs by phreatophytes and other plants
(v) Playa surfaces where discharge occurs by springs
(vi) Combinations of all of these

Source: Motts (1965).

and therefore may appear to be topographically suitable for agriculture, urban development and, especially, airfields. Such is the case in the south-western USA, but the human development of playas has revealed several serious hazards, most notably the development of giant desiccation cracks in the predominantly evaporative environment that can seriously damage roads and runways, sometimes without warning (e.g. Neal *et al.*, 1968). Thus for many playas, it is essential to record during geomorphological surveys, polygonal cracking patterns, surface crust types, and similar phenomena that are indicative of hydrological conditions and potential hazards, in addition to recording sediment types, water-table position, and surface form.

(c) Precipitation in Drylands

In drylands, as elsewhere, a comprehensive understanding of precipitation, its distribution in space and time, and its depth, duration, and frequency characteristics are essential to the proper design of foundations, storm sewers, flood bunds, drainage works, and water-regulating devices within drainage catchments. In general, insufficient attention is paid in the drylands of developing countries either to these matters or to the proper collection and analysis of precipitation and runoff data. Damage from flood events can be

TABLE VI.4a

Classification test results from sands in playas and alluvial plains

Brief description	Depth m	Optimum moisture content	Maximum dry density Mg/m³	CBR* Unsoaked		CBR Soaked		Total water-soluble Salts %	pH value
				Top	Bottom	Top	Bottom		
Fine sand	0.1	9.5	1.720	12	15	5	8	0.05	9.2
Fine sand	0.1	9.0	1.682	11	13	2	2	0.03	8.5
Slightly silty fine sand	1.0	9.5	1.979	8	11	2	2	0.04	8.7
Slightly cemented silty sand with nodules of calcrete	1.9	6.7	2.076	29	37	9	21	0.04	8.5
Calcarenite*	2.0	9.5	2.013	74	87	24	39	0.05	8.7

* For terminology see Fookes & Higginbottom (1975).
Source: Fookes (1976a).

TABLE VI.4b

Some specific salty soil types in playas and alluvial plains, and their engineering significance

Name	Terrain	Groundwater table	Salts	Special significance	Construction techniques
SABKHA	Coastal flat, inundated by sea water either tidally or during exceptional floods	Very near the surface	Thick surface salt crusts from evaporating sea brines. Salts usually include carbonates, sulphates, chlorides and others	Generally aggressive to all types of foundations by salt weathering of stone and concrete and/or sulphate attack on cement bound materials. Evaluate bearing capability	Carefully investigate. Consider tanking concrete foundations; using SR cement. For surfaced roads consider using inert aggregate, capillary break layer or positive cut-off below sub-bases. Use as fill suspect. May not be deleterious to unsurfaced roads
PLAYA	Inland, shallow, centrally draining basin – of any size	Too deep for the capillary moisture zone to reach the ground surface but area will be a temporary lake during floods	None if temporary lake is of salt free water	Non-special. Ground surface may be silt/clay or covered by windblown sands. Evaluate bearing capability	Non-special
SALT PLAYA	As playa but often smaller than a playa	As above but lake of salty water	Surface salt deposits from evaporating temporary salty lake water. Salts usually include chlorides and sometimes nitrates, sulphates and carbonates	Can be slightly to moderately aggressive to all types of foundations by salt weathering and sulphate attack. More severe near water table	As sabkha
SALINA	As playa	Near surface; capillary moisture zone from salty groundwater can reach the surface	Surface crusts from evaporating salty groundwater. Salts include carbonates and many others	Can be slightly to exceptionally aggressive to all types of foundations by salt weathering and sulphate attack	As sabkha

Source: Fookes (1976)

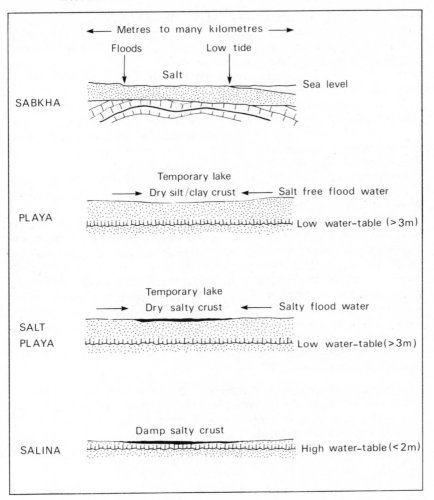

FIG. VI.7 Idealized cross-section of sabkha, playa, salt playa, and salina (after Fookes, 1976b)

catastrophic and where insufficient data are available very conservative designs may result, with high safety factors and correspondingly high costs. One aim of research, therefore, must be to develop primary rainfall/runoff monitoring systems in critical locations, and to formulate good general theoretical prediction models that can be applied locally.

Drylands occur mainly in areas associated with the presence of high-pressure cells around the 30th parallels, where there is subsiding air, relative atmospheric stability, and divergent airflows at low altitudes. These areas are occasionally penetrated also by the rain-bearing low-pressure systems characteristic of the circumpolar westerlies of middle latitudes. Continentality can reinforce the effects of the global circulation pattern and produce areas of extreme aridity (e.g. the Central Sahara). In addition, coastal mountains can cause rain-shadow arid zones, as, for example, in south-western USA. The general result is that desert areas are characterized by low precipitation and relative humidity, and high temperatures, although in some coastal desert regions, such as those around the Arabian Gulf, humidities can be very high during the summer months.

These drylands, however, cannot be distinguished solely in terms of precipitation deficit and most climatic classifications, such as that developed by Meigs and used in Fig. I.1 have employed a moisture or aridity index. Within Meigs's classification, further subdivision is based upon rainfall deficit and the seasonal distribution of rain (Alam, 1972).

Most rainfall in drylands generally occurs as a result of the passage of low-pressure or frontal systems or from small convective cells producing intense storms of short duration. For example, on the Walnut Gulch Research Watershed in south-western USA, Renard (1970) has found that winter rainfall (October to April) is a result of depressions originating in the North Pacific Ocean, whereas the July, August, and September rain is produced by small convective thunderstorms which develop when moist air masses advance from the Gulf of Mexico. A similar situation is described by Kappus et al. (1978) on the Arabian Gulf coast of Iran. Schick and Sharon (1974), however, found that rainfall on the Nahal Yael Research Watershed in Sinai was produced by convective cells developed at the front of a low-pressure belt as a result of chronic atmospheric instability.

Generalizations about desert precipitation are difficult to make, but a number of features of rainfall, are important:

(i) Precipitation can be very varied spatially. Table VI.5 shows the rainfall collected over 10 hectares in Avdat, Israel by 20 rain-gauges during a single storm. The mean value collected was 5.07 mm, the standard deviation 1.75, and the range 7.8 to 2.4 (see also Figs. VI.8 and VI.9).

TABLE VI.5

Rainfall data, Avdat, Negev, Israel, 1 March 1960

	Rain gauge	Rainfall (mm)	Rain gauge	Rainfall (mm)
Automatic	1	7.3	8	4.4
Automatic	2	3.4	9	2.4
Standard		7.8		
Small	1	7.0	10	2.9
	2	6.8	11	4.2
	3	7.3	12	3.0
	4	7.0	13	4.8
	5	5.6	14	5.8
	6	2.2	15	5.0
	7	4.4	16	5.5
			17	4.6

Source: Evenari *et al.*, 1968.

23–24 November 72

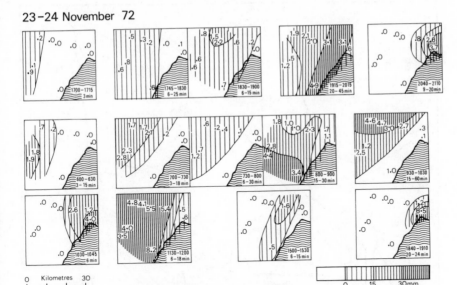

FIG. VI.8 Spatial and temporal variability of rainfall for a storm period in the Nahal Yael area of southern Israel based on studies by D. Sharon (after Schick, 1974a)

(ii) Individual storm precipitation totals can be very high. Table VI.6 lists rainfall totals for individual storms from various drylands of the world. In many cases the single-storm precipitation far exceeds the mean annual precipitation. This results in the typical positively skewed distribution of annual rainfall totals (see Alice Springs, Australia, Fig. VI.10B). Slatyer and Mabbutt (1964) stated that in fact it is not uncommon for the standard deviation of the mean annual rainfall to exceed the mean value.

TABLE VI.6

Extremes of precipitation

Location	Date	Mean annual precipitation (mm)	Storm precipitation
Chicama (Peru)	1925	4	394 mm
Aozou (Central Sahara)	May 1934	30	370 mm/3 days
Swakopmund (South West Africa; Namibia)	1934	15	50 mm
Lima (Peru)	1925	46	1524 mm
Sharjah (Trucial Coast)	1957	107	74 mm/50 min
Tamanrasset (Central Sahara)	Sept. 1950	27	44 mm/3 hr
Bisra (Algeria)	Sept. 1969	148	210 mm/2 days
El Djem (Tunisia)	Sept. 1969	275	319 mm/3 days

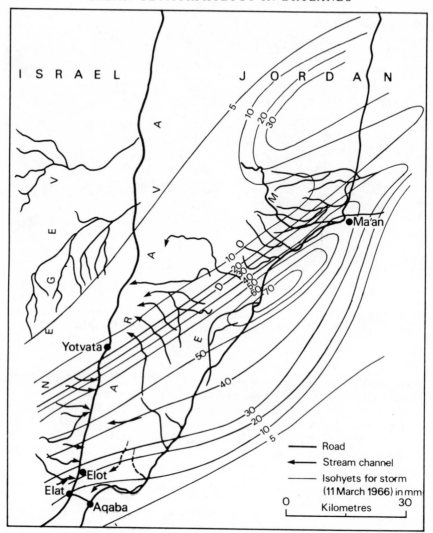

FIG. VI.9 The pattern of rainfall in the area of southern Israel and Jordan affected by the storm of March 1966 (after Schick, 1971)

(iii) Rainfall intensities can be very high. One of the highest recorded rainfall intensities in a very arid area is the 74 mm in less than an hour at Sharjah (Table VI.6); larger Sarahan storms regularly reach intensities of 1 mm/minute.

Superimposed on these general features of dryland precipitation is the possibility of long-term fluctuations. Figure VI.10C displays how there have been periods of time above and below the mean annual precipitation in Tacubaya, Mexico. This can be of great geomorphological importance as shown by Leopold (1951), who suggested that long-term fluctuations in

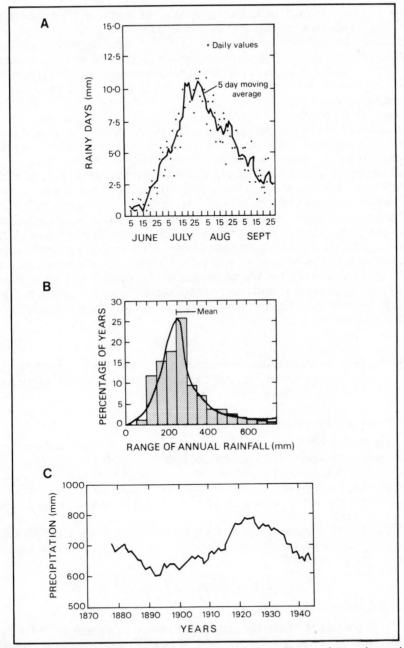

FIG. VI.10 (A) Frequency of historical rainy days during a 73-year observation period and over a 4-month season for Tombstone, Arizona (after Smith and Schreiber, 1973); (B) Annual rainfall distribution 1874–1959 at Alice Springs, Northern Territory, Australia (after Slatyer and Mabbutt, 1964); (C) Long-term fluctuations of precipitation at Tacubaya, Mexico (after Wallén, 1956)

precipitation resulted in increased erosion in southern USA (see Fig. VI.10A). This view is supported by Cooke and Reeves (1976) in southern Arizona, although they demonstrated that human environmental changes such as overgrazing also fundamentally affected soil loss by water erosion. Such fluctuations or trends in precipitation have to be taken into account for both their effect on the geomorphological environment and on the design of engineering structures.

Despite the problems of variability in space and time in precipitation in drylands and the lack of sufficient records, the need for conservative design criteria in engineering works requires that attention is paid to estimating the upper limits of rainfall within a drainage basin or region. Two main techniques are used for estimating the Probable Maximum Precipitation (*PMP*): the maximization and transposition of real or model storms, and the statistical analysis of extreme rainfalls. Myers (1967) outlined the technique of storm (and hence precipitation) maximization. This involves setting maxima to the general synoptic conditions liable to result in a storm which produces 'the theoretically greatest depth of precipitation for a given duration that is physically possible over a particular drainage area at a certain time of year' (Huschke, 1959). The second, statistical procedure for estimating *PMP* uses existing rainfall records to estimate the extreme precipitation. The World Meterological Organization manual (1973) described this technique to derive PMP by:

$$PMP \approx \bar{P} + KSp \qquad (VI.6)$$

where $\bar{P}$ = mean of annual series of 24-hour maximum precipitation,
         Sp = standard deviation of the annual series of 24-hour maximum precipitation,
and K = empirical coefficient which is a function of $\bar{P}$ and rainfall duration.

Kappus *et al.* (1978) compared the two techniques in the development of rainfall frequencies for the Gulf Coast of Iran. No significant difference was found between the two 24-hour *PMP* estimates for point rainfall. This study was extended to include the development of precipitation depth–duration–frequency curves for the area (Fig. VI.11) and it provides a useful model for future research.

Despite such advances there remains a paramount need for reliable observational networks and further fundamental research on desert rainfall and its relation to runoff.

(d) The Generation of Runoff

1. *Hillslopes*

A high proportion of dryland rainfall is absorbed by the ground or lost by evaporation, and the residue available as runoff decreases with annual rainfall towards a minimum of perhaps only 10% in extremely arid areas. Not all low-intensity storms generate runoff, and it is generally thought in this context that precipitation intensity is a more important factor than rainfall total. Published data (Table VI.7) suggest that intensities of at least

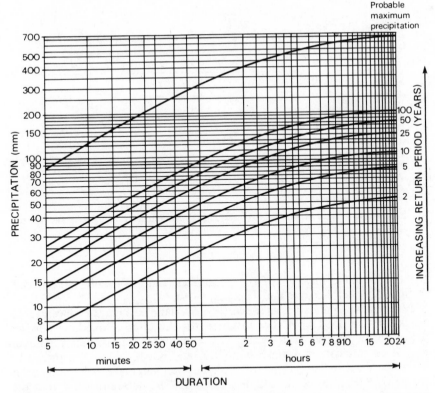

FIG. VI.11 Precipitation depth-duration-frequency curves for part of the Iranian Gulf coast (after Kappus *et al.*, 1978)

1 mm/min and totals of some 10 mm at these intensities are normally required for channelled runoff to occur (see IAHS–AISH, 1979).

The following controls promote maximum and rapid runoff and subsequent cessation of flow from hillslopes and headwater catchments: steep slopes, absence of dense vegetation, immediate antecedent rainfall and soil moisture, the absence of permeable detritus, cemented and compacted mantles, thin soils, and poor water-storage capacity.

On the slopes of the mountain or fan catchments runoff may be either in sheet or in concentrated form. Observations in the field (e.g. Doornkamp *et al.*, 1980) suggest that where the soil surface is compacted it resists infiltration, and where it is thin it quickly becomes saturated so that, during heavy rain, surface detention storage rapidly takes place, puddles become linked by shallow rills, and connected flow fills shallow drainage lines. These lines can overflow to unite and form a sheet of water perhaps 10–30 mm deep which can move quite rapidly, even on low slopes. There appear to be four stages: (i) the visible retardation of rainfall is just apparent; (ii) a third of the surface is covered by water; (iii) the surface is just completely covered; and (iv) there is visible flow at the surface. Runoff is judged to begin at the fourth stage and

TABLE VI.7

Published data on the intensity of rainfall generating runoff in arid areas

Area	Intensity generating runoff	Total
Mojave	34 mm/24 hr	
Negev	1.7 mm/3 min	18 mm
	3.4 mm/10 min	
Ahaggar	25.4 mm/10 min	
C. Australia		12.7 mm
(bankfull Q)		50.8 mm
Piedmont areas		25–50 mm
Nahel Yael	1.5 mm/3 min	8–30 mm
Walnut Gulch		>6 mm
Alice Springs	44.6 mm/50 hr	

Source: Slatyer and Mabbutt (1964); Renard (1970); Schick (1971); Cooke and Warren (1973).

is the acceptable measure of the parameter 'time to runoff' (Rubin and Steinhardt, 1964; Scoging and Thornes, 1979).

Where rills or gullies dissect the surface, overland flow may be concentrated quickly into a flood wave. For instance, the initiation of overland flow on debris-covered hillslopes in southern Israel, requires approximately 3 mm of rain at an intensity of 1 mm/3 min (Yair and Klein, 1973). The threshold value decreases to 2 mm if the soil is already wet and to 0.2–0.4 mm/3 min if the peak intensities occur late in the storm. The data also show that the assumption in humid regions that the runoff coefficient increases with slope angle (0–45°) (e.g. Horton, 1945; Meyer and Monke, 1965) does not hold for dry regions, where the relationship may be inverted because increases in slope angle strengthen the factors that control infiltration (e.g. roughness, coarse grain size) (Evenari *et al.*, 1968; Yair and Klein, 1973; Scoging and Thornes, 1979) (Table VI.8).

TABLE VI.8

Relationship between runoff coefficients and slope angle at experimental plots of Avdat farm, Israel

Slope angle (%)	Runoff coefficient (%)
10.0	27.1
13.5	22.9
17.5	18.5
20.0	14.1

Source: Evenari *et al.*, 1968.

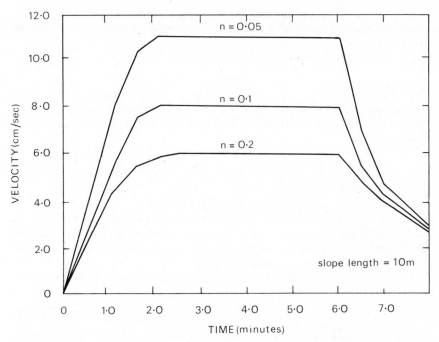

FIG. VI.12 The effect of Manning's *n* on flow velocity on dryland slopes (after Scoging, 1978)

Other characteristics of runoff on hillslopes in drylands include the fact that downslope increase in depth and velocity can be successfully predicted by the Manning equation (see equation VI.12) (e.g. Emmett, 1970; Scoging, 1978). Figure VI.12 provides an example. In addition, the time to initiation of overland flow is typically short, due to surface crust formation and slow infiltration (Scoging, 1978).

These conclusions suggest that the principal component of channelled runoff on hillslopes is overland flow and neither groundwater nor interflow (other than piping – see section (d)) (Evenari *et al.*, 1968; Yair, 1972, 1974; Yair and Klein, 1973). Most of these findings, however, are based on small experimental plots, and it is generally impossible to use them to predict the behaviour of larger areas, since recent studies show that runoff does not occur uniformly over all the slopes, even in conditions of uniform rainfall (e.g. Amerman and McGuiness, 1967; Betson and Marius, 1969; Yair and Lavée, 1975; Yair *et al.*, 1978). This is, of course, explicable in terms of topographic differences in the catchment which are related to spatial differences of soil depth, granulometry, permeability, and storage. Nevertheless, the relationships are not yet fully understood and there are severe problems here for the development of predictive theory (e.g. Osborn and Lane, 1969; Renard, 1970).

At present many estimations are based on Horton's (1945) model for steady state runoff (qs) which at distance X downslope and i (rainfall excess over infiltration) is

$$qs = iX \qquad \text{(VI.7).}$$

Carson and Kirkby (1972) provided a review of the extensive literature on this subject. Infiltration has been shown to be a fundamental control. The most commonly used model of the total rate of infiltration is that derived by Phillip (1957):

$$\mathscr{T} = A + B \cdot t^{1/2} \quad \text{or} \quad F = A_t + 2Bt^{1/2} \qquad \text{(VI.8)}$$

where $\mathscr{T}$ = rate of infiltration,
         F = the total amount infiltrated after time t,
         t = the time elapsed since the beginning of infiltration,
and A and B are constants for the soil and its antecedent moisture content which control the steady rate of flow under a constant potential gradient and the diffusion flow as the soil voids are filled.

A variation of this method is the empirical equation of Swartzendruber and Hillel (1975):

$$i_t = Bt^{-n} + A \qquad \text{(VI.9)}$$

where i_t = infiltration rate at time t,
         A = the final constant infiltrability,
         n = a constant which if equal to 0.5 yields a curve similar to that of Phillip,
and $B = (p - A/t_0^n)$ for $i = p$ (rainfall intensity at $t = t_0$).

A more useful approach, recently fully discussed with respect to soil erosion in the Sonia and Ugijar regions of semi-arid Spain (Scoging and Thornes, 1979) examines the controls of the period following the onset of precipitation during which no excess water is produced because the soil store is unsaturated and because the infiltration rate is flux controlled.

Following Green and Ampt (1911) and Mein and Larsen (1973) the amount of infiltration prior to runoff V is defined by:

$$V_{t_0} = \frac{S_{av} \cdot M_d}{(p/Ks - 1)} \qquad \text{(VI.10)}$$

and the infiltration rate i_t by:

$$i_t = K_s[1 + (S_{av} \cdot M_d / V_t) \qquad \text{(VI.11)}$$

where S_{av} = average capillary function at the wetting front which is a soil dependent parameter, determined for each soil, by integrating the standardized tension-conductivity curve on the wetting phase,
         M_d = the initial soil moisture deficit,
         p = rainfall intensity
         K_s = the saturated conductivity.

Data for drylands indicate that the final infiltration rate is reached in a few minutes, thus allowing an infiltration excess to develop during intense storms (e.g. Thornes, 1976).

Scoging and Thornes (1979) in their Spanish examples used sprinkler tests to show that times to surface flow were much shorter than those reported in the literature (Table VI.9) and that the higher the infiltration rate the quicker runoff occurs due perhaps to a limited availability of storage volume. This has the implication that rainfall intensity may be more relevant in some cases than final infiltrability. In these cases the saturation overland flow model of Kirkby (1978) may be more relevant than the classic Horton overland flow model.

The Spanish work showed that the best-fit curves for their infiltration tests were

$$i_t = A + B/t \qquad \text{(VI.12)}$$

and

$$i_t = At^B \qquad \text{(VI.13)}$$

From this they show that the volume stored (V) to the point at which precipitation intensity equals the final infiltrability A is given by:

$$V = p^B/(p-A) \qquad \text{(VI.14)}$$

and

$$V = p[A/p)^{1/B}] \qquad \text{(VI.15)}$$

and the times to saturated overland flow by:

$$t_0 = B/(p-A) \qquad \text{(VI.16)}$$

and

$$t_0 = (A/p)^{1/B} \qquad \text{(VI.17)}$$

Equation VI.14 is similar to that developed by Kirkby (1976, 1978) which showed that the Phillip equation (VI.8) yielded

$$V = pB/(p-A)^2 \qquad \text{(VI.18)}$$

It can, therefore, also be shown that it is the storms of short maximum intensity that commonly exceed infiltration and storage, and generate slope runoff, indicating that runoff may be more closely related to short-period rainfalls, which should therefore be used in predictive equations.

2. *Runoff in ephemeral stream channels*

Rapid runoff from hillslopes controls the main characteristics of flow generation in ephemeral stream channels. Many authors have shown that ephemeral runoff in small channels characteristically shows a very short time of concentration and rise to peak discharge (e.g. Renard, 1970; Schick, 1970b; Thornes, 1976; Griffiths, 1978b). There is a typical hydrograph form with an instantaneous rise, sharp peak, steep initial recession limb followed

TABLE VI.9

Published values of time to saturation (t_0)

Author	Time $t_0.(s)$	p (mm/h)	Ks (mm/h)	Comments
Rubin, 1966	29.9	2 154.0	478.8	Laboratory, sand
	81.7	1 436.4	478.8	Laboratory, sand
	165.7	1 077.3	478.8	Laboratory, sand
Smith and Cherry, 1973	80.0	127.2	30.0	Computed
	140.0	101.4	30.0	Computed
	170.0	88.8	30.0	Computed
	525.0	51.0	30.0	Computed
Mein and Larsen, 1973	130.0	400·3	50.0	Computed
	240.0	300.2	50.0	Computed
	540.0	200.2	50.0	Computed
	330.0	200.0	50.0	Computed
	480.0	200.0	50.0	Computed
	600.0	200.0	50.0	Computed
Swartzendruber and Hillel, 1975	2 260.0	62.2	52.1	Field tests, loam
	640.8	62.2	40.6	Field tests, loam
Scoging and Thornes, 1979	34.8	44 (mm/ h^{-1}/m^{-2}	168.6	Sprinkler tests until runoff observed. All slopes
	18.89	44	–	Bare slopes, mean
	43.26	44	–	Vegetated slopes, mean
	70.18	44	–	> 50% vegetation, mean
	23.9	44	–	Sprinkler tests with sustained application and runoff collected. Soils initially dry
	14.7	44	–	As above, soils initially wetted by 1st test. 15 min delay

by a *relatively* long, slow fall towards dry conditions (e.g. Fig. VI.13). Duration is usually short, perhaps 1–5 hours, with peak flow occurring in the first 10–30 minutes. Runoff normally only occurs on a few occasions a year, and bankfull discharge is even more unusual. Specific peak discharges can be as high as 17 m^3/sec/km^2, and wave-like rapid advance of the flood is typical.

There are some significant differences between small and large channels. Headwaters and small catchments all exhibit a rapid concentration of overland flow and short times to peak flood. There are several reasons for this. First, slope lengths are generally short, slope angles are steeper than in larger catchments, and valley heads are zones of flow-path concentration. Second, soil cover and available soil moisture storage is low, and times to saturation are short. Third, it is easier for a small catchment to be covered by a single storm. In larger catchments only a part of the system may be functioning in a given storm. Fourth, larger channels may exhibit substantial storage of water in alluvial and slope deposits which delays concentration in long channels and may even give rise to baseflow. For these reasons the storm

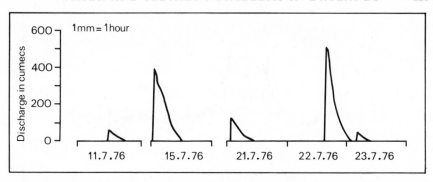

FIG. VI.13 Recorded hydrographs in Mithawan nallah, Pakistan (after Griffiths, 1978b)

hydrographs for larger catchments may display lower and flatter peaks and flatter hydrographs. Where there are steep headwaters flashy peaks may be superimposed on delayed flow or baseflow characteristics (Fig. VI.14). Generally, their duration is longer. Rapid flash floods of small magnitude but possibly catastrophic effect are therefore more likely in medium-small streams. Finally, it is worth noting that there is some evidence that peak discharge per unit area increases with larger catchment size (Renard, 1970).

Regardless of channel size, however, runoff is generally so rapid that the most significant predictors are peak, short-term rainfall intensity, and basin area. Thus rather crude prediction formulae (see section (f)) often estimate hydrograph form and peak discharges rather successfully. In this context, recent work in Arizona suggests that even in arid areas, mean annual flood peak correlates strongly with catchment area for catchments of less than 200 km², with basin slope being of less importance. Wolman and Gerson (1978) believed that arid catchments up to 100 km² behave in a similar fashion to catchments in other climatic zones. Catchments larger than 100 km², however, are less likely to be covered by a single rainstorm, so that runoff per unit area will be less than in more humid catchments; they also argued that large arid catchments retain evidence of large magnitude flows due to the lack of the modifying influence of vegetation and frequent, low-magnitude events, a view in accord with Schick's (1974b) conceptual analysis of river-terrace development in deserts. In addition, since there is such a high correlation between peak short-period (15-minute) rainfall and peak and total discharge (e.g. Renard, 1970), estimates of runoff volume–frequency relations, made from short records of precipitation, can be extrapolated to other similar watersheds with some confidence (Osborn and Hickok, 1968; Renard, 1970).

Finally, it should be noted that in large catchments, snowmelt or frontal rainfall are likely to produce a second family of hydrographs with greater peaks and longer durations than those associated with localized storms. Thus, precipitation conditions must be precisely specified in any attempts to estimate runoff in drylands.

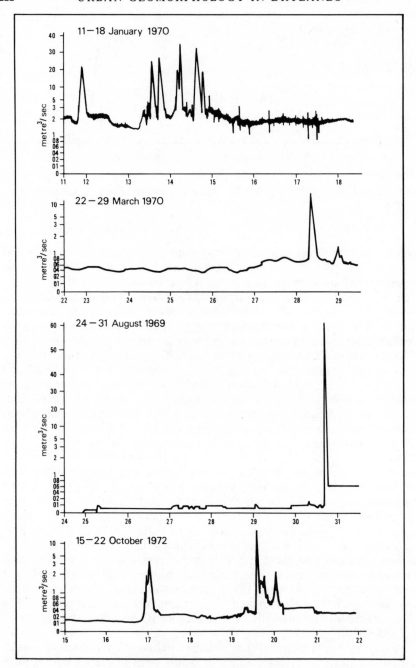

Fig. VI.14 Stage records of discharge for Las Tosquillas, Rio Ugijar, Spain. Storm discharges are superimposed on a baseflow pattern, but show similar features to storm hydrographs for small catchments in more arid areas (after Thornes, 1976)

3. *Piping*

Piping – a term used to described subterranean channels developed by water moving through incoherent sediments – is an important form of subsurface discharge of water and sediment in many semi-arid areas. Because it often poses a subsidence hazard to engineering structures (e.g. Fig. VI.15) it has been extensively studied (e.g. Parker, 1963; Parker *et al.*, 1964; Parker and Jenne, 1967) in North America and elsewhere. Bridge abutments and wing walls, piers, retaining walls, and culvert facings are all regularly damaged by piping and where piping is extensive, highway subsidence or railway distortion can seriously disrupt traffic, and dams may collapse. The dominant type of piping that causes damage to engineering structures is that

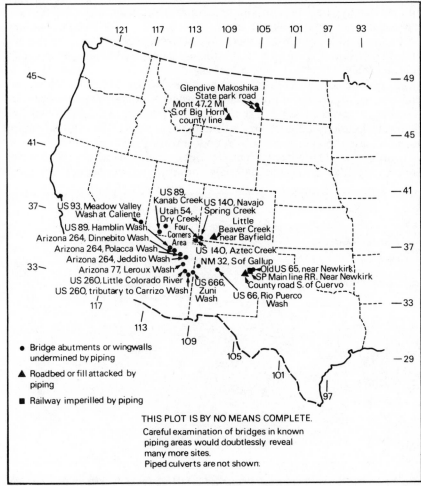

Fig. VI.15 Location of engineering features in the USA known to be imperilled by pipes (after Parker and Jenne, 1967)

of 'stress desiccation' in which surface runoff (including road drainage and irrigation water) drains into a desiccation crack to cause erosion of fine material, the collapse of channel banks and eventually surface settlement that creates an irregular topography (Fig. VI.16).

Other types of piping include the process of particle entrainment under high hydraulic gradients, and the development of sub-surface passages due to variable lithological permeabilities. In some cases large voids do not develop but seepage erosion and 'running sand' cause morphologically similar surface collapse phenomena.

Local processes that assist or are associated with piping include the removal of clay coatings and iron, manganese oxide, calcite and silica cements that bind stable aggregates and the consequent reduction in cohesive strength; clay-mineral swelling; and the physical dispersion of clay minerals and clay aggregates; in addition, secondary breakdown and subsidence of low-density earth materials may occur, following wetting of dryland alluvium, colluvium, and loess in which there may be volume decreases of up to 13%, and surface lowering by up to 5 m – the process of *hydrocompaction* (Cooke *et al.*, 1978).

Piping intensity tends to increase with the Na:Ca-Mg ratio, soluble salt content, montmorillonite and other swelling clay minerals' content, degree of desiccation, rates of surface runoff and denudation, available relief and hydraulic gradients between land surface and gully floors. In addition, piping may increase runoff dramatically by concentrating subterranean flow.

In order to avoid piping problems it is clearly essential that pre-construction surveys carefully define the materials and other environmental conditions

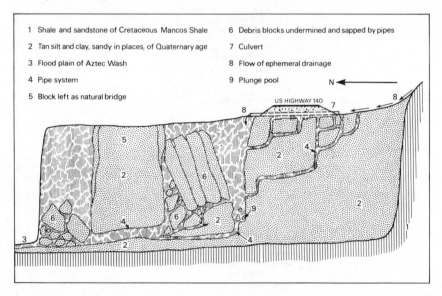

1	Shale and sandstone of Cretaceous Mancos Shale	6	Debris blocks undermined and sapped by pipes
2	Tan silt and clay, sandy in places, of Quaternary age	7	Culvert
3	Flood plain of Aztec Wash	8	Flow of ephemeral drainage
4	Pipe system	9	Plunge pool
5	Block left as natural bridge		

FIG. VI.16 North-south cross-section under US Highway 140 built on Aztec Wash alluvium, in the Four Corners area of south-western Colorado (after Parker and Jenne, 1967)

conducive to piping. Normally it is cheaper to avoid piping areas rather than to apply remedial measures and incur high maintenance costs. Remedial action may include filter or filter drains, catchment drains that outfall in stable channels away from threatened structure, protected abutments, and a careful avoidance of drainage concentration in natural sediments. The processes of piping are extensive but inadequately recognized in drylands, and it is clear that geomorphological surveys can contribute towards an improved understanding of the hazard.

4. *Transit phenomena in ephemeral stream channels*

An evaluation of runoff in ephemeral stream channels would be incomplete without consideration of the changes that take place in such channels during runoff events. Many ephemeral stream channels contain extensive and deep, coarse-grained alluvial deposits that are characterized by high infiltration capacities (transmission losses) and high water-storage potential (Table VI.10). The magnitude of transmission losses into alluvium is variable, and related to flow duration, channel length and width, antecedent moisture, peak discharge, the pattern of flood-wave sequences, the physical properties of the alluvium and suspended sediment load. In the most favourable circumstances, it is possible for the entire available precipitation to be lost within a basin as a primary groundwater recharge source. The effect of transmission losses is to decrease flow downstream (Fig. VI.17), to steepen the flood-wave front (Thornes, 1976), to decrease the flood rise time, to increase relative sediment load, sorting and deposition downstream, and to decrease stream width.

TABLE VI.10

Some published rates of transmission loss and comparisons with watershed point infiltration rates

	Area	Author	Transmission/Infiltration rate	Notes
river beds	Walnut Gulch	Renard, 1970	158 163 m³/km	85–90% prec excess
		Renard and Keppel, 1966	142 542 m³/km	57% runoff
	Todd River, Alice Springs	Slatyer and Mabbutt, 1964	50.8 mm/hr	point infiltration in river bed
	Sierra Nevada, Spain	Butcher and Thornes, 1978	30.00 mm/min	ponded infiltration in river bed
			1.00 mm/min	released irrigation water on to saturated bed
slopes	Nahel Yael	Schick, 1974	0.5–1.5 mm/3 min	point infiltration on watershed slopes
	Negev	Schick, 1971	2.5 mm/hr	point infiltration on watershed slopes

A

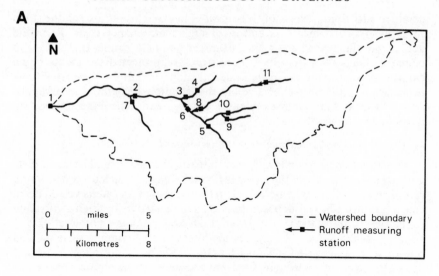

B

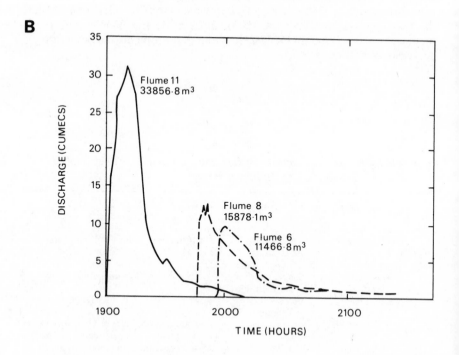

FIG. VI.17 (A) Location of flumes in Walnut Gulch (near Tombstone, Arizona); (B) Runoff hydrographs at three flumes for a flood on 5 August 1968 (after Renard, 1970)

Most transmission losses seem to take place in the early stages of a flood event so that it is common for small, translatory waves, a few centimetres high, to drain rapidly as they pass downstream, and for the succeeding waves to have steeper and steeper fronts as flow progresses downstream (Butcher and Thornes, 1978). These phenomena appear to operate as kinematic waves and it may be that theories of flood routing will be replaced by kinematic-wave routing procedures as knowledge of ephemeral stream channels increases (Renard and Keppel, 1966; Thornes, 1976). The downslope distance reached by floods in ephemeral stream channels varies with the size of the runoff event, especially if the flow is generated mainly in headwater areas (Fig. VI.18).

Other transit phenomena include the effect of desert environmental conditions and transmission losses on velocity, channel morphology, and sediment transport. In general, velocity increases faster downstream within mountain catchments than in equivalent streams in more humid areas. Gradients are usually steeper and runoff more complete from slopes with thinner soils and vegetation, so there is less likelihood of baseflow or of obstruction to flow. Transmission loss, however, means that as soon as flow encounters an alluvial floor, discharge and velocity decrease downstream. This is accompanied by comparable changes in flow depth, channel width (Table VI.11), and by other changes in channel geometry.

The economic significance of transmission losses includes the loss of irrigation water; the analysis of groundwater resources and recharge potential; the estimation of flood hazard and the design of flood prevention structures. Unfortunately, detailed knowledge on these matters comes from only a few areas such as the south-west USA (e.g. Renard, 1970), Israel (Schick, 1970b), Spain (Thornes, 1976) and Pakistan (Griffiths, 1978a). Further research elsewhere is clearly required.

TABLE VI.11

Transmission loss channel reaches, Walnut Gulch experimental watershed, showing the effect on channel width

Channel reach (Fig. VI.17)	Channel length (km)	Average channel width for 'in-bank flow' (m)
7 and 2→1	6.4	66.14
6→2	4.5	48.77
8→6	1.4	20.73
11→8	6.4	13.11
5→6	2.4	19.20
4→3	1.9	10.97
9 and 10→6	4.3	17.37

Source: Renard (1970).

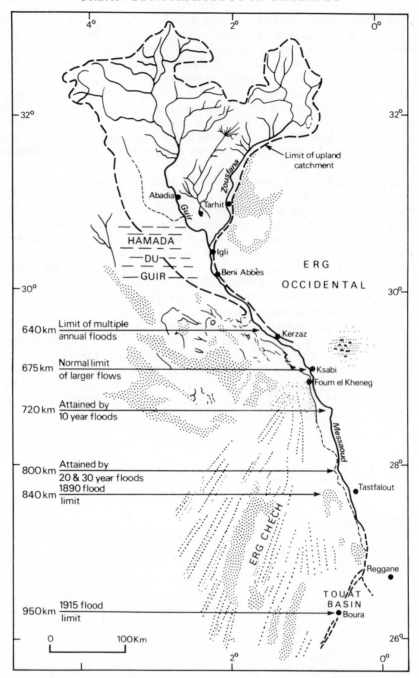

FIG. VI.18 The limits of flooding in the Gur-Saoura-Messaoud catchment of the Sahara (after Vanney, 1960, from Mabbutt, 1977)

TABLE VI.12

Amounts of available sediment produced by burrowing animals, September 1973

Geological formation	Sediment in kg	% of total area	% of total sediment produced
Drorim	121	39.8	20.9
Shivta	392	47.9	67.9
Netser	64	12.3	11.2

Source: Yair (1974).

(e) Sediment Movement

1. *Hillslopes*

Sediment movement on dryland hillslopes is mainly caused by raindrop impact, splash, and overland flow processes including unconcentrated and concentrated surface wash. These processes have been extensively studied and reviewed (e.g. Bennett, 1939; FAO, 1965; Carson and Kirkby, 1972; Cooke and Doornkamp, 1974; Thornes, 1976). Some of the methods available for estimating sediment loss are discussed later. In addition to recent reviews of this field, some research has yielded several conclusions of general interest, and these include:

(i) Studies of desert hillslopes in Sinai by Yair and Klein (1973) showed that there is a linear, positive relationship between the amount of sediment and the volume of runoff, but there is no significant relationship between sediment concentration and volume of runoff. High sediment concentrations were obtained in this study with small runoff amounts, perhaps due to initial sediment flushing and increasing cohesion due to continuous wetting.

(ii) In the same study (Yair and Klein, 1973) it was shown that there is an increase in sediment concentration with decreasing slope angle, perhaps due to the fact that the slope materials in Sinai are more erodible and because slope roughness obscures the effects of slope angle: this relationship is the reverse of that commonly described.

(iii) Yair and Klein (1973) also showed that sediment yield is sporadic and spottily distributed.

(iv) In many drylands burrowing animals, such as isopods, gerbils, porcupines, and gophers, can prepare quantities of soil that are readily removed from the surface even by shallow overland flows. In Yair's (1974) study of sediment yield in the Negev, it appears that as much as 60% of total sediment could be produced or transported to the channel by this means (Table VI.12).

2. *Channels*

The most important aspect of sediment movement, from the dryland engineering and urban development point of view, is to examine the way in which sediment load accumulates and moves in channel systems. In rock gullies, torrential runoff can move all sizes of material so that tributary channels are at times swept clean of debris. At gully mouths, there is a hazard of small-fan and 'delta' development in the main channel which, in turn, may lead to bank erosion, the formation of gravel sheets, lobes, bars, and alterations of bed elevation. Deposits from high-magnitude tributary flows may produce temporary terraces that are steadily removed by smaller-magnitude flows or even swept clean by a further major spate, only to be replaced by a new terrace deposit at an elevation consistent with the scour and fill of the new flow. Major flows may carry huge amounts of sediment which, in the falling stage, is deposited to form elevated valley floors. Thus a common morphological feature (perhaps the best available morphological estimator of possible flow magnitudes in drylands) is a succession of alluvial terraces and bars that represent the gamut of previous flood conditions, evidence that may be used to estimate peak discharge (see section (f)).

Sediment load, both suspended and bedload, characteristically increases downstream as tributaries and slopes supply material. When the effect of transmission loss begins to be felt, the sediment load again rises (relative to water volume) until the stream is supercharged and begins to deposit its load.

Deposition may occur as water is lost during transmission, as the storm itself subsides, and as distance from source increases. Lower reaches of streams are therefore dominated by deposition. As this commonly occurs with bedload deposited first, then suspended load of progressively finer size, a crude particle-size stratification is often evident both downstream and in any one sedimentary section as a result of each flood event. Deposits are usually layered from coarse at the base to fine at the top, and the channel surfaces, when dry, are marked by the sand and silt skins of the final stages of deposition. These skins of finer sediment may significantly influence infiltration properties, especially if the fine materials are not winnowed away by aeolian action, and accumulate to form impermeable bands within the deposits (and even to form perched water-tables). It is therefore important that measurements of these properties are related to the exact geomorphological location of any discharge, velocity, or sediment transport estimates.

The magnitude of sediment movement in drylands has rarely been recorded. Permanent recording stations are desirable, such as those at Walnut Gulch in Arizona (e.g. Renard, 1970) and near Eilat, Israel (e.g. Schick, 1970a and b). Spot measurements using suspended sediment samples are useful, but the problem is to be at the right place, at the right time, with the correct equipment. Some useful reports are summarized in Table VI.13. Implications of these data and other published reports include:

(i) Catchments in close geographical proximity appear to yield similar sediment discharges, and therefore data from an instrumented catchment can be used to give an appreciation of the conditions in neighbouring catchments that are perhaps the subject of terrain evaluation or planning and urban development.

(ii) Streams dominated by longer basin lag times yield less sediment, for a given discharge, than more 'flashy' streams.

TABLE VI.13

Some aspects of sediment yield from drainage basins in drylands

Author	Area	Precipitation peak intensity	Flow	Runoff coefficient	Observed sediment concentration	Load	Comments
Inbar, 1972	Nahal Meshashim	10 mm/hr 120–360 mm in 36 hr	8–340 m/sec 90 min total 10×10^6 m³	0.27–0.56	200–300 ppm minimum	60 000 tons 60 m³/km²	Normal yield 6–8 000 tons/annum Bedload included 7 ton boulders
Schick, 1971	Nahal Yael 1966	–	2.2 m³/sec	–	<10 000 ppm	1.57×10^6 tons	Velocity 1.56 m/sec Deposition area aggraded 1–2 m
	Nahal Roded	–	100 m³/sec	–	131 000 ppm		
Schick, 1970b	Typical Negev values	1.5 mm/3 mins	Rise lasts 2.5–5 min Peak 1.0 m³/sec for 4–8 min	–	4 000–60 000 ppm Headwaters 10 000–15 000 ppm. Fan area 70 000 ppm	0–152 tons/km⁻³ /event uplands 80 tons km⁻² yr	
Schick, 1974a	Muqeibela 5.6 km				Increases downstream 15 000–30 000–80 000 ppm in rare event 30 000–65 000–120 000 ppm mean 25 000–75 000 ppm vertical in flow in rare event 100 000–200 000 ppm		
	Wadi Tweibah 8 km²				100 000–200 000 ppm at 1.6×10^6 ppm reached mudflow consistency		
Cirugeda, 1973	R. Yahr, Spain	–	505 m⁻³sec⁻¹	17%	17% solids	101 m³/sec	
	R. Ugíjar, Spain	–	349 m⁻³sec⁻¹	7%	41% solids	244 m³/sec	
	R. Alcolea, Spain	–	166 m⁻³sec⁻¹	7%	29% solids	66 m³/sec	

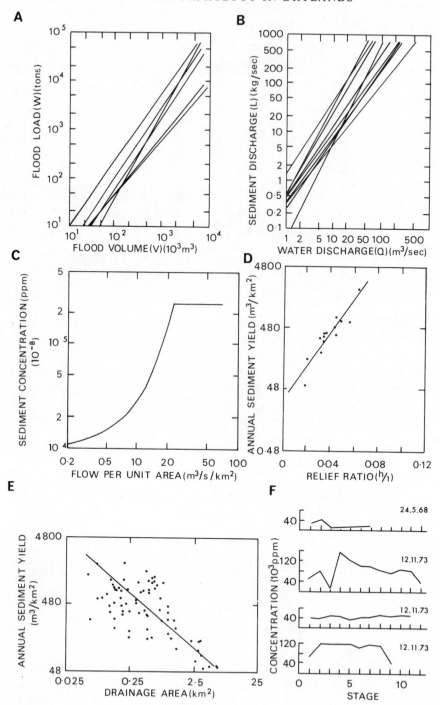

(iii) Flood load is closely related to flood-water volume and water discharge (Fig. VI.19A, B).

(iv) Mean annual sediment yield is highest in semi-arid regions, and diminishes in more arid areas because of lack of precipitation and in more humid areas due to vegetation cover (Langbein and Schumm, 1958).

(v) Annual sediment yield is very variable, and long records are needed to establish a reliable sediment rating curve (e.g. Figure VI.19C). Collection of data should continue until the basic catchment relations are understood.

(vi) The most important morphological indicators of sediment production are relief ratios (Fig. VI.19D) (Manner, 1958; Schumm and Hadley, 1961), drainage basin area (Fig. VI.19E), and drainage density (Schumm and Hadley, 1961). Unfortunately, there have been few attempts to develop predictive equations using these variables, although Lustig's (1965) study is undoubtedly a significant contribution.

(vii) Aggregation within channels seems to occur most frequently where the ratio of channel area to channel length is small.

(viii) There are several types of sediment yield pattern associated with hydrograph rise (Fig. VI.19F). It is important to note that there is a danger in the extrapolation of point suspended load sampling data to predict average values, and spurious results are certain if the overall pattern of sediment behaviour is unknown.

3. Dryland catchment sediment budgets

There have been few attempts to derive complete sediment budgets for dryland catchments. Perhaps the earliest important attempt is that by Leopold et al. (1966). More recently Schick (1977, 1979) has proposed a tentative sediment budget for the Nahal Yael catchment in southern Israel where, from ten years of record for this 0.5 km^2 watershed a mean annual sediment yield of 388 tons/km^2 is estimated, of which dissolved load contributes only 1%, suspended sediment contributes 65%, and the remainder

---

FIG. VI.19 (A) The relationship between flood volume and flood load for various 'streams' in Sinai (after Schick, 1977); (B) Relationship between water discharge and sediment discharge for various streams in Sinai (after Schick, 1977); (C) Estimated generalized sediment rating curve, Nahel Yael Research Watershed (Israel), 1965–6 water year. Based on suspended sediment samples supplemented by bedload observations. For an instantaneous discharge of 2.25 m^3/sec/km^2 the value representing the load concentration is 250 000 ppm or 25% (after Schick, 1977); (D) Relation of mean annual sediment yield to relief ratio for 14 small drainage basins in eastern Wyoming (after Schumm and Hadley, 1961); (E) Relation of mean annual sediment yield to drainage area for 73 small drainage basins in eastern Wyoming (after Schumm and Hadley, 1961); (F) Types of change in sediment concentration with hydrograph rise (after Schick and Sharon, 1974)

is from bedload. Some estimates are available for Spain (ICONA, 1969; Thornes, 1976), ICONA using the Fournier (1960) equations:

$$\text{(A)} \quad y = 6.14\,X\,49.78\,Ht < 6,\,p^2/P < 20 \qquad \text{(VI.19)}$$

$$\text{(B)} \quad y = 27.12\,X - 475.4\,Ht < 6,\,p^2/P < 20 \qquad \text{(VI.20)}$$

$$\text{(C)} \quad y = 52.49\,X\,513.21\,Ht < 6 \qquad \text{(VI.21)}$$

$$\text{(D)} \quad y = 91.78\,X - 737.62\,\text{semi-arid} \qquad \text{(VI.22)}$$

where $X = p^2/P$
          p = rainfall of wettest month, mm
         P = mean annual precipitation, mm
       Ht = mean relief of basin, m
and y = sediment yield,

which yielded approximately 274 tonnes/km^3/yr for the mountain zone and 422 tonnes/km^2/yr in the semi-arid areas. This is similar to the estimate of 380m tonnes/km^2/yr from Colorado (Holeman, 1968) and the estimate by Langbein and Schumm (1958) for semi-arid zones of 265m tonnes/km^2/yr.

These data are in broad agreement with those from America (Wolman and Miller, 1960), Utah (Wolman and Miller, 1960) and New Mexico (Leopold *et al.*, 1966), but somewhat different from those in other climatic environments (see Gregory and Walling, 1973, Table 4.4).

Some data are also available from the Zingg formula as applied by ICONA. They obtained values of 940–1800 m^3/km^2/yr for semi-arid Spain. These are also in agreement with published rates (2000 m^3/km^2/yr, Schumm, 1964; 900 m^3/km^2/yr, Campbell, 1970).

4. *Effects of urbanization on sediment yields*

Much of the previous discussion has been concerned with the nature of sediment movement within watersheds that are, as yet, relatively unaffected by human activity. The imposition of an urban area in a catchment, however, may profoundly change its surface and drainage characteristics and, thus, its sediment yield. Numerous studies in urban areas of temperate climates (e.g. Leopold, 1968; Guy, 1972; Kao, 1974; IAHS, 1974) suggest that, in general, sediment yield increases rapidly during construction phases and declines subsequently to a yield somewhat above that prior to urbanization. It is not yet clear that urbanization in drylands produces similar results, mainly because detailed empirical studies are rare (e.g. Soliman, 1974). Indeed, it is possible that the results in drylands differ substantially from those in temperate areas especially if there is no precipitation during the construction phase. In addition, in drylands surface disruption during construction usually involves the removal of little or no vegetation, and the new surface irregularities may significantly increase surface detention and even infiltration capacity, thus possibly reducing sediment yield from at least small rainfall events. Whatever the case, the uncertain state of present knowledge of the effects of urban development in drylands on sediment yields reinforces the need, discussed in section (g) below, for monitoring sediment yield change during urbanization, for there is no doubt that such changes could strongly alter predictions of sediment yield.

(f) Assessment of the Runoff-Sediment Hazard

1. *Mapping*

The previous discussion has shown that few drylands have been investigated in sufficient detail to provide a complete understanding of their water and sediment systems. The rapid pace of dryland urban development requires that information on these systems be provided quickly and accurately. The purpose of this section is to describe briefly the geomorphological techniques that can be used partly to satisfy this requirement.

The summaries provided above have shown that a whole range of morphological variables are often closely related to channel, slope, water, and sediment phenomena, and one of the most important aspects of a geomorphological contribution is the use of such variables to predict the behaviour of fluvial systems. This is especially important in drylands, where, as has frequently been emphasized, data records are usually short or absent. Morphology, therefore, provides a surrogate for process measurements. The most important variables (the use of which is illustrated in section 3, below) include:

(i) *Morphometric data:* basin area, length, slope; depositional area and slope; channel length, width, slope, and variability; channel 'auto-geometry' and persistent effects of catastrophic events; hillslope length, inclination, limiting and threshold angles; drainage net and pattern characteristics, proportion of gravity, debris and wash-slope areas; and the distribution of characteristic erosional and depositional landform elements.

(ii) *Structure and lithology:* rock types, rock resistance, strength and other mechanical properties, attitudes and bedding characteristics, joint densities, faulting, and evidence of tectonic behaviour.

(iii) *Engineering soils:* mechanical index and strength properties, depth, distribution, sorting, bedding, permeability, infiltration capacities, and antecedent conditions.

(iv) *Vegetation:* type, cover, density, interception value.

(v) *Process:* infiltration, recharge, flood generation and flood characteristics, nature of slope and mass movement, rain impact, splash, sheetwash, piping, rilling and gullying; playa and basin hydrology. Where possible, quantified data to cover peak water and sediment yields, time of concentration, time to peak (lag time), flow duration, transmission loss and storage, in frequency–magnitude–duration terms.

(vi) *Palaeoforms:* indications of channel metamorphosis, abandoned channels, terraces, fossil surfaces, etc.

The most efficient and comprehensive way to obtain this information is by the analysis of existing records, by geodetic and remote-sensing surveys and, above all, by the use of geomorphological mapping techniques as described in Chapter III. Particular attention should be paid to morphological predictors of runoff (e.g. terraces, channel geometry, sediment size, etc.) and to the loci of flood erosion and deposition. The data can be abstracted to produce systematic problem maps, such as those showing fluvial hazards (Fig. VI.20). A specific example is the flood hazard map (Fig. VI.21) produced by the authors for Suez City (Egypt, Ministry of Housing and Reconstruction, 1978; see section (g) 3 below). Of course, the report

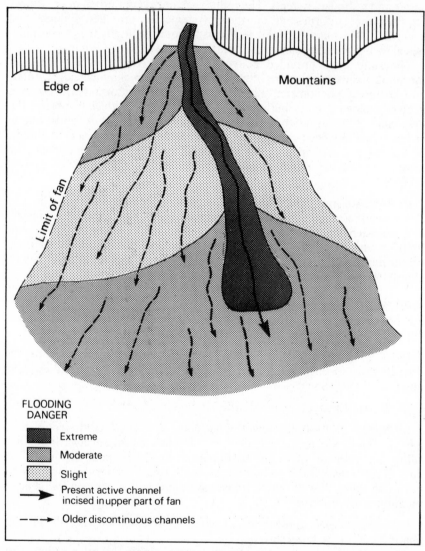

FIG. VI.20 The pattern of flood hazard on a typical alluvial fan in the western United States (after Kesseli and Beaty, 1959)

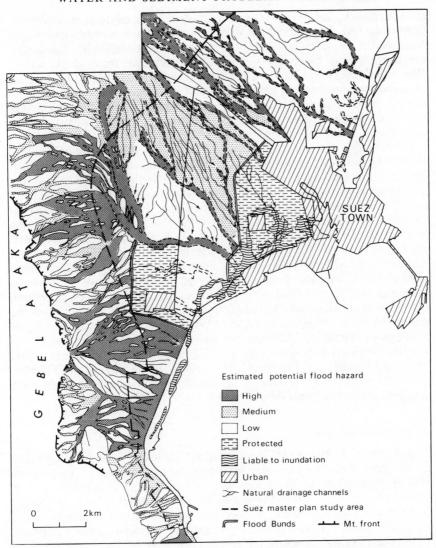

Estimated potential flood hazard

- High
- Medium
- Low
- Protected
- Liable to inundation
- Urban
- Natural drainage channels
- Suez master plan study area
- Flood Bunds
- Mt. front

0   2km

FIG. VI.21 The flood hazard map produced from geomorphological surveys for Sir William Halcrow & Partners on behalf of the Egyptian Ministry of Housing and Reconstruction (1978) by the authors

accompanying such maps must explain the way the data were derived and how they must be interpreted.

2. *Monitoring*

A fundamental and urgent requirement for dryland development is the establishment of precise environmental monitoring systems. Standard methods of monitoring are briefly noted below, but reference to advanced texts is required before any implementation of these methods is attempted. Meteorological data should be derived from fully equipped observation stations, but for specific projects the short-term installation of rain gauges, evaporation pans, piezometers, and infiltrometers (see, for example, Chow, 1964 and Gregory and Walling, 1973) can provide much useful information. Detailed measurements of slope runoff and sediment yield can be made on experimental plots using troughs and collecting tanks. The measurement of runoff in ephemeral stream channels is extremely difficult, due mainly to problems of recording widely spaced and unanticipated events.

Three principal studies of the dynamics of dryland catchments are those at Walnut Gulch (near Tombstone, Arizona), Nahal Yael (near Eilat, Israel); and for catchments near Ugijar (southern Spain). In the Walnut Gulch study, a 150-km^2 watershed is being used for detailed investigations of ephemeral runoff and many data have been published (e.g. Renard, 1970). The use of eleven supercritical flumes throughout the catchment has enabled the monitoring of transmission losses (Fig. VI.17). The Nahal Yael experimental watershed is only 0.7 km^2 in area. It is situated in an extremely arid area and monitored by 11 rainfall recorders, five channel runoff recorders, seven sediment samplers, and three bedload traps – and has begun to provide a stream of original and fundamental observations, some of which have been discussed above (e.g. Schick, 1970a, 1978). The studies in Spain include the use of electronic, remote-sensing stage poles and a comprehensive study of hillslope and channel, infiltration, transmission, flow generation, and flood-routing problems at present only partially summarized in Thornes (1976, 1977) and Butcher and Thornes (1978; Scoging, 1978; Scoging and Thornes, 1979). In addition, a rainfall-channel runoff study involving 18 catchments has been established near Alice Springs, Australia (Chapman, 1964).

3. Prediction of runoff

Roberts and Melickian (1970) stated that 'Ironically, flood protection may be the most urgent requirement for a new desert community'. This means that the prediction of flood and sediment discharge is a prime aim of geomorphological surveys in drylands. However, the selection of a suitable technique for the estimation of flood discharge or soil loss depends on the availability of terrain and precipitation data and emphasizes the importance of the mapping and monitoring methods noted above.

Where data for a sufficiently long period are available, use may be made of the computer simulation models, as outlined in Fleming (1973) to estimate flood events. If the records are long enough but insufficiently complete for computer modelling, standard hydrological techniques such as frequency analysis (Chow, 1964), unit hydrographs (e.g. Sherman, 1932), and co-axial correlation (Andrews, 1962) should be used. These techniques are well documented (e.g. Ward, 1967; Wilson, 1969; Rodda, 1969).

Where records are inadequate or absent, recourse must be made to cruder techniques. These include the use of prediction equations, the use of area as a surrogate for discharge, and estimations of former flood levels from channel geometry.

(*i*) *Predictive equations.* The difficulty of estimating design floods has long been recognized and since Mulvaney (1851) first published his Rational Method numerous formulae have been developed. Chow (1962) listed 103 in use at that time in the USA but only a few are of value in drylands. These include:

US Geological Survey, 1949 (for use in Colorado) (Linsley *et al.*, 1949):

$$Q = 99A^{0.5} \qquad \text{(VI.23)}$$

where Q = discharge in ft^3/sec,
and A = catchment area in sq miles.

Hickock, Keppel, and Rafferty (1959) for use in small semi-arid catchments:

$$\text{(a)} \quad T_L = K_1 \left\{ \frac{A^{0.3}}{S_a \sqrt{D_D}} \right\}^{0.6} \qquad \text{(VI.24)}$$

where T_L = lag time,
         S_a = average slope of catchment,
         D_D = drainage density,
and K_1 = coefficient dependent on units used.

$$\text{(b)} \quad Q_p/V = K_3/T_L \qquad \text{(VI.25)}$$

where Q_p = peak rate of runoff,
         V = total volume of runoff,
and K_3 = coefficient dependent on units used.

(*ii*) *Area as a surrogate for discharge.* The use of catchment area as a surrogate for discharge stems from the usual logarithmic relationship between the two variables, and its validity has been confirmed by analyses of data from several dryland catchments. Generally, within a given environment, discharge attributes – for instance, flood peak, flood volume, or mean annual flow – will increase with catchment size. Specific discharge (runoff per unit area), however, tends to decrease with increase of area (see section (d)). The *magnitude* of the *flood* hazard, as a physical phenomenon, can therefore be assessed although prediction of the actual flood discharge may be impossible. Similar to this type of analysis, is the use of alluvial-fan area as a surrogate for discharge because catchment area and fan area are highly correlated (equation VI.1). It must be emphasized, however, that relationships described in one locality do not necessarily hold for other areas. Thus the technique can only be used to predict flow in ungauged catchments if sufficient data on discharge and area relationships exist for the area under examination.

Much the same is true of the methods of estimating mean runoff in ungauged semi-arid areas described by Moore (1968), in which relations of runoff per unit area to altitude are defined for a hydrologically homogeneous region using a small number of stream flow records from a variety of altitudes and orientations. This relationship is then used to predict mean annual runoff in an ungauged catchment by calculating first, the area of the basin that lies within each altitude zone, second, by multiplying the area of each zone by the appropriate runoff value, and third, by making specific adjustments for local conditions.

(*iii*) *Use of channel geometry*. Field mapping of channel terraces is a useful indication of the stage or magnitude of low-frequency flood events. From the stage height of the flood level, an indication of discharge can be calculated by multiplying the cross-sectional area of the flood by an estimate of flow velocity provided either by Manning's or Chezy's equation:

Manning's equation

$$V = \frac{k R^{2/3} S^{1/2}}{n} \qquad \text{(VI.26)}$$

where V = velocity,
         k = coefficient dependent on units,
         R = hydraulic mean radius,
         S = slope of water surface,
and n = Manning's roughness coefficient.

Chezy's equation

$$V = C\sqrt{(RS)} \qquad \text{(VI.27)}$$

where C = Chezy's coefficient.
         (R and S as for equation VI.26).

The main error involved is the problem of the unknown depth of scour and therefore the true magnitude of the cross-sectional area during the flow conditions.

4. *Prediction of sediment yield*

Sediment yield is dependent on numerous variables, especially climate and relief (e.g. Fournier, 1960). The principal morphological explanatory variables are topographic form (Guy, 1965), relief ratio (Manner, 1958; Schumm and Hadley, 1961), drainage area (Schick, 1970b), drainage density (Hadley and Schumm, 1961), soil erodibility (Anderson and Wallis, 1965), and vegetation cover (Schumm and Hadley, 1961). In addition, relations with stream discharge have been established, so that where a rating curve can be developed from gauged stations it is possible to use the results in ungauged but morphologically similar basins using simple regression and correlation models:

Fournier's equation (1960), for regional patterns:

$$E = ((p^2/P)^{2.65} Cm^{0.46})/1.56 \qquad \text{(VI.28)}$$

where E = mean suspended sediment yield (tons $\text{km}^{-3} \text{yr}^{-1}$),
         p = maximum mean monthly precipitation (mm),
         P = mean annual precipitation (mm),
and Cm = $H \tan \theta$, where H = *mean basin height* (*m*) *and* θ = mean basin slope.

Mean particle size moved by given flow (Miller, 1958):

$$D = 1.0V^2 \qquad\qquad\qquad (VI.29)$$

where D = diameter of mean particle size moved (mm),
and V = water velocity (ft/sec)
(but because of interaction between particles of different sizes, this equation should be treated with reserve).

Walnut Gulch area/sediment yield equation (Renard, 1972) is expressed as:

$$SY = 0.001846A^{-0.1187} \qquad\qquad (VI.30)$$

where SY = sediment yield (acre-feet/acre),
and A = catchment area (acres).
This is only of local applicability; similar empirical equations could be developed elsewhere.

5. *Prediction of soil loss*

In order to predict soil loss from agricultural or construction areas in drylands, a number of important variables need to be considered, notably soil properties (e.g. water-stable aggregates, and soil-erodibility indices, Bryan, 1968); rainfall erosivity (e.g. Wischmeier and Smith, 1958; Stocking, 1977); vegetation cover, type and density (e.g. Meginnis, 1935; Hudson and Jackson, 1959); frequency of cultivation (e.g. Glymph and Holtan, 1969); topography, especially slope angle and length (e.g. Wischmeier and Smith, 1965), and burrowing animal activity (e.g. Yair, 1974). The following is a small selection of suitable equations based on these and related variables:

The Universal Soil Loss equation (e.g. Hudson, 1971):

$$A = R, K, L, S, C, P \qquad\qquad\qquad (VI.31)$$

where A = average annual soil loss (tons/acre),
       R = rainfall factor
       K = soil erodibility factor, expressed as soil loss in tons per acre for each unit of R for the area, assuming continuous fallow land, a 9% slope, 72.6 feet long,

LS = slope length-steepness factor $\left(\dfrac{actual\ length\ \&\ steepness}{72.6\ feet\ and\ 9\%\ slope} \right)$

       C = cropping and management factor (i.e. expected ratio of soil loss from land cropped under specific conditions, to soil loss from tilled fallow),
and P = conservation practice ($=1$ where there are no practices and cultivation is up and down hill, $= <1$ where conservation is employed).

Values assigned to these parameters are discussed and exemplified in Dunne and Leopold (1978).

Boyson (1974), in considering methods for predicting sediment yield in urban areas, defined an 'urban sediment yield equation' as follows:

$$\text{Sheet and rill erosion} + \text{gully erosion} - \text{sediment deposition} \\ - \text{sediment trapped} = \text{offsite sediment yield} \quad \text{(VI.32)}$$

Given certain assumptions, Boyson showed that this could be re-written as:

$$RKLSCP - \text{sediment trapped} = \text{offsite sediment yield} \quad \text{(VI.33)}$$

Soil transport from unvegetated fields in relation to slope:

Zingg (1949)	$S\alpha x^{1.6}\beta \tan^{1.4}$	(VI.34)
Musgrave (1947)	$S\alpha x^{1.35}\beta \tan^{1.35}$	(VI.35)
Kirkby (1969)	$Sx^{1.73}\beta \tan^{1.35}$	(VI.36)

where S = soil transport (cm^3/cm/yr),
        x = slope length (m),
and $\tan$ = slope gradient.

(g) Human response to Runoff Hazards

1. *The record*

In 1919, local floods in Cairo followed 43 mm of rain, and trams were buried in mud up to their window-sills. In 1952, an alluvial fan flood at Desert Hot Springs, California, caused serious damage to property. In 1914, 1938, 1952, 1963, 1969, and 1978 (and on many other occasions) serious problems, loss of life and property damage accompanied heavy winter rainfall in Los Angeles, California. On the first day of Passover 1963, Eilat was struck by a sudden storm that cut roads, exposed foundations and where water entered pre-fabricated houses through the doors inhabitants were compelled to make holes in the walls facing downslope in order to let the water out (Schick, 1971). In 1965, inhabitants of Denver (on the S. Platte River) experienced 350 mm of rain in three hours which caused extensive damage. In 1966, an unexpected flood at Ma'an in southern Jordan killed 70, injured 250, destroyed half the buildings, and left 3,000 homeless (Schick, 1971).

Unforgettable experiences? Things of the past? It is certain that the first problem to be overcome is the poor perception of the hazard by planners, engineers, environmental managers, and the population as a whole, and the first sensible response is to alert them effectively to the hazards of drylands. Perhaps because of the 'inexorable, monotonous dryness' (Schick, 1971), the dryland urban population does not expect, and is usually not prepared for, these hazards. Much dryland urban development is done in ignorance of the potential environmental threats of water and the sediment in it. The danger is particularly acute when planning and development are carried out by aliens from temperate lands. A first priority, therefore, of urban planning in drylands should be that all personnel concerned should be provided with technical briefing, and inhabitants should be advised on the flood risk of an area.

2. *Flood warning*

As discussed above, desert floods are natural and inevitable, but there is usually little time between the occurrence of precipitation and the arrival of a flood wave. Even in very dry areas, floods may occur annually. For example, on one catchment in the Negev, Schick (1970a) estimated that channel flow will occur every year, a major flow on average every three years, and a widespread destructive event on average every ten years. Thus, a flood warning system (Barnes, 1975) is commonly needed on channels that pass through or near urban and other developments.

Although some areas of the world do have sophisticated flash-flood warning systems (as in parts of the USA) it is not a sensible recommendation to suggest, or expect, that such systems will be introduced into most dryland areas. The basic networks of data collection are simply not available. The only early-warning storm system known to the authors in an extremely arid area is in the Sinai, and employs remote sensing of rainfall-runoff, synchronized central recorders, and automatic telephone dialling of four predetermined numbers (Fig. VI.22; Porath and Schick, 1974). There are also attempts to link the warning to controlled traffic lights etc. It is clearly desirable in drylands to develop flood warning systems/guidelines at least at the military and/or police level; to produce a flood-forecasting handbook of local relevance; to instal basic precipitation-runoff monitoring systems, and to survey those aspects of landforms and deposits of relevance to flood studies.

3. *Alternatives to warning and monitoring systems*

In many drylands lack of data prevents the establishment of adequate flood warning systems, and prohibits the development of precise design criteria and factors of safety. In this situation, planning, management, and design must normally rely on the following: (i) the use of predictive equations (see section (f)), and (ii) the use of geomorphological indicators of *relative flooding probability*. The morphological character of the catchments, the morphological indicators of erosion thresholds, scour, erosion, and deposition, and surveys of damage in previous events can be usefully employed to map the areas of relative flood danger (see Fig. VI.21), as suggested by Kesseli and Beaty (1959), discussed by Schick (1970b, 1971), and employed in the author's study of Suez (Egypt, Ministry of Housing and Reconstruction, 1978). In the latter study the following procedure was adopted: detailed mapping of drainage channels using aerial photographs and field surveys, including recording of erosional and depositional forms, sediment characteristics, the form, slope, and catchment size of major channels and damage caused by previous events. Using this information, the area was qualitatively divided

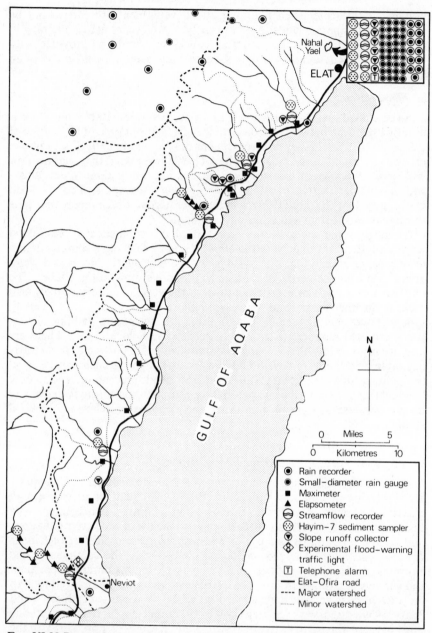

FIG. VI.22 Research layout for flood warning systems on the Elat-Ofira Road, Sinai (after Schick and Sharon, 1974)

into high-, medium-, and low-risk areas, and the basis provided for sensible land zonation and flood-control planning.

This approach is useful for two main purposes. First, for land-use zonation in preparing a master plan, in which the safest areas only are developed – such a use is sensible, efficient, and costs little. Second, as a guide to the adoption of sensible control measures in areas where high flood risks cannot be avoided. As Schick (1971, p. 151) declared:

Geomorphic evaluation of the terrain – in itself a relatively simple and inexpensive procedure – will point to preferences which must be considered together with other factors which operate in the selection of sites for desert installations such as dwellings and industrial plants, ports and airfields, roads and railways. The effectiveness of a terrain flood hazard evaluation is immeasurably increased if it can be correlated with exact data, of which there is at present very little. However, even without local basic knowledge as to rain intensities, peak flows, fluvial parameters and other factors, just an intelligent attempt to try as little as possible to control the flood and as much as possible avoid it offers considerable benefits. This means, above all, to resist the natural tendency of engineers to copy methods used in regions where most of their work is done – the non-arid regions.

There is thus a priority here: in the absence of detailed design data, proposed development schemes should be preceded by predictive and field geomor- phological mapping surveys, and these should be analysed to give relative flood hazard maps, land-use zoning maps and design warnings. The approach is efficient and cheap, and will remain suitable until adequate remote-sensing and flood warning systems are installed.

4. *Avoidance and control*

In humid lands, runoff and sediment problems have been recognized and management has responded for many years. A common response, for example, has been to develop high-cost flood protection works, river training, and drainage improvement structures. Recently, there has been substantial progress towards the codification of management practices in such areas, especially with respect to erosion and sediment control during construction, and to repair and maintenance of damaged surfaces. In the USA, for example, many local authorities have adopted practices that commonly include minimizing the area of land exposed during development; minimizing the time of exposure; using vegetation and/or mulches to protect temporarily exposed surfaces; installing sediment basins to trap sediment from land undergoing development; providing proper facilities to conduct increased runoff; installing permanent erosion control structures as soon as possible; preserving natural vegetation wherever possible; and, of the greatest importance, planning for sediment control (as far as possible at the site of its production), and adopting a development plan that seeks to minimize erosion potential (e.g. US Dept Housing and Urban Development, 1970; Foote, 1972; Kao, 1974).

Such codification may well be relevant to urbanization in drylands, and indeed such practices are not unknown in deserts. But as yet no such codification has been developed exclusively to meet dryland requirements. The following comments are designed to contribute towards this task.

Recent research in drylands is increasingly supporting the view that intervention with conventional engineering solutions is not always beneficial and that the use of 'more natural' designs based on, and in sympathy with, the tendencies of the landscape itself may be cheaper and more effective, at least until more brutal controls can be based on a well-documented understanding of the environment. This sympathetic 'geomorphological' design approach might be employed in the following ways:

(*i*) *Avoidance*. As noted above, all areas thought to be, or known from previous events to be locations of high flood risk, debris flow, piping, or other hazards can be avoided by sensible, *informed* land-use planning.

(*ii*) *Location*. Locate highways, etc., in low-risk areas and especially in such a way as to minimize the number of channel crossings. Thus several authors (e.g. Schick, 1971; Fookes, 1976b; Griffiths, 1978a) recommend that the most sensible place to cross an alluvial fan is near the apex of the fan, with due regard to channel movement and entrenchment, because variable, shifting channels, and flood areas are thus largely avoided (Fig. VI.23). Alternative routes include those close to the playa margin (but here flooding and saline ground may be hazards), or at an intermediate position where infiltration diminishes channel flow and major channel crossings are minimized. Crossings of well-established and entrenched channels are preferable to crossing braided or unentrenched flow lines.

(*iii*) *Use of existing features*. A general principle of great value is that established natural drainage lines are the most efficient and cheapest means of disposing of runoff and that they should be disturbed as little as possible. Low-cost plans can be developed that include a minimum of flood bunds and channel stabilization techniques using strong points and protected channel banks.

(*iv*) *Minimum disturbance*. Roads and other structures are likely to disturb the 'equilibrium' character of the desert surface. For example, where a road crosses a channel at right angles, the road in cross-section is horizontal whereas the channel is sloping, so that the upper edge of the road is likely to be above channel grade. The commonest solution to this problem is to build culverts, drainage ditches, dykes, or diversion bunds. But such a solution is relatively expensive and not always successful: systems that concentrate flow can cause increased flow velocities, erosion, entrenchment, and undermining of structures. Experience in the Middle East suggests that on low-cost roads, bridgeless crossings are often both less expensive and more effective than culverts. In designing such crossings, it may be possible to allow the road surface to be slightly below channel grade so that the road is covered by a thin veneer of protective sediment during floods and the only remedial work involves the removal of a thin sediment cover after the flood. Traffic disruption is likely to be short-lived. Similarly, sediment traps and other deep ditches should be avoided, as these often quickly fill up.

Thus, a final priority for future research in the field of water and sediment problems in drylands should be the development of a design handbook that uses 'natural environmentally sympathetic' designs based on information from field surveys. Such low-cost methods are often preferable to large-scale

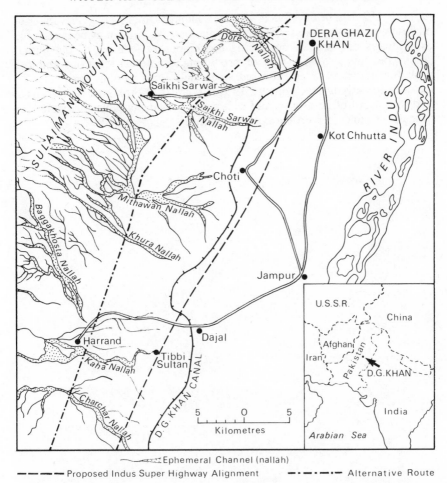

—————— Ephemeral Channel (nallah)

— — — — Proposed Indus Super Highway Alignment   —·—·—·— Alternative Route

FIG. VI.23 The proposed locations of the Super-Indus highway in relation to the Sulaiman Mountains, Pakistan (after Griffiths, 1978a)

interference with natural desert systems which are as yet poorly understood, and where long-term consequences of interference are not known.

(h) Summary

The movement of water and sediment in fluvial systems of drylands poses serious, commonly underestimated and certainly under-studied problems for the engineer and the planner. Precisely monitored fluvial data are available from only a small number of experimental catchments, but these reveal that fluvial activities in drylands, like the fluvial landscape it creates, are in many ways distinctively different from those in temperate areas. Thus, models for prediction need to take this distinctiveness into account, but at the same time they often have to be developed without a sound basis of locally relevant data on fluvial processes. As a result, geomorphological surveys may be invaluable as a means not only of describing contemporary features of the fluvial landscape, but also as a way of predicting hazardous locations and the probable nature of water and sediment yield, and guiding the choice of location for urban development. In this context, morphometric variables, and the record of landforms and sediments provide suitable surrogates for process variables; nevertheless, more monitoring is required.

VII

PROBLEMS OF SAND AND DUST MOVEMENT BY WIND IN DRYLANDS

(a) Introduction

The movement of sand and dust by wind occurs in many environments, but it is most pronounced, and poses the most serious problems in drylands. If buildings, cultivated areas, pipelines, and transportation networks in deserts are to avoid being attacked and perhaps buried by sand and dust, extensive control measures may be necessary. Urban areas in drylands may present obstacles to the natural and pre-existing patterns of aeolian debris movement. But, equally important, the nucleation of desert settlements dependent on localized water-supply and limited suitable agricultural land causes pressure on land to be concentrated around settlements so that the desert ecosystem is likely to be most disturbed in the immediate neighbourhood of urban areas. As a result, vegetation may be destroyed and soil structure damaged, thus promoting sand and dust movement and leading to increases in the magnitude and extent of aeolian problems. In this way, the causes of sand and dust problems are often intimately related to the problems of desertification that have received much recent attention (e.g. Paylore and Haney, 1976; United Nations, 1977).

Avoidance or amelioration of the undesirable consequences of sand and dust movement requires an understanding of the nature of that movement, and identification of actions that accelerate the process and of practices that reduce its occurrence. Thus, the objectives of this chapter are to assess the nature, intensity, and distribution of sand and dust movement by wind in drylands, and the problems arising from aeolian activity; to establish a methodology for assessing and monitoring sand and dust movement; to examine how sand and dust movement; and wind erosion hazards, can best be inferred or predicted from readily available information; and to evaluate measures for controlling sand and dust movement.

(b) The Nature of Sand and Dust Movement by Wind

Sand and dust movement, as a geomorphological process, is determined by the interaction of five main factors: the velocity and turbulence of the wind, on the one hand, and the inherent roughness, cohesion, and grain size of the surface on the other. Thus it is the interaction between the *erosivity* of the wind and the *erodibility* of the surface that ultimately determine whether or not sand and dust will move. Vegetation adds a complicating factor, since it reduces wind speed near to the ground surface.

Sand and dust are distinct, and problems arising from their movement are also often quite different. Sand has a range of grain sizes from 0.08 to 2.0 mm diameter; dust has a grain-size distribution with diameters less than 0.08 mm (Bagnold, 1941); dust usually has a lower specific gravity than sand (quartzose sand has a specific gravity of 2.65 gcm^{-1}); dust particles are usually irregular in shape (often platy) thus encouraging contact between grains and increasing their cohesion; sand, on the other hand, is virtually cohesionless.

Dust travels as suspension load in the wind, and may be lifted to great heights by turbulence. Sand movement, however, is chiefly by saltation, and is concentrated near to the ground in the moving 'saltation curtain'. In flight sand grains extract momentum from the airstream so that grain impact with the sand bed causes rebound or burial of the grain in a crater, both of which tend to eject more grains into the airstream. The force of bombardment of saltating grains rolls other, impacted grains (some of which may be too massive to saltate) along the surface as creep load. Sand movement is initiated on a loose, dry sand surface at the fluid threshold and movement is sustained at the slightly lower impact threshold (Fig. VII.1). Dust, in contrast, usually forms an aerodynamically smooth, highly cohesive surface, so that only high-velocity winds are capable of entraining it; although in the presence of saltating sand, dust may be entrained by ballistic impact. As a result of sorting processes that segregate sand from dust, superficial aeolian bedforms are invariably built of sand-sized particles (Cooke and Warren, 1973).

The detailed mechanics of sand and dust movement are not reviewed here as they have been thoroughly summarized elsewhere, notably in Bagnold's (1941) classic work, Chepil and Woodruff's (1963) seminal summary of wind erosion research, and in recent geomorphological surveys (Cooke and Warren, 1973; Mabbutt, 1977).

(c) Problems of Sand and Dust Movement

Problems arising from sand and dust movement can be classified and examined in terms of three fundamental aeolian processes: deflation, transport, and deposition.

1. Deflation problems

Deflation, the removal of sand and dust by wind from desert surfaces, is primarily a problem because it leads to the depletion of some of the most important soil constituents – silt, clay, and organic matter – leaving behind coarser particles, lower levels of fertility, and a reduced ability to retain water. In addition, once lost, surface soil in drylands does not regenerate quickly, so that the deflation loss may be fairly permanent. Other effects of deflation include scour and undermining of footings to telegraph poles etc., and scouring beneath pipelines, railway sleepers, and even roads – effects that can lead to collapse of the structures. Deflation is a natural process, but it usually leads quite quickly to the establishment of wind-stable surfaces,

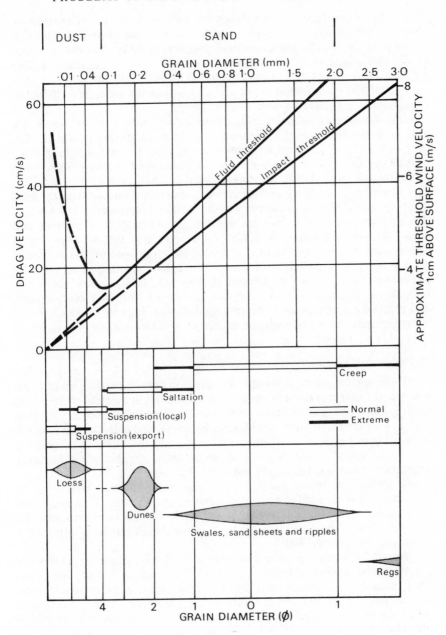

FIG. VII.1 Relationship between grain size, fluid and impact threshold wind velocities, characteristic modes of aeolian transport, and resulting size-grading of aeolian sand formations (after Bagnold, 1941 and Folk, 1971; from Mabbutt, 1977)

such as those comprising stone pavement, saline crusts, or soil aggregates. Only where stability cannot easily be achieved, as in mobile dune fields or on unpaved roads, is deflation a continuing natural problem. But deflation can be initiated where surface stability is disrupted by human interference. The Arizona Department of Transportation (1975) reported, for example, that the annual loss by deflation of silt and clay was in the order of 5–50 kg per vehicle mile on unpaved roads.

Such interference is most widespread in agricultural areas which are usually on the outskirts of settlements, and in areas of urban construction: thus man-induced deflation problems ensuing from the destruction of stone pavements and surface crusts etc., or the removal of vegetation are often urban-related in drylands. Of course, as the urban area grows and local agriculture expands to try and accommodate demand, pressure on the land increases and deflation problems may be further exacerbated. The form of this pressure varies, but may include the sedentarization of nomads and consequent concentration of herds and overgrazing, the development of commercial ranching and rainfed, irrigated cropping systems, the drilling of deep wells, and installation of pumps to replace traditional methods of groundwater extraction, thus causing shallow wells to dry up and the water-table to fall. Reduction of soil moisture is common in these circumstances, and is an important factor in promoting deflation, the rate of which varies approximately inversely with the square of effective surface soil moisture (Chepil *et al.*, 1963a).

2. *Problems of transport*

(i) Abrasion problems. Although the saltation curtain reaches up to about a metre, the maximum abrasive effect of sand is felt at a height of 20–25 cm (Sharp, 1964). Abrasion effectiveness is determined by numerous variables, such as particle shape, size, orientation, hardness, and surface texture of the material under attack; sizes, hardness, mass, and sharpness of projectiles; and environmental factors such as topography, neighbouring particles, vegetation, and wind velocity (Cooke and Warren, 1973).

Soil clods and other more or less indurated soil materials may be disintegrated or abraded as a result of saltating grain impact, thus promoting the impoverishment of soil structure, and rendering soil even more erodible (Chepil, 1946). Abrasion can also be extremely injurious to certain plants (Lyles and Woodruff, 1960).

Of greater direct consequence in urban areas is abrasion of structures and equipment by aeolian material. Abrasion heights may in fact be higher over hard, man-made surfaces, and the abrasion problem consequently more severe, because saltation heights and the rate of sand movement are increased over such surfaces as roads and runways. Sand abrasion can have an erosive effect on building materials: faces at angles greater than 55° to wind direction tend to be pitted, whereas more acutely inclined faces become fluted and grooved. Telephone and telegraph poles, and fences may be sand-blasted at their bases. Glass loses its transparency, first becoming pitted and then frosted: even the relatively high car windscreen may suffer (especially, of course, when the car is in motion). Paintwork is easily damaged. Maintenance and replacement costs for mechanical equipment are increased. Equipment such as generators and pumps may suffer from worn piston rings, scored cylinders, damaged bearings etc. Air filters require frequent changes. As a result of a storm on 27 December 1953, a single insurance company in California had 1200 claims for wind damage to windscreens and bodywork totalling $165 000 (US) (Clements *et al.*, 1963).

(ii) Visibility and other problems. Dust storms are common desert phenomena, and they may vary in size from those covering as much as 2500 × 600 km (Idso, 1976) down to diminutive dust devils (e.g. Smith and Leslie, 1976). Long-distance transport of dust has been reviewed by Goudie (1978). Problems associated with dust storms include the spread of disease through pathogen transport, suffocation of cattle, development of static electricity, interruption of radio, telephone and telegraph services, disruption of transport, the damaging of property, and harm to human health (Idso, 1976; Morales, 1977, 1979). But the problem that has received most attention is that of visibility reduction, a problem of greatest importance to the transport industry. Flights into and out of airports affected by blowing dust may be delayed (and in places there may be a danger of runway skidding), as at Sharjah and Bahrain in the Arabian Gulf (e.g. Houseman, 1961). Clements *et al.* (1963) described sand and dust storms in the Mojave Desert (California) that reduced visibility to such an extent that road traffic had to be halted.

3. Depositional problems

(i) Dust. Problems associated with dust deposition within drylands include the burial and killing of young plants, the rendering of roads impassable, infiltration of dust into houses creating problems of sanitation and housekeeping, and contamination of food and drinking water. For example, Clements *et al.* (1963) reported that dust between relay contacts and abrasion of switches causes particular problems for a Californian telephone company.

(ii) Sand. The commonest depositional problems are associated with the bulk transport of sand *en masse* in bedforms. Dunes encroach upon and often completely bury urban obstacles of all types, including roads, railways, runways, pipelines, and cultivated gardens. Larger aeolian features than dunes, the *draa,* have a similar effect but if they move, they move much more slowly.

In settlements, a common outcome of sand encroachment is land abandonment, reduction in the intensity of land use, failure of communications and depopulation. The town of In Salah, in the Algerian Sahara, provides a significant example: the inhabitants are fighting an endless war against encroaching sand that threatens to overwhelm their palm-trees, and which they attempt to ward off with palm-frond fences. Where control schemes have been implemented, their management is often poor or absent; sand fences are often wrongly constructed: they may be too low, or incorrectly spaced, or too close to the oasis. Ultimately, as elsewhere in this region of the Sahara on the fringes of the sand seas, the settlement may be abandoned as clearing costs become prohibitive. Similar abandoned oases have been described in Iran (Iran, Dept. of Environment, 1977).

In fabricating structures within urban areas the engineer and designer must not only consider wind loads for stability and safety of the buildings and inhabitants, but also the movement of sand and dust by winds. The debris will abrade the structures, and interactions between wind and buildings can produce local debris accumulations and debris infiltration into buildings. Areas of wind funnelling and vortices can result in unpleasant wind abrasion conditions for people's comfort, vegetation, or traffic. Shelter in airflow separation zones will promote the growth of sand accumulations, blocking roads and passageways, burying vegetation, and encroaching on living areas.

Communication lines are vulnerable to sand movement and can be especially inconvenienced by migrating dunes: on roads, for instance, detours may be necessary around dunes, or costly sand-removal measures may be required; and heavy-duty four-wheel drive vehicles are often essential. Some roads and railroads may be designed to be self-cleansing, but others, especially where they cross active sand fields, may be continuously in jeopardy, as, for example, the road from Kharga to Asyuit in the New Valley, Egypt, the new road from Al Ain to Abu Dhabi (UAE), the railroad out of Plaster City, California, and the first section of the proposed trans-Saharan railroad. In the first case the new road north of Al Kharga was built on a dam about as high as the local barchan dunes, which led initially to sand storage behind the dam (and temporary relief to the road), but dunes soon began to climb over the drifts, advance on the road and partially bury telephone lines. Control measures, such as the dam, commonly fail because of a lack of understanding of the processes involved. One particularly common error is to assume that by flattening dunes the problems ensuing from their movement will be eliminated: this is rarely so, because the winds will soon recreate aerodynamically suitable surface forms similar to the original dunes.

Pipelines and similar features pose especially difficult problems, because deep burial by sand makes maintenance and inspection difficult; and unsupported pipes on active dunes can be left high above the ground as dunes move on, putting them under torsion and possibly causing fracture. Such was the fate of part of the phosphate conveyor-belt at El Aaiun (Spanish Sahara) (De Benito, 1974).

4. *Geographical extent of the problems*

The problems of sand and dust movement can and do occur in most drylands, but the problems are most serious in areas of active sand dunes and those areas of stabilized dunes that are destabilized by human activity, and these are particularly severe in the 'sand seas' (McKee, 1979). Unanticipated problems are most likely to arise in the latter areas from developments that fail to appreciate the consequences of disturbing vegetation and surface covers on fixed dunes (Fig. VII.2). Dust problems are fairly ubiquitous in drylands, and even beyond them, but they are concentrated in areas of particularly strong winds and surface instability: major source areas are the Sahara, the southern coast of the Mediterranean Sea, north-east Sudan, the Arabian Peninsula, the lower Volga and North Caucasus in the USSR, parts of Argentina, Afghanistan, the Sonoran-Mojave Desert, and the western

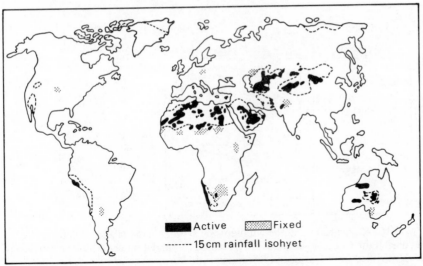

FIG. VII.2 Geographical extent of active and fixed sand dunes in drylands (after Cooke and Warren, 1973)

Great Plains of the USA. (Rapp, 1974; Idso, 1976; Goudie, 1978; Morales, 1979).

(d) Assessment of the Problems

1. *Introduction*

The mutual relationship between wind erosivity and surface erodibility determines where and when sand and dust movement occur, and such movement may be initiated from stable conditions by a change, positive or negative, in one or more of the variables listed in Fig. VII.3.

(i) Erosivity. Theoretical and experimental studies have shown that wind erosion is related to a large number of atmospheric properties in an extremely complicated manner. In general erosivity will vary spatially and temporally according to the distribution of regional and local (diurnal) wind regime. The most important wind variables are mean wind velocity; wind direction; frequency; period and intensity of gusts (governed by the vertical wind profile and the transient and steady drag); and the vertical turbulence exchange (controlled by temperature stratification and surface roughness). An assessment of erosivity therefore depends on the availability of accurate wind records.

Regrettably, such records are not universally available for drylands. As a result, data extrapolation and interpolation must be relied upon in judging the erosivity of many dryland sites. This is possible, but accuracy is limited: horizontal data extrapolation or interpolation beyond 50 km is unlikely to be more than 75% accurate (Lyles, 1976). Conclusions based on short records of wind speeds are at best tentative, can be fallacious, and may require special

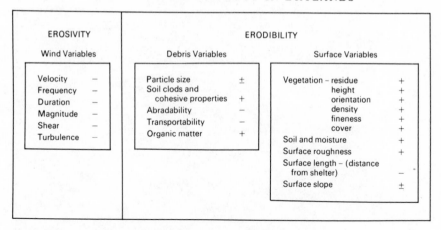

FIG. VII.3 Key variables in the wind-erosion system. Wind erosion will normally be reduced if the values of variables are increased ($+$) and if other variables are reduced ($-$) (modified from Cooke and Doornkamp, 1974)

manipulation (Wallington, 1968; Brookfield, 1970). Although attempts are made to standardize anemometer heights to 10 m, many observations are not made at this height and therefore require adjustment: often on-site inspections of anemometer locations are required and field estimates of surface roughness should be made (e.g. Johnson, 1965). The time base for wind data is important. If, for example, instantaneous wind speeds are cubed over an hour and averaged, the result may be considerably higher than that from cubing average annual wind speed. Incorrect conclusions can easily be drawn from station comparisons using data recorded at different time-base periods.

Despite these problems, there have been several attempts to predict erosivity, and the frequency of wind erosion and dust-blowing events (e.g. Dubief, 1952, in the Sahara; Zingg, 1949, 1950; Chepil *et al.*, 1962; Skidmore, 1965; and Skidmore and Woodruff, 1968, in the Great Plains of the USA).

(ii) Erodibility. The two primary factors controlling erodibility are vegetation and surface type. The character of plant cover – its composition, seasonal variation etc. – plays an important part in determining the erodibility of the surface. In particular, it tends to reduce erodibility in protecting the surface from wind erosion by displacing the velocity profile away from the surface, and by increasing the cohesion of particles through inputs of organic matter and salt linings along roots. Where plant cover is sparse or absent, the soil-surface conditions determine erodibility. Table VII.1 presents a brief classification of the major soil-surface types commonly found in drylands and their relative erodibilities (for a detailed review, see Petrov, 1976). Of these types, the sand deserts (see Fig VI.2) have the greatest quantities of material in motion at any one time because their 'soils' have only single-grained structure and minimal aggregation of particles.

(iii) Sources of information on erosivity and erodibility include the published

TABLE VII.1

Soil-surface types and their relative erodibilities

Soil-surface type	Arabic name	Local names	Amount of sand and/or dust movement
Sand Deserts	Erg	tamakhak (Arabic) kum, barchan (Turkish) edeien (Berber) sha-mo (Chinese) elisun (Mongolian)	Large amounts of sand and dust in motion when unvegetated. Fixed sand sheets and dunes become unstable when vegetation cover is degraded
Sand-Pebble Deserts	Serirs	ergs, azrirs (Arabic) gobis (Mongolian) regs (Persian)	Quantity of sand and dust in motion depends on the development of the pebble pavement. Intermediate amounts moved when undisturbed. Large quantities moved when pebbles and/or vegetation removed from surface
Pebble Deserts	Regs	hamada (Arabic)	Little or no sand and dust movement in undisturbed state; intermediate amounts when pavement disturbed or removed
Gravelly Hamada			Intermediate between pebble desert and rocky hamada
Rocky Hamada			Little or no movement of sand and dust
Solonchak	Sebkha	sabkha, chott (Arabic) sor, shor (Turkish) kevir, kebir (Persian)	Dust movement increases as surface becomes desiccated when groundwater table drops. Breakdown of crusts results in large quantities of dust being suspended. Floodouts and playas are the most important source of dust

Sand sheets and dune fields developed on thick loose arenaceous sediments

Poorly developed pebble pavement with variable quantities of sand filling the space between pebbles

Well-developed multi-layered pebble pavement. Weakly cemented, poorly sorted gravel. Plant cover very sparse

Weak removal of weathering products. In lower horizons the weathering products are coarsely fragmented while upper horizons contain poorly rounded gravel. Virtually devoid of fine material

Saline depression with groundwater table close to surface. Sometimes saline crusts on surface

Source: after Petrov (1976).

literature, remote-sensing imagery, fieldwork, laboratory work and, less directly, predictive equations. Each of these sources is briefly reviewed below with examples given of the type of information each can yield. Naturally, the different sources are commonly used together to enhance the quality of the overall assessment.

2. Literature

Although topographical, vegetational, pedological, and geological maps may provide data relevant to assessing sand and dust movement, the most important published source is that of meteorological records. In any project it is unlikely that there will be time to collect sufficient meteorological data in the field, yet records of long duration are the only way to demonstrate adequately the effects of strong winds of short duration. Analysis of meteorological data usually requires the construction of diagrammatic generalizations, such as wind roses. The latter indicate frequency and strength of all winds. But a better approach is to calculate sand and dust roses for effective (i.e. sand and dust transporting) winds. Bagnold (1953) gave the following equation for calculating sand- (and dust-) movement roses:

$$Q = \frac{1.0 \times 10^{-4}}{\log{(100_z)^3}} t\,(V-16)^3 \qquad (VII.1)$$

where Q = sand movement in tonnes per metre,
         t = the number of hours that the wind of V km/hr blows, and 16 km/hr is the threshold velocity,
and z = height above ground in metres.

Q is summed for effective winds of all speeds for each direction to give a sand-movement rose.

Skidmore (1965) and Skidmore and Woodruff (1968) assessed the direction and relative magnitudes of wind erosion forces by computing wind-erosion force vectors, r_j, for each of 16 principal directions:

$$\mathbf{r}_j = \sum_{i=1}^{n} (\overline{V-V_t})_i^3 F_i \qquad (VII.2)$$

$$j = 0 \to 15$$

where $(V-V_t)_i^3$ = the cubed mean windspeed above threshold within the ith speed group,
         F_i = the percentage of total observations that occur in the ith speed group and direction under consideration,
and j = the direction (values 0 to 15 numbered in an anticlockwise direction, starting with east).

The sum of the vector magnitudes gives the total magnitude of wind erosion forces for the location:

$$F_i = \sum_{j=0}^{15} \sum_{i=1}^{n} (\overline{V-V_t})_{ij}^3 F_{ij} \qquad \text{(VII.3)}$$

The relative erosion vector is given by r'_j such that

$$\sum_{0}^{15} r'_j = 1$$

$$r'_j = \frac{\sum_{i=1}^{n} (\overline{V-V_t})_i^3 F_i}{\sum_{j=0}^{15} \sum_{i=1}^{n} (\overline{V-V_t})_{ij}^3 F_{ij}} \qquad \text{(VII.4)}$$

The preponderance of wind erosion forces can be calculated. A preponderance of 1 indicates no preferred wind erosion direction, while a preponderance of 2 indicates erosion forces twice as great parallel with the direction line as normal to it: where preponderance is large, windbreaks oriented normal to the maximum erosion forces will perform efficiently. Thus, preponderance is useful in determining the correct configurations of windbreaks and fences to control erosion. The vector and its sum provide an excellent index of erosivity because they take account of the capacity of the wind to erode, the prevailing wind direction, and the directional distribution of wind-erosion forces. Vectors indicate how the factors vary throughout the year, the time and place when erosion hazard is greatest, and thus, the time and place when the need for protection is greatest, and the correct orientation of barriers to reduce the problems.

3. Remote-sensing imagery

(i) Satellite photographs. Many publications report the use of satellite imagery applications in dune environments. For example, Vinogradov (1976) used the spectral reflectivity coefficient to monitor areas of aeolian land degradation. Such problem areas are especially obvious where homogeneous pasture surfaces are crossed by administrative boundaries between different modes of land use. The advantages of LANDSAT imagery for studying dune characteristics in relation to other environmental features can be summarized as follows: (i) it allows direct comparison of areas because approximately the same scale prevails on all images, (ii) it permits the recognition of major trends and lineations, (iii) it provides mosaics from false-colour prints that show sand patterns over large areas and prove useful as base maps, and (iv) it furnishes colour images that are a much better guide to the presence of sand either bare, lightly vegetated, or in dune form, than any of the single black and white bands of ERTS-1 (McKee and Breed, 1973).

Useful studies of sand and dust problems using LANDSAT imagery include those by Seevers and Drew (1973, 1974) on the Sand Hills of Nebraska, on dunes in central Australia by Simonett et al. (1969), by Mainguet (1976) on sandflow in Saharan and Australian ergs, Romanova (1964, 1968) on the Karakum Desert, and by Bowden et al. (1974) on the dust plumes of the Los Angeles area.

(ii) Aerial photographs. Where good-quality, well-indexed aerial photographs exist for drylands, the analysis of the problems of sand and dust movement is more rapid, efficient, and accurate (e.g. Stone, 1961; Stewart, 1968). Such imagery is especially useful in studying erodibility and land degradation, active dune systems, fixed dunes, rate of dune movement, and in specialized mapping techniques.

(a) Erodibility and land degradation. Once landforms and soil-surface types are identified on aerial photographs, certain information can be inferred concerning erodibility. For instance, dead vegetation can be distinguished from living vegetation, different lithologies and, often, soils can be recognized – all invaluable data for the assessment of erodibility.

(b) Active dune systems as indicators of wind regime. Although the characteristic appearance of desert landscapes on aerial photographs has been reviewed (Davis and Neal, 1963), little has been published on the recognition of dune landscapes. The trend of dunes can sometimes be inferred from dune images on aerial photographs in the way outlined on Fig. VII.4. Parameters describing dune geometry which are of use in determining aspects of dune and sand movements can be measured from photographs by photogrammetric techniques (for example, rate of dune movement is proportional to dune height). In areas where wind data are inadequate or absent, indirect evidence of wind regime can be obtained from the trends of *active* dunes. Such dunes can serve as reliable indicators of dominant wind and sand-movement direction (e.g. Monod, 1958; King, 1960; Brookfield, 1970; Wilson, 1971; Higgins *et al.*, 1974). Petrov's (1976) classification of dune forms and their corresponding wind regimes is shown in Table VII.2 and Fig. VII.5. For each of the five wind regimes, the dune forms are examined in relation to the soil-surface type and the depth of groundwater-table. The table does not include all dune types (e.g. obstacle dunes) but it is the most comprehensive available at present.

(c) Fixed dunes, formed in the past and stabilized by vegetation and/or surface crusts, were not necessarily formed by present winds and should in no circumstances be used to infer contemporary wind regime. Urban development on fixed dune fields can create serious problems if the surface is destabilized, so that the recognition and monitoring of fixed dunes is important: they are generally recognized by soil profiles accordant to dune surfaces and a vegetation cover (Churchwood, 1963a,b), modification of dune form and pattern by non-aeolian agencies (Smith, 1963), and discordance of trend with present effective winds. Ridges of eluviated dunes are rounded and low, the flanks are gentle and can be gullied, and the inter-dune hollows are partially infilled. The fixed dune field may also have its own drainage system.

(d) The rate of dune movement can be measured if successive aerial photographs are available for the same area (e.g. Hunting Surveys Ltd, 1977). For example, planimetric outlines of barchans plotted on sequential aerial photograph flights are used to compute horizontal displacement. As a general rule, the rate of movement of a barchan is inversely proportional to

BARCHAN

(A) Aerial Photograph

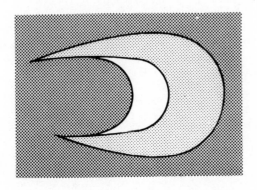

(B) Interpretation

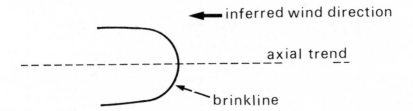

inferred wind direction

axial trend

brinkline

FIG. VII.4 The assessment of dune trend from aerial photographs. The brinkline, at the top of the slipface, is easily distinguished by the light-coloured sand of the slipface contrasting with the darker sand of the windward slope. This tonal difference results from the slipface being unrippled (avalanching processes of sand sorting rather than formation of ripples) so that it reflects more light than the rippled windward slope, or because the two faces are at different angles to the light. Inter-dune areas with a clay surface largely clear of loose sand show up dark on aerial photographs. Inter-dune areas which are littered with gravel, or covered with salt crusts, show up light because they reflect more light

TABLE VII.2

Basic types of aeolian relief of sands in deserts of the world

Wind regime	Character of underlying surface			
	Plains consisting of compact deposits, with deep groundwaters, lacking vegetation	*Plains consisting of loose sands, with deep groundwaters, lacking vegetation*	*Solonchak plains, with shallow groundwaters and halophilous vegetation*	*Sand plains, with deep groundwaters and psammophilous vegetation*
Dominance of one direction of active wind	1. Isolated barchans and barchan chains, more or less stable. Fields of parallel barchan chains. Complex barchan ridges	2. Isolated barchan chains. Fields of parallel chains. Complex asymmetric barchan ridges	3. Monticule bars, asymmetric coppice mounds extending in direction of prevailing wind	4. Small, medium, and large sand ridges, fixed by psammophilous vegetation
Dominance of two opposing active winds	5. Isolated barchans and barchan chains. Fields of parallel barchan chains	6. Isolated barchan chains. Fields of parallel barchan chains	7. Heaped sands – fine, medium, and coarse. Small sand ridges	8. Small, medium, and large sand ridges, fixed by vegetation
Dominance of two perpendicular wind directions	9. Frequently crossed, isolated barchans and barchan chains undergoing transformation	10. Honeycomb barchan sands	11. Heaped coppice sands, small, medium, and large, depending on size of shrubs	12. Honeycomb, hummocky, and analogous relief forms of fixed sands, small, medium, and large
Dominance of two winds blowing at an acute angle	13. Asymmetric barchans and barchan chains. Linear ridges. Seifs. Whalebacks	14. Asymmetric barchan chains. Seifs. Whalebacks	15. Monticule bars. Asymmetric coppice mounds	16. Small, medium, and large sand ridges and sand hillocks, fixed by vegetation
Variously angled, without dominance of one particular direction of active wind	17. Systems of pyramidal dunes. Isolated dunes	18. Isolated pyramidal dunes, rarely groups	19. Heaped sands, small, medium, and coarse, depending on size of shrubs	20. Sand hillocks and honeycomb sands – large, medium, and small. Network ridges. Fixed pyramidal sands

Source: Petrov (1976).

its mass. Thus, some measure of dune size should be given as a factor used to modify recorded rates of advance to determine rate of sand movement. Dune height, determined either from aerial photographs by stereoscopic methods or from field measurements, is usually used. Lettau and Lettau (1969) have used calculations of the rate of movement of barchan dunes to compute the rate of bulk transport of sand: dune volume was calculated from aerial photographs and bulk transport was computed by multiplying dune volume by the annual rate of dune advance.

(e) *Specialized mapping techniques*. Geomorphological maps normally require aerial photographs as a basis. These maps might classify aeolian hazard areas – actual and potential – according to erosion, transportation, and depositional tolerances in urban areas or areas to be urbanized, and in terms of hazard severity. Such a classification (Table VII.3) was developed in a study of a new airport site in Dubai (Halcrow Middle East, 1977a) (see also Chapter III).

A final comment on aerial photographs in studying aeolian problems is important: interpretation of aeolian phenomena on air photographs is often difficult; training and experience are required before it is possible to extract relevant data consistently and at an acceptably high level of accuracy; and 'field checks' are essential.

4. Fieldwork

(i) *Erodibility* can be assessed in the field in various ways, chiefly through using experimental plots or methods of soil classification. The experimental plot has the advantage that the environmental conditions can be controlled to a large extent (e.g. by using a portable wind tunnel); the main disadvantage is the influence of the plot itself on the empirical results, a problem that makes extrapolation of the results over wider areas difficult (e.g. Butterfield, 1973). Soils can be classified in the field according to their erodibility (e.g. Lyles, 1976). Classification may be made on the basis of depth, structure, and texture of the surface horizon and the depth of non-erodible sub-surface clay and gravel horizons in determining the depth to which erosion may continue and the amount of material available for drift once erosion is initiated (Condon and Stannard, 1954).

(ii) *Dust monitoring*. Instruments used for dust monitoring have been thoroughly discussed by the Task Committee (1970): techniques include settling jars (Hendrickson, 1968) and similar sedimentation collectors (Fairweather *et al.*, 1965; Smith and Twiss, 1965; Chatfield, 1967; Salem and Sowelim, 1967; Arizona Department of Transportation, 1974, 1975; Morales, 1979); and those involving filtration, impingement, electrostatic precipitation, thermal precipitation, and centrifuging.

Counting dust particles is usually done on a standard biological microscope. Particle-size distribution is determined by sieving, optical microscopy, sedimentation, elutriation, centrifugal classification, or by using the electron

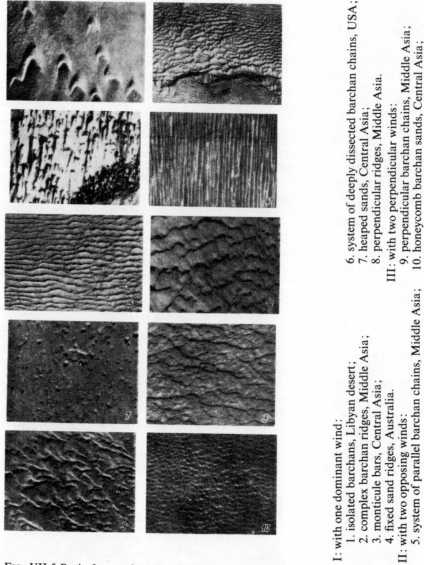

I: with one dominant wind:
1. isolated barchans, Libyan desert;
2. complex barchan ridges, Middle Asia;
3. monticule bars, Central Asia;
4. fixed sand ridges, Australia.
II: with two opposing winds:
5. system of parallel barchan chains, Middle Asia;
6. system of deeply dissected barchan chains, USA;
7. heaped sands, Central Asia;
8. perpendicular ridges, Middle Asia.
III: with two perpendicular winds:
9. perpendicular barchan chains, Middle Asia;
10. honeycomb barchan sands, Central Asia;

FIG. VII.5 Basic forms of aeolian relief of sands and their corresponding types of wind regime (after Petrov, 1976)

microscope. Methods of chemical analysis are standard and numerous (see Stern, 1968).

Chepil and Woodruff (1957) examined the relationship between measured dust concentration and visibility and found that (at 2 m height)

$$C = \frac{56.0}{V^{1.25}} \tag{VII.5}$$

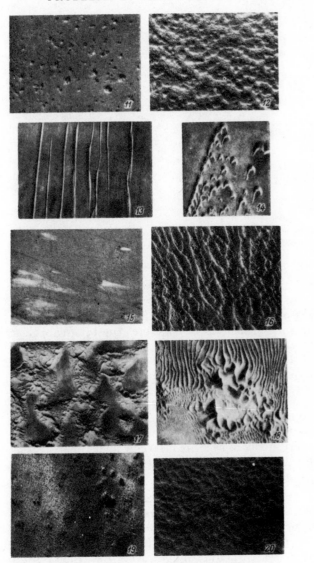

11. heaped sands, Central Asia;
12. fixed honeycomb sands, Middle Asia.
IV: with two winds blowing at an acute angle:
13. linear sand ridges, Central Asia;
14. wedge-shaped arrangement of isolated barchans;
15. coppice mounds, Middle Asia;
16. sand ridges parallel to wind direction, Middle Asia.

V: with several mutually opposed winds:
17. pyramidal sands, Sahara;
18. pyramidal dune, Central Asia;
19. heaped sand, Central Asia;
20. fixed sands with honeycomb depressions, Middle Asia. (See also Table VII.2)

where C = concentration of dust in milligrams per cubic metre,
and V = horizontal visibility in kilometres.

Chepil (1957a,b) and Chepil and Woodruff (1957) also analysed the sorting, size, and composition of dust particles collected during dust storms.

Synoptic meteorological studies of tropical dust storm conditions at the surface and upper levels of atmospheric disturbance are required to determine the weather patterns that are conducive to long-range transport and the frequency of dust-storm occurrence. Standard weather observations, used to study the frequency distribution of dust storms in space and time, could

TABLE VII.3

A classification of dunes according to their potential hazard as a source of mobile sands for a site in Dubai

	Vegetation		
	No vegetation	Sufficient to produce surface roughness	Sufficient to be a major stabilizing influence
Dune morphology	High hazard	Moderate hazard	Low hazard
Angular crest Steep slip face Active 'blowouts' Exceptionally high	Class 1	Class 2a	Class 3a
Rounded/broad 　crests No active slip face Comparatively low	Moderate hazard Class 2b	Low hazard Class 3b	Little hazard Class 4

Source: Halcrow Middle East (1977a).

reveal the sources and sinks of the dust. Goudie (1978) has listed the most important dust-bearing winds (Table VII.4). Widespread dust storms of great duration occur in the warm sectors of low-latitude depressions, whereas localized wall-like dust clouds usually mark the passage of fronts.

Dust haze conditions have been studied by monitoring radon-222 (Carlson and Prospero, 1972). Radon 222 is a radioactive (half life 3.82 days) inert gas produced by the uranium 238 decay series. The emanation rate of radon for the ocean is about 100 times less than that over land, so that oceanic air parcels have relatively low radon concentrations. Radon concentrations within areas of dense dust-produced haze are markedly higher than in clear areas or above a trapping temperature inversion.

(iii) Sand monitoring. Measurement of total sand flow can only be approximate because the presence of any collecting device inevitably interferes with the airstream. The particle collector must minimize airflow disturbance, eddy production, and deflection effects by presenting a thin section to the wind and possessing a fairly open structure and smooth walls to prevent internal back pressure. Several examples of horizontal and vertical collectors are shown in Fig. VII.6 (Bagnold, 1941; Belly, 1964).

In addition to measuring sand movement empirically, the measurement of near-surface wind velocities is also important, especially in terms of predictive equations. Techniques commonly used include pitot tubes (Svasek and Terwindt, 1974) and cup anemometers. Turbulence has been measured with hot wires, strain-gauge anemometers (Knott, pers. comm.), bivanes and directional vanes (Sethuraman and Brown, 1976). Various flow visualization techniques have also been used, such as introducing a tracer airflow through, for instance, exploding smoke bombs (Sharp, 1966), meteorological balloons tracked with balloon theodolites (Knott, pers. comm.), and fluorescent sand

TABLE VII.4

Dust-bearing winds

Africa	Khamsin (Egypt)
	Dachami (South Sahara)
	Ghibli (Tripolitania)
	Gobar (Ethiopia)
	Haboob (Sudan)
	Harmattan (West Africa)
	Sahel (Morocco)
	Chili (Tunisia)
	Chidili (South Algeria)
	Leste (Madeira)
Asia	Belat (S. Arabia)
	Buran (S-E. Russia), Kara Buran (Turkestan)
	Shimal (The Gulf)
	Scirocco (Arabia, Palestine, Mesopotamia)
	Scihaitan (Baluchistan)
	Scistan (Iran)
	(India)
	Andhi (India)
Americas	Pampero Sucio (Argentina)
	(Mexico)
	Palouser (Idaho, Montana)
Australia	Brickfielder
Europe	Calina (Spain)
	Kassava (Hungary)
	Scirocco (S. Europe)
	(S. Russian steppe)

Source: Goudie (1978).

grains – counted with ultraviolet lamps (Ingle, 1966). Measurement of dust movement can be studied, in addition to using air photographs, with reference to surveyed stakes (e.g. McKee and Douglass, 1971).

5. *Laboratory analyses*

(i) Geometric and dynamic similarity in wind tunnels. The advantage of using models in wind tunnels to study sand and dust movement is that it permits control and systematic variation of parameters governing aeolian activity, such as wind velocity, surface roughness, etc. In order to use experimental measurements to supplement field data, the model and its environment should obviously simulate the field conditions as closely as possible (Rim, 1958). Similarity depends primarily on matching the appropriate Reynolds number (a measure of the balance between viscous and inertial forces resisting flow) and the geometrical boundary conditions. Reynolds numbers for full-scale structures are large and to obtain them in the laboratory the model must be relatively large; the size is, however, limited by its blockage effect on air flow which may distort velocity distributions round the model. Complete accounts of similarity requirements, together with assessments of the relative success of simulations have been given by several

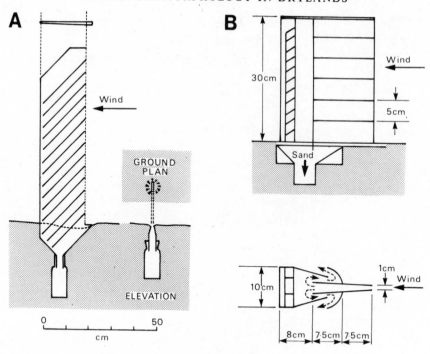

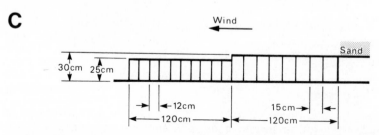

FIG. VII.6 Horizontal (A, B) and vertical (C) sand collectors (after Bagnold, 1941 and Belly, 1964)

authors (Cermak, 1958, 1971; Plate and Cermak, 1963; Cermak *et al.*, 1966; MacVehil *et al.*, 1967).

(ii) Velocity profiles and turbulence in wind tunnels. Velocity profiles can be modelled using two techniques of profile generation – 'roughness' and 'graded blockage' methods (Lloyd, 1966; Lawson, 1968). Turbulence spectra can be produced with turbulence grids and roughness elements (Comte-Bellot and Corrsin, 1966; Counihan, 1969, 1970, 1971). Hot-wire anemometry is the

main tool used for laboratory measurement of turbulence (Hinze, 1959; Corrsin, 1963; Bradshaw, 1975). Surface stress can be measured by Preston tubes. Smoke and neutral density helium-filled bubbles can be used to visualize the flow field. These methods will not be considered in detail here.

(*iii*) *Experimental results* from wind-tunnel studies are very extensive, and include work on the threshold velocities of particle movement (Greeley *et al.*, 1977) and Chepil's (1950) work on erodibility. Whitney (1978) used a wind tunnel to carry out wind-blast tests in which electron-microscopy surveillance was used to monitor changes on the surface of test specimens. Dietrich (1977) wind-blasted rocks of varying hardness for two years with dust particles in a wind of 28 km/hr, and discovered that the rate of erosion is directly related to the bond strength of the material being wind blasted, and that the mass of the abrasive is more important than its hardness. Architectural problems associated with sand and dust movement are important in drylands, and these can be investigated in wind tunnels. For example, Duchemin (1958) carried out experimental studies in wind tunnels aimed at solving design problems for buildings in arid areas. Such experiments led to the definition and quantification of aerodynamic effects for different building configurations (Fig. VII.7): areas of high velocity cause discomfort by the flying sand and dust, and wear by abrasion; areas of shelter suffer from sand drifting and deposition.

6. *Predictive equations*

In many ways, the processes and factors controlling the movement of sand and dust, or rather our understanding of them, can best be summarized in a series of predictive equations, equations that are invaluable in attempts to manage the problems. The equations are based on the relations between erodibility and erosivity. Some of the factors are represented by their true value and some by an index describing their behaviour relative to arbitrarily established standard conditions. The equations can be grouped into two broad categories: (i) complete equations, that contain parameters describing all the groupings of factors; and (ii) partial equations that estimate sand and dust movement with parameters describing only some of the groupings of factors affecting soil erosion.

(*i*) *Complete equations*
Universal equations for measuring wind erosion. An early version of a universal equation was developed by the US Agricultural Research Service (Niles, 1961) and developed by Woodruff and Siddoway (1965). Standard conditions for the equation are those for the vicinity of Garden City, Kansas.

$$E = f(I', K', C', L', V) \qquad \text{(VII.6)}$$

where E = computed soil loss (tons/acre/yr),
   I' = soil and knoll erodibility index,
   K' = soil ridge roughness factor,
   C' = climatic factor,
   L' = field length along prevailing wind erosion direction,
and V = equivalent quantity of vegetation cover.

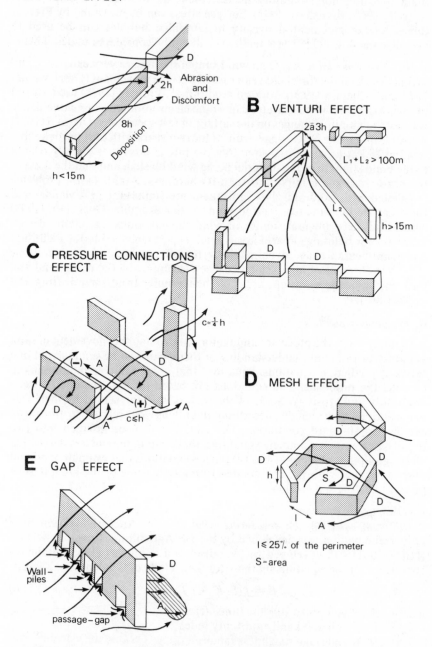

FIG. VII.7 Aerodynamic effects of building configuration on deposition of sand (D) and abrasion and discomfort resulting from flying sand and dust (A) (modified from Gandemer, 1977)

F   CORNER EFFECT

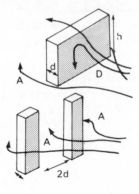

G   WAKE EFFECT

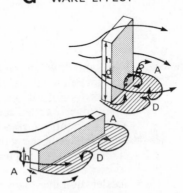

H   CHANNEL EFFECT

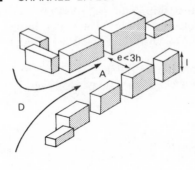

I   TOWER IN AN OLD
        SETTLEMENT

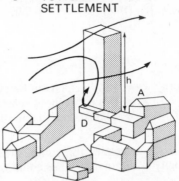

The manual resolution of this equation is laborious because the elements are
not multiplicative – erosion can only be predicted by using nomograms and
tables. A computer solution has been published (Skidmore *et al.*, 1970a,b).
Application of the equation to large areas is difficult because success depends

on precise data that may not be available in many drylands. But the equation is a comprehensive summary of the wind erosion system, it is a predictive tool for determining the potential amount of wind erosion under existing conditions, and it is a useful guide for determining control measures necessary to reduce erosion to an acceptable level.

(ii) Partial equations cannot be entirely accurate, but they can be useful in areas where the values of neglected factors do not change. As a consequence, they are of little relevance for erosion assessment over wide areas.

Bagnold (1941) derived an equation for predicting *the rate of sand movement*:

$$q = C\sqrt{\left(\frac{d}{D}\right)\frac{p}{g}} V'^3_*$$ (VII.7)

where q = total sand flow (gm/cm/sec),
        C = a coefficient to take account of sand grading (with values of 1.5 for nearly uniform sand, 1.8 for naturally graded dune sand, 2.8 for poorly sorted sand with a wide range of grain size, and 3.5 for a pebbly surface),
        D = the diameter of the standard 0.25 mm sand,
        d = diameter of the sand under consideration,
        p = density of air,
        g = acceleration due to gravity,
and V' = velocity gradient of the air.

Figure VII.8A illustrates this relationship, and Fig. VII.8B shows the sensitivity of the formula to variations in C and d. The equation has been used successfully by many workers (e.g. Kadib, 1963, 1964; Belly, 1964; Tsoar, 1974; Howard *et al.*, 1977).

The most difficult parameter to measure in the field is V_* and it is not a value recorded in standard meteorological observations. However, it can be shown that (Bagnold, 1941, p. 69):

$$V'_* = \alpha(V - V_t)$$ (VII.8)

where V = wind velocity at height z,
        V_t = is the threshold velocity at height k',
and α = a constant equal to $(0.174/\log{}^z/_{k}')^3$.

Thus the necessity for complicated measurements of atmospheric shear velocity is eliminated. Thus equation VII.7 can be written as:

$$q = \alpha C\sqrt{\left(\frac{d}{D}\right)\frac{p}{g}} (V - V_t)^3$$ (VII.9)

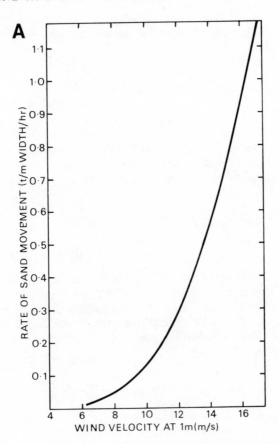

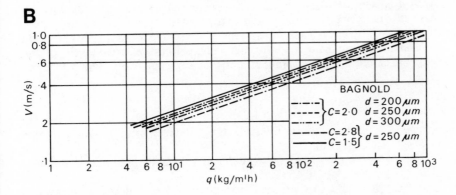

FIG. VII.8 (A) Relationship between rate of sand movement and wind velocity at a standard height of one metre ($c=1.8$, $d = 0.25$ mm); (B) Sensitivity of Bagnold's predictive equation (after Bagnold, 1941)

Taking 10 m as the standard height for measurement of wind velocity, a predictive equation can be written for use with standard meteorological data:

$$q = \alpha D \sqrt{\left(\frac{d}{D}\right)} \, \frac{p}{g}(V_{10} - V_t)^3 \qquad \text{(VII.10)}$$

where V_{10} = wind velocity at 10 m.

Values of C, d, k', and V_t will vary from site to site. They can be estimated from velocity profile measurements in the field or in a wind tunnel, or standard values can be taken. For uniform sand of 0.25 mm diameter, C is 1.5, k' is 0.3 cm, and V_t is 2.5 m sec^{-1}. For an average dune sand, C is 1.8, k' is 1.0 cm and V_t is 4.0 m sec^{-1}, C is 1.8 and d/D is 1.0, equation VII.7 becomes:

$$q = 4.39 \times 10^{-10}(V_{10} - 4.00)^3 \qquad \text{(VII.10a)}$$

The curve on Fig. VII.9 gives the rate of movement of an average dune sand with mean grain diameter of 0.25 mm.

Many other empirical relationships have been proposed for calculating the rate of sand movement (for a review see Cooke and Warren, 1973).

The rate of barchan dune advance was also predicted successfully by Bagnold (1941):

$$C = \frac{q}{\gamma h} \qquad \text{(VII.11)}$$

where q = rate of sandflow,
γ = a packing factor,
and h = dune height.

Dust blowing, it has been found experimentally, varies directly with the cube of wind velocity and inversely to the square of the effective moisture at the soil surface. On the assumption that the average moisture content of the soil surface is proportional to the effective precipitation, the quantity of dust blowing can be calculated from climatic data (Chepil et al., 1962, 1963a):

$$C = \frac{V^3}{(P-E)^2} \qquad \text{(VII.12)}$$

where C = the climatic index in the universal wind erosion equation (VII.6),
V = average annual wind velocity (in miles per hour) at standard height of 10 m,
and P-E = effective precipitation or moisture index (Thornthwaite, 1948).

The equation is standardized using the annual average value of 2.9 for Garden City, Kansas:

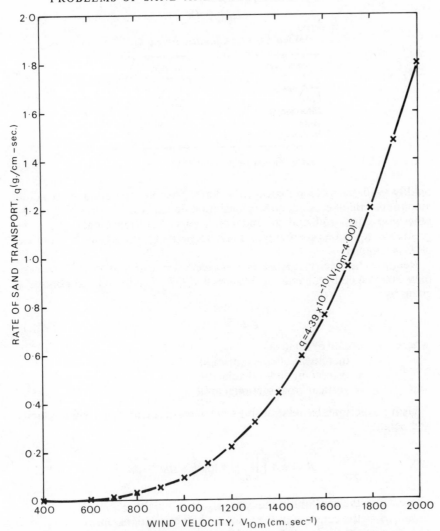

FIG. VII.9 Predictive equation for use with wind velocity measurements at a standard of 10 m

$$C' = \frac{100V^3}{2.9(P-E)^2} \% \qquad \text{(VII.13)}$$

Areas of potential deflation and deposition of dust can be located by calculating the climatic index and comparing the computed value against standard values (Table VII.5) (Yaalon and Ganor, 1966). The predictive equation for dust is derived by computing a regression between the climatic index and the annual number of dust storms (Chepil *et al.*, 1963b).

The main advantage of the climatic index is that it is simple and uses

TABLE VII.5

Wind Erosion Climatic Index, C

Wind erosion	Climatic index (%)
Very low	0–17
Low	18–35
Intermediate	36–71
High	72–150
Very high	>150

Source: Yaalon and Ganor (1966).

readily available meteorological data. But it gives no indication of the actual amount or rate of erosion, only information on the relative erodibility under otherwise equal soil and ground conditions. Additional data and other predictive equations are required if it is necessary to predict actual quantities of dust removed.

Relations between *measured dust concentrations and visibility* were successfully established by Chepil and Woodruff (1957). The rate of soil removal is given by:

$$R = \sum (C_\Delta u_\Delta) A \qquad \text{(VII.14)}$$

where R = rate of removal,
C_Δ = increment of concentration,
u_Δ = increment of wind velocity,
and A = vertical cross-sectional area.

Using experimental relationships between concentration, wind velocity, and height:

$$R = aA \int \left[\frac{1}{m} h^{-n} \log h + B h^{-n} \right] dh \qquad \text{(VII.15)}$$

where a = the constant in the concentration–height relationship,
B = the constant in the velocity–height relationship,
n = the constant slope of the concentration–height relationship,
m = the constant slope of the velocity–height relationship,
and h = the height above ground
Following a complex series of assumptions and substitutions it can be shown that:

$$R = \frac{56}{V_d^{1.25}} \left[\frac{2.33}{m} + B \right] \qquad \text{(VII.16)}$$

where V_d = the visibility in kilometres.

Other examples of such an equation are provided by Fuchs (1964) and Robinson (1968).

(e) Control Measures

1. *Avoidance, removal, or control*

A sound general rule in environmental management is to avoid hazard areas and surface disturbance as far as possible: avoidance is often more effective and cheaper than control both in the short and the long term. Similarly, if surface disturbance is inevitable, in general the less the disturbance, the better. Such rules are extremely pertinent to aeolian problems especially in the context of large and active dune fields, where the scale of natural forces is considerable and control measures, even expensive and extensive control measures, are likely to be of only temporary value.

The best means of avoidance lies in sensible site selection, a procedure that requires prior knowledge of existing or potential problems and therefore requires antecedent environmental surveys. Even when the general location of urban development is determined by other factors, the site may be designed sensibly to avoid or to minimize the impact of aeolian problems. Al Ain, in eastern Abu Dhabi, for example, has to be located for political and hydrological reasons near to the old oasis but the new city plan successfully seems to minimize (although not eliminate) the potentially serious sand movement problems associated with a nearby, extensive and active dune field.

Closely allied to the notion of avoidance in the eyes of some developers is that of removing the material by earthmoving equipment. An extreme and rarely practicable solution is to remove all the moving sand and thus eliminate the problem entirely – often an expensive solution, but possible if only small dunes are involved. Alternatively, where sand is continuously causing problems by advancing across roads, etc., clearance provides a temporary solution, but one that is generally both continuous and expensive in the long term. As indicated earlier, the solution of removing dunes by flattening them is unlikely to succeed because air-flow working on the newly flattened surface is likely soon to reconstitute the mobile bedforms.

2. *Vegetational stabilization*

Permanent stabilization of mobile sand and dust can often only be achieved effectively through the development of a vegetation cover. This solution usually requires the use of a combination of mechanical, chemical, and botanical methods at least until the vegetation has become firmly established.

(i) Environmental conditions. Attempts at vegetational stabilization should consider the interrelationships between the following habitat factors – character of substrate, thickness of sand deposit, degree and nature of salinization; water storage capacity, nutrients, and structure of the substrate or soil; quantity and quality of water available for the plants (such as precipitation regime, soil moisture, air humidity); depth of water-table and its chemistry; type of movement and rate of displacement of moving sand and dust; exposure to predominant wind direction and solar radiation.

The main types of vegetational stabilization conditions in arid regions may be grouped as follows:

Sands of semi-arid regions (with annual precipitation exceeding 200–250 mm; potential evaporation 1500–1600 mm) where mesophyllous psammophytes can grow. Within these areas it may be possible to use a wide variety of sand-binding plants. Vegetational stabilization is usually successful, with plantings showing a high survival rate. Surface cover develops rapidly so that barren sands may be stabilized in 5–10 years.

Sands of arid deserts (annual precipitation 100–250 mm and evaporation 2300 mm) where xerophyllous psammophytes (shrubs and grasses) survive, vegetational stabilization is possible but less easily attained. Young plants must be protected from being blown away or buried. The variety of sand-binding plants is more limited. Woody species cannot grow; the main plants are shrubs, semi-shrubs, and perennial and annual grasses. Sowing and planting fail in the periodic droughts. The rate of surface cover development is much slower.

Sands of extremely arid deserts (annual precipitation less than 100 mm and evaporation 2800 mm) where there is insufficient precipitation to ensure the protection of sands by herbaceous vegetation. Vegetational stabilization is impossible in these unfavourable conditions (low rainfall and high air and soil temperatures). Vegetational stabilization over limited areas must resort to physico-chemical amelioration, with mechanical protection provided by surface stabilization practices.

There is a rich literature on the selection of appropriate plants. A world-wide survey of plant selections is given by Kaul (1970).

(ii) Natural recovery. Protection of an area (e.g. by enclosure) where relics of the original vegetation are still present, may result in the spontaneous redevelopment of the vegetation cover, although the speed and effectiveness of response will vary greatly with local conditions. Much depends on the stage of deterioration of vegetation and soil and also the respective sizes of degraded and undegraded areas adjacent to the recovery zone. In general, recovery is slower the more arid is the climate, the shallower the soil and the more degraded the vegetation. Recovery is generally restricted to areas of shallower groundwater conditions where the problems of deflation are minor.

(iii) Artificial recovery involves the transformation of the natural ecosystem by planting with species that may or may not belong to the native vegetation. Some of the measures used here are related to those designed to conserve soil from wind erosion, such as contour ploughing, cover cropping, strip cropping, and the use of fertilizers and mulches (e.g. Cooke and Doornkamp, 1974). For example, the use of adequate and appropriate fertilizers in preparing sandy areas for re-vegetation cannot be overemphasized. In sand the nutrient content is commonly low, so that high initial amounts of fertilizer are often necessary to maintain a high plant growth rate – nothing is gained by sowing a complex seed mixture in a soil with insufficient nutrients; at the same time, the high permeability, low moisture capacity and lower adsorptive power of sand makes the washing out of fertilizers a serious potential problem.

Mulches control erosion by sheltering the erodible surface (Chepil *et al.*,

1963b; see Cooke and Doornkamp, 1974, p. 58). Here, the effectiveness of plant residues depends on their resistance to decay, and their density and orientation – the more erect, and the finer and denser the residue, the greater the control of erosion. Different types of cutting instrument maintain different amounts of residue. On loose sand, little natural anchorage of mulch occurs, so that artificial stabilizing is required. For this purpose, both rapid-curing cutback asphalt and rapid-setting asphalt emulsion, among other materials, are effective anchoring agents.

3. Surface stabilization

In some circumstances, especially in extremely arid areas, vegetational stabilization of moving sands is impossible, so that sand fixation must depend on special surface coverings or man-made obstacles.

(i) Water is only a good surface stabilizer if the surface is kept wet – if the surface dries out, as it quickly does in areas of high evaporation, the protection is lost. Permanent wetting requires frequent spraying, but minerals and sediment precipitated from irrigation water might form a protective cement in time. The method is expensive, but often an excellent temporary expedient.

(ii) Gravel, stones, and crushed rock greater than 2 mm in diameter are stable under aeolian conditions and provide an excellent means of stabilizing a sand surface (e.g. Chepil et al., 1963c). Even dune sand (where no traffic is involved) can be stabilized with fine, medium, or coarse gravel spread uniformly over the surface. It is often a problem, however, actually to spread material mechanically over unstable sand surfaces. Cost of transport of these bulky materials is an important factor in this method.

(iii) Oil, despite its ugly appearance, has been used successfully to stabilize large areas at low cost. Where vegetation is ultimately to stabilize the surface, the oil must not restrict plant growth by toxic effects or prevent water penetration. Data on the environmental effects of oil stabilization are virtually non-existent. Three types of oil are commonly used: low-gravity asphaltic oil (as used in road construction); high gravity deep-penetrating waxy oil; and crude oil (Kerr and Nigra, 1952; FAO/DANIDA, 1974).

(iv) Chemical sprays are widely used to stabilize surfaces. They usually require special equipment and trained personnel; amounts depend on soil structure, slope, spraying technique, and degree of stabilization required; sometimes seeds and fertilizers may be applied at the same time. In principle, the spray soaks a few millimetres into the surface where the water base evaporates and the particles remaining bind sand grains together. Germination and growth of plants are supported by water penetrating through the pores of the stabilized layer. Evaporative losses are often reduced by the stabilized layer. But germination may be retarded, especially where chemicals are highly concentrated. Local destruction of the chemical crust has to be prevented or the crust repaired by respraying, because rapid deflation may begin which can undermine and rapidly break up the crust. Many chemical stabilizers are only temporary, disintegrating after a year or so, and they are expensive, so they are normally only used in conjunction with other methods, especially vegetation stabilization. The following list of chemical stabilization products is not comprehensive, but includes the main products:

Wood cellulose fibres go into a suspension with water fertilizer, grass seed, and an anchoring agent (such as asphalt or resin emulsion) to produce a sprayable slurry. *Resin-in-water emulsion* forms a highly stable film on sandy soils, but it disintegrates rapidly on fine-textured soils; seeds in the emulsion have their germination restricted, so that seeds should be broadcast before spraying (Rostler and Vallerga, 1960; Rostler and Kunkel, 1964). *Latex-in-water emulsions* produce a rubbery surface film that does not penetrate the surface easily; in order to control wind erosion effectively the cover must be continuous which, if achieved, can inhibit water penetration; to avoid breaking the emulsion it should be diluted with neutral water in the ratio of at least 1 : 1. *Gelatinized starch solution* is relatively unstable, decomposing quickly; but it is fairly cheap.

(v) Effectiveness and cost of materials. Costs and effectiveness of stabilizers are not easily evaluated because so much depends on local geomorphological and economic circumstances. For example, on level terrain it may be possible to use standard agricultural equipment and boom sprayers, whereas in dune country hand spraying may be necessary or four-wheel-drive equipment required. For further evaluation see Chepil *et al.* (1960, 1963c), Lyle *et al.* (1969) and Armbrust and Dickerson (1971). The Arizona Department of Transportation (1974, 1975) provided a useful experimental review of stabilizers ('dustproofers') of potential value in controlling dust and erosion along dryland highways.

4. Fences

(i) Theoretical and laboratory modelling. The effectiveness of any fence depends on the relations between free wind velocities, sheltered area velocities, and fluid threshold velocities. The fully protected zone of any fence is reduced as wind velocity increases. The most effective fences are semi-permeable for, although velocity reductions are smaller than for impermeable fences, they restrict diffusion and eddying effects and their influence extends further downwind.

The aerodynamic action of a windbreak is simple in principle. The windbreak exerts a drag force on the wind field, causing a net loss of momentum, and thus providing a shelter effect (Hagen, 1976). 'Bleed flow' through the windbreak is reduced and windbreak drag increases as porosity of the windbreak is reduced (e.g. Jensen, 1954; Baltaxe, 1967). Below a certain porosity, a region of large-scale flow separation occurs in the lee of the windbreak – the lower the porosity, the stronger the turbulent eddying in the separation zone (Castro, 1971; Raine, 1974; Mulhearn and Bradley, 1977). It is therefore necessary to distinguish between mean wind reduction and wind protection: while a less porous windbreak gives greater reduction in mean windspeed, the greater turbulence in its wake may make it less effective overall for wind protection.

(ii) Vegetational windbreaks and shelterbelts. Where irrigation water is available or groundwater is reasonably close to the surface, windbreaks of living plants can be used to control erosion. Studies of windbreaks for erosion control are widely dispersed in the literature (Caborn, 1957; Read, 1961; Stoeckeler, 1962; van Eimern *et al.,* 1964; Baltaxe, 1967; FAO, 1969; Bhimaya, 1976; Tinus, 1976). Windbreaks aid wind-erosion control in two main ways – they decrease surface wind shear stress; and they trap moving sand and dust. Optimizing control depends on manipulation of design variables such as porosity, porosity distribution, height, width, shape, and resilience. Orientation is also important – the preponderance of wind forces (Skidmore, 1965; Skidmore and Woodruff, 1968) can be used to ascertain the most efficient orientation (at right angles to the predominant wind-erosion force).

Windbreaks can be used to improve significantly visibility and air quality. As Honda's (1974) studies showed, plants can be efficient dust traps, with efficiency being inversely proportional to porosity and the nature of the relationship depending on leaf and dust characteristics. As low porosity is advantageous for trapping sand, but consequent regions of high turbulence in the lee of the barrier accentuate dust blowing, clearly a compromise is necessary in selecting the optimum porosity.

Windbreaks do have limitations; they afford competition to adjacent crops for moisture and nutrients (e.g. Woodruff *et al.,* 1959); they are expensive, shelter limited areas, take time to mature, require water, and are 'fixed'.

(iii) Diversion fences may be constructed from diverse materials, such as wood panels, metal plates, stone walls, or earthworks, and must clearly be of sufficient strength to withstand the wind. The fences are normally set up in either a single slant, or a 'V' pattern at an acute angle to wind direction. The more acute the angle, the longer the lifespan, but the less the area protected. Fence height can be kept to a minimum since most sand is transported at no more than 30 cm above the surface. Surfaces around the fences should be stabilized to avoid erosion, otherwise there is a possibility of fence collapse by undermining. Diversion fences are relatively expensive, usually temporary, and often require clearing of lee accumulations.

(iv) Impounding fences. Figure VII.10 illustrates some types of impounding fence. Good, relatively inexpensive fences may be constructed from local vegetation, such as palm fronds. Indeed, the use of so-called Lubian squares – a grid of approximately 2 m-spaced brushwood lines covering a large area to windward of a feature to be protected, has been particularly successful (Fookes, pers. comm.). The effect of such fences is to reduce the transport capacity of the wind both in front of *and* behind the barrier (Fig. VII.11A). The volume of sand accumulated around a fence is proportional to fence height (Kerr and Nigra, 1952). The first fence erected is likely to fill rapidly, but the next increment of fence would increase the effective life of the fence by about four times; the third increment would increase effective life by nine times (Table VII.6). The initial height, and the stages at which height must be increased depend directly on the rate of sand transport – a factor of value in both predicting fence height and longevity, and one that can be determined

A

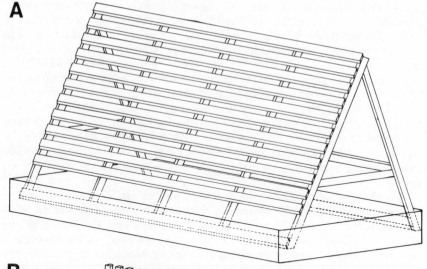

B

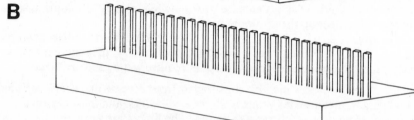

C

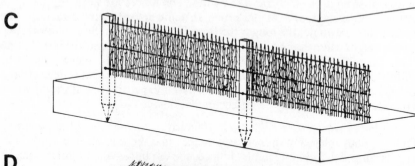

D

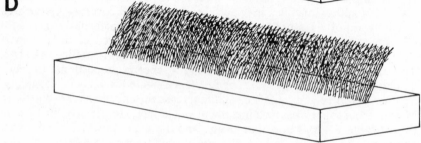

fairly easily. The protective effect of a fence ends when it is buried and the sand accumulation has a streamlined shape, so that wind crosses it without loss of transport capacity.

Fences have to be installed at some distance from the object to be protected and at right angles to the prevailing wind direction. A three-fence system used in conjunction with surface stabilization between the last fence and the protected object has been shown to be reasonably effective and long-lasting at low costs (Fig. VII.11B). Sand accumulates mainly and first around fence 1; as the effectiveness of the first fence declines, sand accumulates increasingly at fence 2; fence 1 then needs to be raised. Accumulation at fence 3 calls for raising at both fences 1 and 2. Height of fence 1 can be increased until its leeward accumulation covers fence 2; then another fence could be constructed to windward.

Savage (1963), using slat fencing and brush fencing on coastal dunes in a temperate climate, found that the effective height of the fence was increased by raising the base 0.3 m off the ground, for the fencing filled more slowly than fencing in contact with the ground. He proposed two methods of dune building, and the results and methods are described fully in Savage and Woodhouse (1968). Manohar and Bruun (1970) observed a similar sequence of dune development around fences.

(v) Architectural control. New housing and settlement should be designed to reduce or prevent the problems arising from sand and dust movement by, for instance, paying attention to layout and orientation of houses (Duchemin, 1958). Walls, doors, and windows should preferably face directions that will minimize sand and dust penetration. Air-filtration systems may be desirable; outer storm doors, window shutters, and weather stripping can be fitted to buildings.

Building configuration can relieve the problems of sand and dust movement. The bar effect (Fig. VII.7A) can be minimized by ensuring that buildings are sufficiently separated so that areas are self-cleaning without unnecessary discomfort. Uncomfortable zones associated with tunnel phenomena can be avoided by reducing the heights of buildings and infilling the open space between the converging arms, encouraging the wind to pass over the convergence rather than through it (Fig. VII.7B). Pressure connection effects (Fig. VII.7C) and associated transverse currents are prevented by ensuring that rows of buildings are not built perpendicular to the wind, or if they are built that way, they are separated by distances at least equal to the height of the buildings. Gap effects (Fig. VII.7E) can be minimized by restricted building heights. Channelling effects (Fig. VII.7F)

FIG. VII.10 Types of sand impounding fences: (A) The A-style snow fence constructed in movable sections; (B) Picket-type fence. Each picket is movable. Easy establishment and maintenance; (C) Brush barrier constructed by weaving brush into set wires. This fence is immovable and must be reconstructed each time it is necessary to raise the dune; (D) Brush barrier, probably the cheapest fence to construct where brush is available. It must be rebuilt each time the dune is raised (after McLaughlin and Brown, 1942)

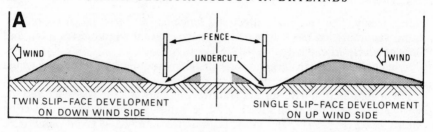

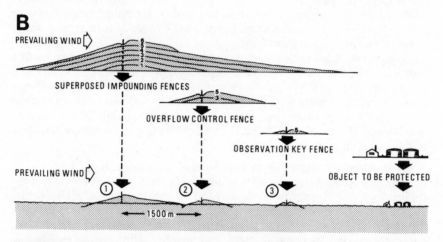

FIG. VII.11 (A) Impounding effect of porous fence; (B) Three fence system to protect extensive areas, such as villages, shops, yards, industrial plants etc. (after Kerr and Nigra, 1952)

TABLE VII.6

Effective life schedule of an impounding fence north of Dhahran Stabilizer

Fence increment	Total height (M)	Observed effective life of barrier (Years)	Estimated effective life of barrier (Years)
1	1.5	$\frac{1}{2}$	
2	3.4	2	
3	4.9		$4\frac{1}{2}$
4	6.8		8
5	7.6		$12\frac{1}{2}$
6	9.1		18
7	10.6		$24\frac{1}{2}$
8	12.1		32
9	13.7		$40\frac{1}{2}$
10	15.2		50
20	30.4		200

Source: Kerr and Nigra (1952).

can be reduced by careful building orientation. Air pockets produced by mesh configuration (Fig. VII.7D) can be used to great advantage in the designs of buildings. The relative dimensions of the mesh will determine whether the wind passes over the mesh or blows into it. Gandemer (1977) proposed that the nondimensional parameter

$$S/h_{m'}^{2}$$

where S = the surface area of the mesh,
and h_m = the mean height,

governs whether the flow passes over or into the mesh. A closed mesh with leeward opening and

$$S/h^2_m \leqslant 20,$$

with an overall pyramidal or hemispherical shape would minimize the problems of sand and dust blowing and deposition. The wind would pass over the shape and around its sides. Many of the control measures already outlined have to be used in conjunction with the manipulation of building configuration because bluff bodies in any configuration will cause zones of localized shelter, with associated depositional or wind-accelerational problems, where abrasive and comfort problems will be important.

(vi) Dune stabilization and destruction. Flattening dunes is not a solution to sand and dune movement problems. The flat sand surface is unstable, and sand dunes soon begin to redevelop. Transposing dunes is costly and it, too, usually provides only temporary amelioration. Destruction of dunes by trenching – that is by cutting longitudinal and transverse trenches with a bulldozer across a dune to destroy its symmetry – is also expensive and temporary. Dunes located a considerable distance from a threatened object (at least 20 times dune height) can best be stabilized by fences and/or surface treatment.

Fences set up to windward of a dune cause local deposition. Once the shelter effect of the fences has diminished sufficiently, the wind will erode more sand from the windward side of the dune than it would have done if the fence had not caused deposition on its leeward side: as a result, the dune will migrate faster, and will diminish in size. Complete dune dissipation is possible in this way, if the area between the fence and the dune is 'paved' to prevent sand from being eroded.

When the windward side of a dune is stabilized with oil, chemicals, etc. the dune will not migrate, but sand arriving at the foot of the windward slope will be carried to leeward by the higher velocities over the stable surface. The crescent-shaped outline of a barchan, for instance, will turn into a streamlined oval as sand is deposited on the leeward side, so that the stabilized dune will occupy several times the area of the original dune (Fig. VII.12) and there will still be through transport of sand over the dune. Because of the considerable leeward growth of the dune this material can only be applied when the original migratory dune is at a distance of about three times the width of the dune from the object that is protected. When the migrating dune is too close

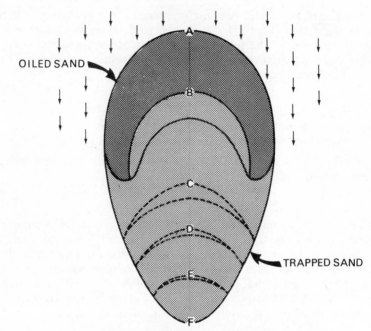

PLAN VIEW OF PROGRESSIVE GROWTH STAGES

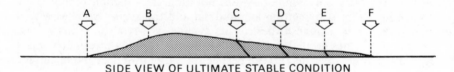

SIDE VIEW OF ULTIMATE STABLE CONDITION

FIG. VII.12 Dune stabilization by oiling (after Kerr and Nigra, 1952)

to the object to be protected, the windward slope is stabilized in the same way and sand-impounding fences are set up to windward. The fences will reduce sand supply to the foot of the dune and thus prevent leeside deposition.

Dune stabilization by vegetation cover is expensive and requires good management. Vegetation may be used to transform transverse or parabolic dunes into smaller and smaller parabolic dunes by blowouts (Fig. VII.13). For vegetation cover to be successful, the supply of sand needs to be cut off to prevent burial and sand-blasting effects; and the planting surface needs to be stabilized.

(vii) Conclusions. Sand and dust control in drylands differs fundamentally from that in humid regions because the areas involved are often more extensive, the problems more severe, the economics more precarious, and the scope for manipulation, especially through vegetation, is more limited. In general, several separate solutions, used in harmony, are likely to succeed

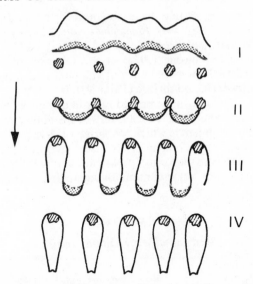

FIG. VII.13 Vegetational transformation of transverse dunes to parabolic dunes: (I) moment of approach of transverse dunes to shrubs; (II) vegetation stabilizing part of the transverse dunes, making possible the movement of their free parts through the gaps between shrubs and the formation of parabolic dunes; (III) subsequent stage in the transformation of parabolic dunes and their elongation; (IV) stage of formation of dunes parallel to the wind. Dots – slip-faces; dashed lines – lines of dune crests (Petrov, 1976 after de Martonne)

better than a single panacea. Effective vegetation stabilization usually requires the combination of mechanical, chemical, and biological methods at least in the initial phase of vegetation growth. Once vegetation is established, other defences may be unnecessary. Successful control of sand and dust depends largely on adequate appraisal of the problem before development and then sound management practices including maintenance and surveillance.

(viii) Case study: dust abatement at Agnew, Western Australia. Proposed nickel mining operations by Australian Selection (Pty.) Ltd. will cause major wind erosion problems at Agnew and the new town which is to be built on a sand plain covered with 10-m high north-west/south-east trending transverse dunes.

The predominant effective winds in the area blow from the north-west and the south-west quadrant. The most critical period for surface instability is from October to February when strong westerly winds are most likely and the rainfall is least effective owing to the high evaporation. Seasonal distribution and total rainfall are extremely variable. The long-term annual average is 200 mm, but the value for any one year can range from twice that amount to virtually zero. Open shrub woodland of varying density is characteristic of the area. Trees are up to 10 m high with an understorey of *spinifex*. A number of shrub and tree species may have potential for replanting programmes: *Acacia aneura, A. kempeana, Eremophila* spp., *Eucalyptus striatocalyx, E. gongylocarpa,* and *E. oldfieldii.*

Marshall and Churchwood (1974) of the Division of Land Resources Management, CSIRO, prepared a list of guidelines on actions necessary to maintain stability of the soil surface (Table VII.7). The preventive measures, they suggested, should be implemented prior to any developmental work.

Primarily the benefits from reducing surface instability are reflected in the more pleasant environment in which people have to live and work. Additional benefits include the improvement of the visual environment; the provision, at low cost, of amenity in the form of parklands, nature trails, and recreational areas; and the scientific value of maintaining undisturbed habitats.

TABLE VII.7

Recommended preventative measures

Regional Stability

1. Establish shelterbelts and clumps of natural vegetation

detailed mapping of vegetation to identify areas and individual trees and shrubs to be retained

roads, buildings, etc. planned to avoid retained areas

2. Retain natural vegetation in a perimeter buffer zone and sanctuary area 1 km wide

north–south orientation (90° to prevailing wind) for maximum efficiency

retained areas fenced prior to clearing and construction work

vehicles and stock prohibited in fenced buffer zone

establish nursery of mainly native plants

Local Stability

1. Surface stabilization

water stabilization; mulches of wood chips, crushed rock; experiments to evaluate petro-chemical stabilizers; plant introduced species on dunes

2. Fences

horizontally lathed fences, 30–50% porosity, round houses, and solid obstacles; shelterbelts, trees, and hedges planted on dunes, town perimeter, etc.

3. Architectural design

houses not regularly aligned but staggered to provide effective reduction in wind speed; concentric or curved roads (no long straight sections running east–west) with sealed surfaces; development of gardens and planting of street trees

Landscape Management

1. No development on unstable landscape units

dune crests and flanks fenced and ground cover improved by planting, irrigation, and fertilization; access and use restricted to gravel paths

2. Constant surveillance

control fencing, shelterbelts, buffer zones, etc.

supervise clearing, construction, and planting

control public access

detect and treat unstable or eroded areas in incipient stage

3. Monitor mining activities and their effects

tailings dumps should be north or south of the town;

unsatisfactory substrate for plant growth should be stabilized with rock or chemical stabilizers.

excessive water usage should be monitored to prevent desiccation and salinization

Implementation

1. Penalty clauses for damage to vegetation

contracts for clearing and construction

local bye-laws

2. Co-ordination of clearing and building

minimize bare unprotected surfaces

Source: Marshall and Churchwood (1974).

(f) Summary

This chapter has provided a brief review of the problems caused by aeolian processes to settlements in drylands, the physical causes that lie behind them, and the range of management practices that can help to solve them. A number of suggestions have been made that have proved successful in studying and controlling sand and dust movements in drylands. But there are dangers in seeking universal solutions: the first recommendation is to seek solutions only in the context of local circumstances; the second is to assess those circumstances in the light of the well-established principles of wind and sediment movement.

If sand and dust movement is an actual or potential problem in an area of urban development, considerable economies can be achieved if the hazard is adequately surveyed, and the survey results incorporated in the urban design. Areas of substantial active sand dunes should be avoided. Where this is not possible, objects or areas liable to suffer as a result of sand and dust movement are often best concentrated in a single area which can then be comprehensively protected. For example, communication lines of various sorts can all follow the same route, with the route being adequately protected.

The justification for an aeolian control project should be clearly related to scientifically based understanding of the project and its effects on adjacent areas. Why is the area eroding? What problems arise from the erosion and what are the alternative solutions? How will the project affect the natural ecosystem (the sand budget especially)? To answer these and related fundamental questions, geomorphological mapping is an extremely valuable preliminary, by identifying aeolian landform types and their dynamics, and helping to initiate a spatial classification of the hazard. It should be followed by an effective well-conceived monitoring scheme to quantify the dynamic elements of the system.

In evaluating alternative solutions in the light of the preliminary scientific evidence, cost-benefit analysis is desirable, but it is extremely difficult because of so many intangible benefits (e.g. beauty) and so many hidden costs. In addition, projects should be monitored before, during, and after implementation, so that data on the aeolian system and its response are to hand to inform management practices and similar projects in adjacent areas. Finally, in order to protect the protection measures and to ensure that their full benefits are realized, land-use controls (such as sand-dune protection ordinances, and access restrictions) are normally essential.

REFERENCES

ABU-LUGHOD, J. L., 1961, Migrant ajustment to city life: the Egyptian case, *Am. J. Sociol.*, **57**, 22–32.

ABU-LUGHOD, J. L., 1965, Urbanization in Egypt: present state and future prospects, *Economic Development and Cultural Change*, **13**, 313–43.

AITCHISON, G. D., and K. GRANT, 1967, The PUCE programme of terrain description, evaluation and interpretation for engineering purposes, *Proc. 4th Reg. Conf. for Africa on Soil Mech. and Foundn. Engng.* (Cape Town), **1**, 1–8.

AKILI, W., and E. H. FLETCHER, 1978, Ground conditions for housing foundations in the Dahran region, Eastern Province, Saudi Arabia, *Proc. Int. Ass. Housing Sci. Conf.*, **2**, 532–46.

ALAM, F. C. K., 1972, Distribution of precipitation in mountainous areas of West Pakistan, *Geilo Symposium on the Distribution of Precipitation in Mountainous Areas* (Norway), 31 July–5 Aug.

ALLAN, J. A., 1978, Better resolution from LANDSAT, *Nature*, **273**, 189–90.

ALLISON, T. R., 1977, The availability and evaluation of building materials in the Gulf, in *Construction Problems and Finance for Middle East Developments* (Middle East Construction, London), 134–42.

ALLUM, J. A. E., 1966, *Photogeology and Regional Mapping* (Pergamon Press, Oxford).

ALTER, J. C., 1930, Mud floods in Utah, *Mon. Weath. Rev.* **58**, 319–21.

AMERICAN SOCIETY OF PHOTOGRAMMETRY, 1960, *Manual of Photographic Interpretation* (Washington, DC).

AMERICAN SOCIETY FOR TESTING MATERIALS (ASTM), 1971, Standard method of test of soundness of aggregates by use of sodium sulfate or magnesium sulfate, Standard C88.

AMERMAN, C. R., and J. L. McGUINESS, 1967, Plot and small watershed runoff: its relation to larger areas, *Trans. Am. Soc. Agric. Engrs.*, **10**, 464–6.

ANDERSON, H. W., and J. R. WALLIS, 1965, Some interpretations of sediment sources and causes, Pacific coast basins in Oregon and California, *US Dept Agric. Misc. Publ.*, **970**, 22–30.

ANDREWS, F. M., 1962, Some aspects of the hydrology of the Thames Basin, *Proc. Instn. Civ. Engrs.*, **21**, 55–90.

ANON, 1972, The preparation of maps and plans in terms of engineering geology, *Q. J. Engng. Geol.*, **5** (4), 295–382.

ANSTEY, R. L., 1965, Physical characteristics of alluvial fans, *US Army Laboratories Tech. Rep.*, **ES–20** (Natick, Mass.)

ANTEVS, E., 1952, Arroyo-cutting and filling, *J. Geol.*, **60**, 375–85.

ARIZONA BUREAU OF GEOLOGY AND MINERAL TECHNOLOGY, 1978, *Geological Investigation Series* **1**, A–J (Studies of the McDowell Mountains area, Maricopa Co., Ariz.).

ARIZONA DEPARTMENT OF TRANSPORTATION, 1974, 1975, *Soil Erosion and Dust Control on Arizona Highways* (Arizona Dept. Transportation, Phoenix), 4 vols.

ARMBRUST, D. V., and J. D. DICKERSON, 1971, Temporary wind erosion control: cost and effectiveness of 34 commercial materials, *J. Soil Wat. Conserv.*, **26**, 154–7.

ASSOCIATION OF ENGINEERING GEOLOGISTS, 1965, *Geology and Urban Development* (AEG, Los Angeles Section).

ASTM–see American Society for Testing Materials.

BAGNOLD, R. A., 1941, *The Physics of Blown Sand and Desert Dunes* (Methuen, London).

BAGNOLD, R. A., 1953, Forme des dunes de sable et régime des vents, in *Actions Eoliennes, Centre National de Recherche Sci., Paris, Coll. Int.*, **35**, 23–32.

BAHRAIN, MINISTRY OF WORKS, POWER AND WATER, 1976, *Bahrain Surface Materials Resources Survey* (eds. D. Brunsden, J. C. Doornkamp and D. K. C. Jones), Consultancy Rep., Feb. 1976, vols. 1–6.

BAILEY, R. W., C. L. FORSLING, and R. J. BECRAFT, 1934, Floods and accelerated erosion in northern Utah, *US Dept. Agric. Misc. Publ.*, **196**.

BALTAXE, R., 1967, Air flow patterns in the lee of model windbreaks, *Arch. Mete. Geophys. Bioklim. Serie B: Allgemeine und Biologische Klimatologie*, **15**, 287–312.

BARDINET, C., M. ALBUISSON, and J. M. MONGET, 1978, Télédétection de l'évolution saisonnière des paysages de la zone de N'Djamena-Kousseri, in *Atlas du Tchad* (Inst. de Géographie, Paris).

BARNES, H. H., 1975, Inventory of flash-floods – methods for computing; protection, forecasting networks; economical consequences, disasters, *Hydrol. Sci. Bull.* **20**, 35–50.

BAUDET, J., P. FUMET, P. MASSON and R. JONEAUX, 1959, Low-cost roads, *Proc. 11th Congr. Permanent Int. Ass. Roads Congresses*, Rio de Janeiro, Section 1, Question IV, Paper 62.

BAWDEN, M. G., 1967, Applications of aerial photography in land-systems mapping, *Photogramm. Record.* **5**, 461–4.

BEATY, C. B., 1963, Origin of alluvial fans, White Mountains, California and Nevada, *Ann. Ass. Am. Geogr.*, **53**, 516–35.

BEAUMONT, P., 1968, Salt weathering on the margin of the Great Kavir, Iran, *Geol. Soc. Am. Bull.*, **79**, 1683–4.

BEAUMONT, P., 1972, Alluvial fans along the foothills of the Elburz Mountains, Iran, *Palaeogeography, Palaeoclimatology, Palaeoecology*, **12**, 251–73.

BECKETT, P. H. T., and R. WEBSTER, 1971, The development of a system for terrain evaluation over large areas, *R. Engrs., J.* **85**, 243–58.

BECKETT, P. H. T., R. WEBSTER, G. M. MCNEIL and C. W. MITCHELL, 1972, Terrain evaluation by means of a data bank, *Geog. J.*, **138**, 430–56.

BELLY, P. Y., 1964, Sand movement by wind, *US Army Corps of Engineers, Coastal Engineering Research Center, Techni. Memo.*, **1** (Washington, DC).

BENNETT, H. H., 1939, *Soil Conservation* (McGraw-Hill, New York).

BERGER, M. (ed.), 1974, *The New Metropolis in the Arab World* (Octagon Books, New York).

BERRY, B. J. L., 1973, *The Human Consequences of Urbanization: Divergent Paths in the Urban Experience of the Twentieth Century* (Macmillan, London).

BERRY, B. J. L., and J. D. KASANDA, 1977, The social consequences of Third-World urbanization: the culture of poverty revisited, in B. J. L. Berry and J. D. Kasanda, *Contemporary Urban Ecology* (Macmillan, New York, and London), Ch. 18.

BETSON, R. P., and J. B. MARIUS, 1969, Source areas of storm runoff, *Wat. Resour.*, **5** (3), 574–82.

BETZ, F. Jr., 1975, *Environmental Geology* (Dowden, Hutchinson & Ross, Penn.).

BHIMAYA, C. P., 1976, Shelterbelts – functions and uses, in *Conservation in Arid and Semi-arid Zones*, FAO Conservation Guide, **3**, (FAO. Rome), 17–28.

BIRKS, J. S., and C. A. SINCLAIR, 1980, Well-lubricated Emirate economy, *Geographical Magazine*, **L111**, 470–80.

BIROT, P., 1968, *The Cycle of Erosion in Different Climates* (Batsford, London).

BLACKWELDER, E., 1928, Mudflow as a geologic agent in semi-arid mountains, *Geol. Soc. Am. Bull.*, **39**, 465–80.

BLACKWELDER, E., 1940, Crystallization of salt as a factor in rock weathering, *Geol. Soc. Am. Bull.,* **51**, 1956.

BLIGHT, G. E., 1976, Migration of subgrade salts damages thin pavements, *Transp. Eng. J., Proc. Am. Soc. Civ. Engrs.,* **102**, No. TE4, 779–91.

BLIGHT, G. E., J. A. STEWART, and P. F. THERON, 1974, Effect of soluble salt on performance of asphalt, *Proc. 2nd Conf. on Asphalt Pavement in S. Africa,* Durban, sect. 3, 1–13.

BLYTH, F. G. H., and M. H. DeFREITAS, 1974, *A Geology for Engineers* (Arnold, London).

BORCHERT, J. R., 1967, American metropolitan evolution, *Geog. Rev.,* **57**, 301–23.

BOWDEN, L. W., J. R. HUNING, C. F. HUTCHINSON and C. W. JOHNSON, 1974, Satellite photography presents first comprehensive view of local wind: the Santa Ana, *Science,* **184**, 1077–78.

BOYSON, S. M., 1974, Predicting sediment yield in urban areas, in D.T.Y. Kao (ed.), *National Symposium on Urban Rainfall and Sediment Control* (University of Kentucky, Lexington), 199–203.

BRADSHAW, P., 1975, *An Introduction to Turbulence and its Measurement* (Pergamon Press, Oxford).

BRIDGES, E. M., and J. C. DOORNKAMP, 1963, Morphological mapping and the study of soil patterns, *Geography,* **48**, 175–81.

BRINK, A. B. A., and A. A. B. WILLIAMS, 1964, Soil engineering mapping for roads in South Africa, *CSIR Pretoria Res. Rep.,* **227**.

BRINK, A.B.A., J. A. MABBUTT, R. WEBSTER and P. H. T. BECKETT, 1966, *Report of the Working Group on Land Classification and Data Storage* (MEXE, Christchurch, England), Rep. **940**.

BRINK, A. B. A., and T. C. PARTRIDGE, 1967, Kyalami land system: an example of physiographic classification for the storage of terrain data, *Proc. 4th Reg. Conf. for Africa on Soil Mech. and Foundn. Engng* (Cape Town), **1**, 9–14.

BRINK, A. B. A., T. C. PARTRIDGE, and G. B. MATHEWS, 1970, Airphoto interpretation in terrain evaluation, *Photo Interpretation,* **5**, 15–30.

BRINK, A. B. A., T. C. PARTRIDGE, R. WEBSTER and A. A. B. WILLIAMS, 1968, Land classification and data storage for the engineering use of natural materials, *Proc. 4th Conf. Austr. Road Res. Board,* 1624–47.

BRITISH STANDARDS INSTITUTION, 1951, *Single-sized roadstone and chippings,* BS **63**.

BRITISH STANDARDS INSTITUTION, 1967, *Gravel aggregates for surface treatment on roads,* BS **1984**.

BRITISH STANDARDS INSTITUTION, 1973, *Sampling and testing lightweight aggregates for concrete,* BS **3681**, part 2.

BRITISH STANDARDS INSTITUTION, 1972, *The structural use of concrete, Part 1. Design materials and workmanship, Code of Practice,* **110**.

BRITISH STANDARDS INSTITUTION, 1973, *Specification for foamed or expanded blast furnace slag; lightweight aggregate for concrete,* BS **877**, part 2.

BRITISH STANDARDS INSTITUTION, 1973, *Aggregates from natural sources for concrete (including granolithic),* BS **882**.

BRITISH STANDARDS INSTITUTION, 1975, *Methods of sampling and testing of mineral aggregates, sand and fillers,* BS **812**, parts 1–4.

BRITISH STANDARDS INSTITUTION, 1976, *Lightweight aggregates for concrete,* BS **3797**, part 2.

BROOKFIELD, M., 1970, Dune trend and wind regime in Central Australia, *Zeit. für Geomorph.,* Suppl. **10**, 121–58.

BRUNSDEN, D., J. C. DOORNKAMP, P. G. FOOKES, D. K. C. JONES and J. M. M. KELLY, 1975, Large-scale geomorphological mapping and highway engineering design, *Q. J. Engng. Geol.,* **8**, 227–53.

BRUNSDEN, D., J. C. DOORNKAMP and D. K. C. JONES, 1976, Report on a geomorphological interpretation of the available 1:15 000 monochromatic aerial photography, in *State of Dubai Aggregate Survey* (Wimpey Laboratories Ltd, Hayes, Middx).

BRUNSDEN, D., J. C. DOORNKAMP and D. K. C. JONES, 1979, The Bahrain Surface Materials Resources Survey and its application to planning, *Geogr. J.,* **145**, 1–35.

BRYAN, K., 1925, Date of channel trenching (arroyo cutting) in the arid Southwest, *Science,* **62**, 338–44.

BRYAN, R. B., 1968, The development, use and efficiency of indices of soil erodibility, *Geoderma,* **2**(1), 5–26.

BUCKLEY, H. E., 1951, *Crystal Growth* (Wiley, New York).

BULL, W. B., 1964, Alluvial fans and near-surface subsidence in Western Fresno County, California, *US Geol. Surv. Prof. Paper,* **437-A**.

BULL, W. B., 1968, Alluvial fans, *J. Geol. Educat.,* **16**, 101–6.

BULL, W. B., 1977, The alluvial-fan environment, *Prog. Phys. Geogr.,* **1**, 222–70.

BULL, W. B., and K. M. SCOTT, 1974, Impact of mining gravel from urban streambeds in the southwestern United States, *Geology,* **2**, 171–4.

BUTCHER, G. C., and J. B. THORNES, 1978, Spatial variability in runoff processes in an ephemeral channel, *Zeit. für Geomorph.,* Supp. **29**, 83–92.

BUTLER, B. E., 1960, Riverine deposition during arid phases, *Aust. J. Sci.,* **22**, 451–2.

BUTTERFIELD, G. R., 1973, The susceptibility of High Country soils to erosion by wind, *Proc. Calibration Control Seminar,* Mussey Univ., N.Z.

BUTZER, K. L., 1974, Accelerated soil erosion: a problem of man-land relationships, in I. R. Manners and M. W. Mikesell (eds.), *Perspectives on Environment* (Ass. Am. Geog., Washington, DC), 57–78.

CABORN, J. M., 1957, Shelterbelts and microclimate, *Bull. Forestry Commission (Lond.),* **29**.

CALIFORNIA DIVISION OF MINES AND GEOLOGY, 1973, *Urban Geology – master plan for California,* Bull. **198**.

CAMPBELL, I. A., 1970, Erosion rates in the Steveville Badlands, Alberta, *Can. Geogr.,* **14**, 206–16.

CARLSON, T. N. and J. M. PROSPERO, 1972, The large-scale movement of Saharan air outbreaks over the northern equatorial Atlantic, *J. Appl. Met.,* **11**, 283–97.

CARSON, M. A., 1971, *The Mechanics of Erosion* (Pion, London).

CARSON, M. A. and M. J. KIRKBY, 1972, *Hillslope Form and Process* (CUP, Cambridge).

CARTER, L. J., 1974, Off-road vehicles: a compromise plan for the California desert, *Science,* **183**, 396–9.

CASTRO, I. P., 1971, Wake characteristics of two dimensional perforated plates normal to an air stream, *J. Fluid Mech.,* **46**, 599–609.

CEMENT AND CONCRETE ASSOCIATION, 1970, Impurities in aggregates for concrete, *Advisory Note,* **18**.

CERMAK, J. E., 1958, Wind tunnel for the study of turbulence in the atmospheric surface layer, *Fluid Dynamics and Diffusion Lab.,* Rep. **CER 58** (Colorado State University, Fort Collins, Colo.).

CERMAK, J. E., 1971, laboratory simulation of the atmospheric boundary layer, *AIAA Journal,* **9**, 1746–54.

CERMAK, J. E., V. A. SANDBORN, E. J. PLATE and G. H. BINDER, 1966, Simulation of atmospheric motion by wind tunnel flows, *Fluid Dynamics and Diffusion Lab.* Tech. Rep. **CER 66**, (Colorado State University, Fort Collins, Colo.).

CHANDLER, R. J., 1977, The application of soil mechanics methods to the study of slopes, in J. R. Hails (ed.), *Applied Geomorphology,* (Elsevier, Amsterdam), 157–82.

CHAPMAN, T. G., 1964, Design and initial instrumentation of a rainfall-runoff experiment in the Alice Springs area, *CSIRO Div. Land Res. Tech. Memo.*, **64/5**.

CHATFIELD, E. J., 1967, A battery operated sequential air concentration and deposition sampler, *Atmospheric Environment*, **1**, 509–13.

CHEN, C-S., 1973, Population growth and urbanization in China, 1953–1970, *Geogr. Rev.*, **63**, 55–72.

CHEPIL, W. S., 1946, Dynamics of wind erosion: IV. The translocating and abrasive action of the wind, *Soil Sci.*, **61**, 167–77.

CHEPIL, W. S., 1950, Properties of soil which influence wind erosion: the governing principle of surface roughness, *Soil Sci.*, **69**, 149–62.

CHEPIL, W. S., 1957a, Sedimentary characteristics of duststorms: I. sorting of wind-eroded soil material, *Am. J. Sci.*, **255**, 12–22.

CHEPIL, W. S., 1957b, Sedimentary characteristics of duststorms: III. Composition of suspended dust, *Am. J. Sci.*, **255**, 206–13.

CHEPIL, W. S., F. H. SIDDOWAY and D. V. ARMBRUST, 1962, Climatic factor for estimating wind erodibility of farm fields, *J. Soil Wat. Conserv.*, **17**, 162–5.

CHEPIL, W. S., F. H. SIDDOWAY and D. V. ARMBRUST, 1963a, Climatic index of wind erosion conditions in the Great Plains, *Proc. Soil Sci. Soc. Am.*, **27**, 449–52.

CHEPIL, W. S., and N. P. WOODRUFF, 1957, Sedimentary characteristics of duststorms: II. Visibility and dust concentration, *Am. J. Sci.*, **255**, 104–14.

CHEPIL, W. S., and N. P. WOODRUFF, 1963, The physics of wind erosion and its control, *Adv. Agron.*, **15**, 211–302.

CHEPIL, W. S., N. P. WOODRUFF, F. H. SIDDOWAY and D. V. ARMBRUST, 1963b, Mulches for wind and water erosion control, *US Dept. of Agric.*, *ARS*, **41–84**.

CHEPIL, W. S., N. P. WOODRUFF, F. H. SIDDOWAY, D. W. FRYREAR and D. V. ARMBRUST, 1963c, Vegetative and nonvegetative materials to control wind and water erosion, *Proc. Soil Sci. Soc. Am.*, **27**, 86–9.

CHEPIL, W. S., N. P. WOODRUFF, F. M. SIDDOWAY and L. LYLES, 1960, Anchoring vegetative mulches, *Agric. Engng.*, **41**, 754–5, 759.

CHOW, V. T., 1962, Hydrologic determination of waterway areas for designs of drainage structures in small basins, *Engng. Stn. Bull.*, **462** (University of Illinois, Urbana).

CHOW, V. T. (ed.), 1964, *Handbook of Applied Hydrology* (McGraw-Hill, New York).

CHRISTIAN, C. S., 1957, The concept of land units and land systems, *Proc. 9th Pacific Sci. Congr.*, **20**, 74–81.

CHRISTIAN, C. S., J. N. JENNINGS and C. R. TWIDALE, 1957, Geomorphology, in B. T. Dickson (ed.), *Guide Book to Research Data for Arid Zone Development* (UNESCO, Paris), 51–65.

CHRISTIAN, C. S., and G. A. STEWART, 1968, Methodology of integrated surveys, *Aerial Surveys and Integrated Studies, Proc. Toulouse Conf.* (UNESCO, Paris), 233–80.

CHURCHWOOD, H. M., 1963a, Soil studies at Swan Hill, Victoria, Australia. II. Dune moulding and parna formation, *Aust. J. Soil Res.*, **1**, 103–16.

CHURCHWOOD, H. M., 1963b, Soil studies at Swan Hill, Victoria, Australia. IV. Groundsurface history and its expression in the array of soils, *Aust. J. Soil Res.*, **1**, 242–55.

CIRUGEDA, J., 1973, Informe relativa las crecidas de Octubre de 1973 en el dureste: estudio de caudales, *Centro de Estudios Hidrograficos*, Madrid.

CLEMENTS, T., R. O. STONE, J. F. MANN and J. L. EYMANN, 1963, A study of windborne sand and dust in desert areas, *US Army Laboratories, Earth Sciences Division, Tech, Rep.*, **ES-8** (Natick, Mass.).

CLEVELAND, G. B., 1971, *Regional Landslide Prediction, California* (Resources Agency, Sacramento).

CLOUDSLEY-THOMPSON, J. L., and M. J. CHADWICK, 1964, *Life in Deserts* (Foulis, London).

COATES, D. R. (ed.), 1971, *Environmental Geomorphology* (Publications in Geomorphology, State University of New York, Binghamton).

COATES, D. R. (ed.), 1976a, *Urban Geomorphology* (Geological Society of America, Colo.), Special Paper, **174**.

COATES, D. R. (ed.), 1976b, *Geomorphology and Engineering* (Dowden, Hutchinson & Ross, Stroudsburg).

COLE, D. C. H., and J. G. LEWIS, 1960, Progress report on the effect of soluble salts on stability of compacted soils, *Proc. 3rd. Aust. – N.Z. Conf. Soil Mech. Foundn. Engng.*, Sydney 22–26 August, 29–31.

COLE, W. F., 1959, Some aspects of the weathering of terracotta roofing tiles, *Aust. J. App. Sci.*, **10**, 346–63.

COLORADO GEOLOGICAL SURVEY, 1969, The governor's conference on environmental geology, *Colorado Geological Survey*, Special Publication, **1**.

COMTE-BELLOT, G., and S. CORRSIN, 1966, The use of a contraction to improve the isotropy of grid-generated turbulence, *J. Fluid Mech.*, **25**, 657–82.

CONDON, R. W., and M. E. STANNARD, 1954, Erosion in Western New South Wales, *J. Soil Conserv. Serv. N.S.W.*, **10**, 17–26.

COOKE, R. U., 1974, The rainfall context of arroyo initiation in southern Arizona, *Zeit. für Geomorph.*, Suppl., **21**, 63–75.

COOKE, R. U., 1977, Applied geomorphological studies in deserts: a review of examples, in J. R. Hails (ed.) *Applied Geomorphology* (Elsevier, Amsterdam), 183–225.

COOKE, R. U., 1979, Laboratory simulation of salt weathering processes, *Earth Surface Processes*, **4**, 347–59.

COOKE, R. U., and J. C. DOORNKAMP, 1974, *Geomorphology in Environmental Management* (OUP, Oxford).

COOKE, R. U., A. S. GOUDIE and J. C. DOORNKAMP, 1978, Middle East – review and bibliography of geomorphological contributions, *Q. J. Engng. Geol.*, **11**, 9–18.

COOKE, R. U., and P. MASON, 1973, Desert knolls pediment and associated landforms in the Mojave Desert, California, *Revue Géomorph. Dyn*, **22**, 49–60.

COOKE, R. U., and R. W. REEVES, 1976, *Arroyos and Environmental Change in the American South-West* (Clarendon Press, Oxford).

COOKE, R. U., and I. J. SMALLEY, 1968, Salt weathering in deserts, *Nature*, **220**, 1226–7.

COOKE, R. U., and A. WARREN, 1973, *Geomorphology in Deserts* (Batsford, London).

CORRSIN, S., 1963, Turbulence: experimental methods, *Handbuck der Physik*, vol. VIII/2 (Springer Verlag), 525–90.

COSTELLO, V. F., 1977, *Urbanization in the Middle East* (CUP, Cambridge).

COUNIHAN., J., 1969, An improved method of simulating an atmospheric boundary layer in a wind tunnel, *Atmospheric Environment*, **3**, 197–214.

COUNIHAN, J., 1970, Further measurements in a simulated atmospheric boundary layer, *Atmospheric Environment*, **4**, 259–75.

COUNIHAN, J., 1971, Wind tunnel determination of the roughness length, *Atmospheric Environment*, **5**, 637–42.

COUNTY OF LOS ANGELES BOARD OF SUPERVISORS, 1971, *Environmental Development Guide* (Los Angeles County, Regional Planning Commission).

COX, J. B., 1970, A review of the geotechnical characteristics of the soils in the Adelaide City area, *Inst. Engrs. (Aust.) Soils Symp.*, 1970, 49–63.

CROFTS, R., 1973, Slope categories in environmental management (Department of Geography, University College London), unpublished paper.

CURRY, R. R., 1966, Observations on Alpine mudflows in the Ten-Mile Range, central Colorado, *Bull. Geol. Soc. Am.*, **77**, 771–6.

CURTIS, L. F., J. C. DOORNKAMP and K. J. GREGORY, 1965, The description of relief in the field study of soils, *J. Soil Sci.*, **16**(1), 16–30.

DAVIS, C. K., and J. T. NEAL, 1963, Descriptions and airphoto characteristics of desert landforms, *Photogramm. Engng.*, **29**, 621–31.

DAVIS, K., 1965, The urbanization of the human population, *Sci. Am.*, **213**, 40–53.

DAVIS, K., 1969, *World Urbanization 1950–1970, Vol. 1: Basic Data for Cities, Countries and Regions* (Institute of International Studies, University of California, Berkeley).

DAVIS, K., 1972, *World Urbanization 1950–1970, Vol. 11: Analysis of Trends, Relationships and Development* (Institute of International Studies, University of California, Berkeley).

DEARMAN, W. R., and P. G. FOOKES, 1974. Engineering geological mapping for the civil engineering practice in the United Kingdom, *Q. J. Engng. Geol.*, **7**, 223–56.

DE BENITO, G. A., 1974, Sand dune stabilisation at El Alaiun, West Sahara, *Int. J. Biomet.*, **18**, 142–4.

DELER, J.-P., 1970, Croissance accélérée et formes de sous developpment urbaine à Lima, *Cah. d'Outre-mer.*, **23**, 73–94.

DEMEK, J., (ed.), 1972, *Manual of Detailed Geomorphological Mapping* (Academia, Prague).

DEMEK, J., and C. EMBLETON, (eds.), 1978, *Guide to Medium-scale Geomorphological Mapping* (IGU, Stuttgart).

DENNY, C. S., 1967, Fans and pediments, *Am. J. Sci.*, **265**, 81–105.

DETWYLER, T. R., and M. G. MARCUS, (eds.), 1972, *Urbanization and Environment* (Duxbury, California).

DIETRICH, R. V., 1977, Impact abrasion of harder by softer materials, *J. Geol.*, **85**, 242–6.

DIRECTORATE OF OVERSEAS SURVEYS, 1968, The land resources of Lesotho, *Land Resources Division, Land Resources Study*, **3**.

DITTON, R. B., and T. L. GOODALE, 1972, *Environmental Impact Analysis: Philosophy and Methods* (University of Wisconsin Sea Grant Program, Madison).

DOORNKAMP, J. C., 1971, Geomorphological mapping, in S. M. Ominde (ed.), *Studies in East African Geography and Development* (Heinemann, London), 9–28.

DOORNKAMP, J. C., D. BRUNSDEN, D. K. C. JONES, R. U. COOKE and P. R. BUSH, 1979, Rapid geomorphological assessments for engineering, *Q. J. Engng. Geol.*, **12**, 189–204.

DOORNKAMP, J. C., D. BRUNSDEN and D. K. C. JONES, (eds.), 1980, *Geology, Geomorphology and Pedology of Bahrain* (GeoBooks, Norwich).

DOORNKAMP, J. C., and P. D. TYSON, 1973, A note on the areal distribution of suspended sediment yield in South Africa, *J. Hydrol.*, **20**, 335–40.

DORSSER, H. J. van, and A. I. SALOME, 1973, Different methods of detailed geomorphological mapping, *Koninklijk Nederlands Aardijkskunding Genootschap Geografie Tijdschrift*, **7**(1), 71–7.

DOWLING, J. W. F., 1968, Land evaluation for engineering purposes in northern Nigeria, in G. A. Stewart (ed.) *Land Evaluation* (Macmillan, Melbourne), 147–59.

DUBIEF, J., 1952, Le vent et le deplacement du sable au Sahara, *Inst. Recherche Sahariennes Travaux*, **7**, 187–90.

DUCE, J. T., 1918, The effect of cattle on the erosion of canyon bottoms, *Science*, **47**, 450–2.

DUCHEMIN, G. J., 1958, Essai sur la protection des constructions contre l'ensablement à Port-Etienne (Mauritanie), *Bull. Inst. Français d'Afrique Noire*, Ser. A, **20**, 675–86.

DUDAL, R., and M. BATISSE, 1978, The soil map of the world, *Nature and Resources*, **14**, 2–6.

DUNNE, T., and L. B. LEOPOLD, 1978, *Water in Environmental Planning* (Freeman, San Francisco).

DWYER, D. J., 1975, *People and Housing in Third World Cities: perspectives on the problems of spontaneous settlements* (Longman, London and New York).

EASTAFF, D. J., C. J. BEGGS and M. D. MCELHINNEY, 1978, Middle East – geotechnical data collection, *Q. J. Engng. Geol.*, **11**, 51–63.

ECKEL, E. B., 1958, Landslides and engineering practice, *Highway Research Board, Special Report.*, **29**.

EGYPT, MINISTRY OF HOUSING AND RECONSTRUCTION, 1976, *Suez Master Plan* (Prepared by Sir William Halcrow and Partners, Robert Matthew, Johnson-Marshall and Partners, Economic Consultants Ltd and Hamed Kaddah and Associates), 3 vols.

EGYPT, MINISTRY OF HOUSING AND RECONSTRUCTION, 1978, *Suez Area Subsurface Investigation* (Prepared by Sir William Halcrow and Partners, Robert Matthew, Johnson-Marshall and Partners, Economic Consultants Ltd and Hamed Kaddah and Associates, with contributions by D. Brunsden, R. Cooke, J. C. Doornkamp, D. K. C. Jones and P. R. Bush), 3 vols.

EHLEN, J., 1976, Photo analysis of a desert area, *US Army Engineer Topographic Labs.* (Fort Belvoir, Virginia), Tech. Rep. **ETL-0068**.

ELLIS, C. I., 1973, Arabian salt bearing soil (sabkha) as an engineering material, *Transport and Road Research Lab. (UK), Report*, **LR 523**.

ELLIS, C. I., and R. B. C. RUSSELL, 1973, The use of salt-laden soils (sabkha) for low cost roads, *Dept. of Environment, Transport and Road Research Lab.*, Paper **PA 78/74**.

EMMETT, W. W., 1970, The hydraulics of overland flow on hillslopes, *US Geol. Surv. Prof. Paper.*, **622-A**.

EMMETT, W. W., 1974, Channel aggradation in western United States as indicated by observations at Vigil Network sites, *Zeit. für Geomorph.*, Suppl. **21**, 56–62.

ERIKSSON, E., 1958, The chemical climate and saline soils in the arid zone, in *Climatology, Reviews of Research, Arid Zone Research*, **10**, 147–88.

EVANS, I. S., 1970, Salt crystallization and rock weathering: a review, *Revue Géomorph. Dyn.*, **19**, 153–77.

EVENARI, M., L. SHANAN and N. TADMOR, 1968, Runoff farming in the desert: 1. Experimental layout, *Agron. J.*, **60**, 29–32.

FAO, 1965, Soil erosion by wetter-zone measures for its control on cultivated lands, *FAO Agric. Dev. Paper*, **81**.

FAO, 1969, Report on the FAO study tour on shelterbelts and windbreaks in the USSR, Parts I and II, *FAO*, **TA 2561** (UN Development Programme, Rome).

FAO/DANIDA, 1974, *Report on the FAO/DANIDA Inter-regional Training Centre on Heathland and Sand Dune Afforestation*, Denmark and Libya, 26 Aug. – 21 Sept., 1973 (FAO, Rome).

FAIRWEATHER, J. H., A. F. SIDLOW, and W. L. FAITH, 1965, Particle size distribution of settled dust, *J. Air Pollution Control Ass.*, **15**, 345–7.

FEZER, F., 1971, Photo interpretation applied to geomorphology – a review, *Photogrammetria*, **27**, 7–53.

FISHER, W. B., and H. BOWEN-JONES, 1974, Development surveys in the Middle East, *Geogr. J.*, **140**, 454–66.

FLEMING, G., 1973, *Computer Simulation in Hydrology* (Elsevier, New York, Oxford, and Amsterdam).

FLETCHER, J. E., K. HARRIS, H. B. PETERSON, and V. N. CHANDLER, 1954, Piping, *Trans. Am. Geophys. Un.*, **35**, 258–63.

FOOKES, P. G., 1976a, Middle East – inherent ground problems, *Engng. Group Geol. Soc. Conf. Engng. Probl.* Nov., 7–23.

FOOKES, P. G., 1976b, Road geotechnics in hot deserts, *J. Inst. Highway Engrs.*, **23**, 11–23.

FOOKES, P. G., 1976c, Plain man's guide to cracking in the Middle East, *Concrete*, **10**, 20–2.

FOOKES, P. G., 1977, Natural construction materials in the Gulf, in *Construction Problems and Finance for Middle East Developments* (Middle East Construction, London), 143–64.

FOOKES, P. G., and R. BEST, 1969, Consolidation characteristics of some Late Pleistocene periglacial metastable soils of East Kent, *Q. J. Geol. Soc., Lon.* **2**, 103–28.

FOOKES, P. G., and L. COLLIS, 1975a, Problems in the Middle East, *Concrete*, **9**(7), 12–17.

FOOKES, P. G., and L. COLLIS, 1975b, Aggregates and the Middle East, *Concrete*, **9** (11), 14–19.

FOOKES, P. G., and L. COLLIS, 1976, Cracking and the Middle East, *Concrete*, **10** (2), 14–19.

FOOKES, P. G., and W. J. FRENCH, 1977, Soluble salt damage to surfaced roads in the Middle East, *J. Inst. Highway Engrs.*, **24** (12), 10–20.

FOOKES, P. G., and I. E. HIGGINBOTTOM, 1975, The classification and description of near-shore carbonate sediments for engineering purposes, *Geotechnique*, **25**, 406–11.

FOOKES, P. G., and I. E. HIGGINBOTTOM, 1980, Some problems of construction aggregates in desert areas, with particular reference to the Arabian peninsula, *Proc. Inst. Civ. Engrs.*, **68**, 39–67, 69–90.

FOOKES, P. G., and J. L. KNILL, 1969, The application of engineering geology in the regional development of northern and central Iran, *Engng. Geol.*, **3**, 81–120.

FOOKES, P. G., and A. POOLE, 1978, Selection and durability of rock and concrete materials for breakwaters and sea defence works, *Engng. Group Geol. Soc. Conf. on Coastal Engng. Geol.*, Southampton, Sept, unpublished paper.

FOOTE, L., 1972, *Soil Erosion and Water Pollution Prevention, NACE Action Guide Series*, **XII** (National Association of County Engineers, Washington, DC).

FOURNIER, F. 1960, Debit solide des cours d'eau. Essai d'estimation de la perte en terre subie par l'ensemble du globe terrestre, *Int. Ass. Sci. Hydrol., Publ.* **53**, 19–22.

FRÄNZLE, O., 1966, Geomorphological mapping, *Nature and Resources* (UNESCO), **II** (4), 14–16.

FRENCH, W. J., and A. B. POOLE, 1974, Deleterious reactions between dolomites from Bahrain and cement paste, *Cement and Concrete Research*, **4**.

FRENCH, W. J., and A. B. POOLE, 1976, Alkali aggressive aggregates and the Middle East, *Concrete*, **10**, 18–20.

FUCHS, N. A., 1964, *The Mechanics of Aerosols* (Macmillan, New York).

FULTON, F. S., 1964, *Concrete Technology* (Portland Cement Institute, Johannesburg), 3rd edn.

GANDEMER, J., 1977, Wind environment around buildings: aerodynamic concepts, *Proc. 4th Int. Conf. on Wind Effects on Buildings and Structures*, 423–32.

GELLERT, J. F., and E. SCHOLZ, 1964, *Katalog des Inhaltes von geomorphologischen. Detailkarten aus verschieden europäischen Ländern* (Institute für Geographie, Potsdam).

GEOGRAPHICAL RESEARCH INSTITUTE OF THE HUNGARIAN ACADEMY OF SCIENCES, 1963, *Legend of the Detailed Geomorphological Maps of Hungary* (Budapest).

GILEWSKA, S., 1967, Different methods of showing the relief on the detailed geomorphological maps, *Zeit. für Geomorph.*, **11**, 481–90.

GILLULY, J., A. C. WATERS, and A. O. WOODFORD, 1968, *Principles of Geology* (Freeman, San Francisco), 3rd edn.

GLENNIE, K. W., 1970, *Desert Sedimentary Environments* (Elsevier, Amsterdam).

GLYMPH, L. M., and H. N. HOLTAN, 1969, Land treatment in agricultural watershed hydrology research, in W. L. Moore and C. W. Morgan (eds.) *Effects of Watershed Change on Streamflow* (Texas University Press, Austin), 44–68.

GOUDIE, A. S. 1973, *Duricrusts in Tropical and Subtropical Landscapes* (OUP, Oxford).

GOUDIE, A. S., 1974, Further experimental investigation of rock weathering by salt and other mechanical processes, *Zeit. für Geomorph.*, Suppl. **21**, 1–12.

GOUDIE, A. S., 1977, Sodium sulphate weathering and the disintegration of Mohenjo-Daro, Pakistan, *Earth Surface Processes*, **2**, 75–86.

GOUDIE, A. S., 1978, Dust storms and their geomorphological implications, *J. Arid Environments*, **1**, 291–310.

GOUDIE, A. S., R. U. COOKE and I. S. EVANS, 1970, Experimental investigation of rock weathering by salts, *Area*, **4**, 42–8.

GOUDIE, A. S., and J. WILKINSON, 1978, *The Warm Desert Environment* (CUP, Cambridge).

GRANT, K., 1968, A terrain evaluation system for engineering, *CSIRO Austr. Div. Soil Mech. Tech. Paper*, **2**.

GRANT, K., 1972, Terrain classification for engineering purposes of the Melbourne area, Victoria, *CSIRO Austr. Div. Appl. Geomech. Paper*, **11**.

GRANT, K., and G. D. LODWICK, 1968, Storage and retrieval of information in a terrain classification system, *Proc. Conf. Aust. Road Res. Board*, **4** (2), 1667–76.

GREELEY, R., B. R. WHITE, J. B. POLLACK, J. D. IVERSEN and R. N. LEACH, 1977, Dust storms on Mars: considerations and simulations, *NASA Tech. Memo.* **78423** (Washington, DC).

GREEN, W. H., and G. A. AMPT, 1911, Studies on soil physics. I. The flow of air and water through soils, *J. Agric. Sci.*, **4**, 1–24.

GREGORY, K., and D. WALLING, 1973, *Drainage Basin Form and Process* (Edward Arnold, London).

GRIFFITHS, J. S., 1978a, Estimation of low frequency event discharge from hill torrents in the area of Dera Ghazi Khan, *Indus Super Highway Board*, Jan. unpublished report.

GRIFFITHS, J. S., 1978b, Flood assessment in ungauged semi-arid catchments as a branch of applied geomorphology, *Geography Dept., King's College, London, Occasional Paper*, **8**.

GRUNERT, J., 1978, Fossil landslides in the central Sahara, in E. M. van Zinderen Bakker and S. A. Coetzee (eds.), *Palaeoecology of Africa and the Surrounding Islands*, 10–11.

GRUNERT, J., and H. HAGEDORN, 1976, Beobachtungen an Schichtstufen der Nubischen Serie (Zentral Sahara), *Zeit. für Geomorph.*, Suppl., **24**, 99–110.

GULICK, J., 1967, Baghdad: portrait of a city in physical and cultural change, *J. Am. Inst. Plann.*, **33**, 246–55.

GUY, H. P., 1965, Residential construction and sedimentation at Kensington, Md., in *Proc. Federal Inter-Agency Sedimentation Conf.*, (Jackson. Miss., 1963), *US Dept. Agric. Misc. Publ.*, **970**, 30–7.

GUY, H. P., 1970, Sediment problems in urban areas, *US Geol. Surv. Circular*, **601-E**.

GUY, H. P., 1972, Urban sedimentation in perspective, *Proc. Am. Soc. Civ. Engrs, J. Hyd. Div.*, Dec. 209–16.

HAANTJENS, H. A., 1968, The relevance for engineering of principles, limitations and developments in land system surveys in New Guinea, *Proc. Conf. Aust. Road Res. Board*, **4**, 1593–1612.

HAARS, W., H. HAGEDORN, D. BUSCHE and H. FORSTER, 1974, Zur Geomorphologie des Shir-Kuh Massivs (central-Iran), *Marburger Geogr. Schri.*, **62**, 39–48.

HADLEY, R. F., and S. A. SCHUMM, 1961, Hydrology of the Upper Cheyenne River Basin: B. Sediment sources and drainage-basin characteristics in Upper Cheyanne River Basin, *US Geol. Surv. Water-Supply Paper*, **1531**, 137–98.

HAGEDORN, H., D. BUSCHE, J. GRUNERT, K. SCHAFFER, E. SCHULZ and A. SKOWRONEK, 1978, Bericht über geomissen Schaftliche Unter sachungen am Westrand des Murzuk-Bechens (zentrale Sahara), *Zeit. für Geomorph.*, Suppl., **30**, 20–38.

HAGEN, L. J., 1976, Windbreak design for optimum wind erosion control, in R. W. Tinus (ed.), *Proc. Symp. on Shelterbelts on the Great Plains* (Denver, Colorado), *Great Plains Agric. Counc. Publ.*, **78**, 31–6.

HAILS, J. R. (ed.), 1977, *Applied Geomorphology* (Elsevier, Amsterdam).

HALCROW MIDDLE EAST, 1977a, *Report on Geomorphological Investigations in Dubai, April 1977* (prepared by J. C. Doornkamp, D. Brunsden, D. K. C. Jones, P. Bush, M. J. Gibbons and R. U. Cooke), 3 vols.

HALCROW MIDDLE EAST, 1977b, *Khor Khwair Reconnaissance Geomorphological Investigation* (report prepared by D. K. C. Jones, D. Brunsden, J. C. Doornkamp, P. R. Bush and M. J. Gibbons).

HALCROW, SIR W. K. and PARTNERS, 1975, *Special Report on Concrete Quality and Related Factors in Suez* (Arab Rep. of Egypt Min. of Housing and Reconstruction), Nov. 1975.

HALL, P., 1974, *Urban and Regional Planning* (Penguin, Harmondsworth).

HARRIS, C. D., 1971, Urbanization and population growth in the Soviet Union, 1959–1970, *Geogr. Rev.*, **61**, 102–24.

HARRIS, W. D., 1971, *The Growth of Latin American Cities* (Ohio University Press, Athens, Ohio).

HARRISON, D. (ed.), 1970, *Specification 1970* (Architectural Press, London).

HARRISON, J. V., and N. L. FALCON, 1937a, The Saidmarreh landslip, southwest Iran, *Geogr. J.*, **89**, 42–7.

HARRISON, J. V., and N. L. FALCON, 1937b, An ancient landslip at Saidmarreh in south-western Iran, *J. Geol.*, **46**, 269–309.

HATHERLY, L. W., and M. WOOD, 1957, The seasonal variations in subgrade soil moisture and temperature with depth in Baghdad, Iraq, *Proc. 4th Inst. Conf. Soil Mech. Foundn. Engng* (London), **2**.

HAYES, C. J., 1971, Landslides and related phenomena pertaining to highway construction in Oklahoma, *Ann. Oklahoma Acad. Sci.*, **2**, 47–57.

HICKOK, R. B., R. Y. KEPPEL, and B. R. RAFFERTY, 1959, Hydrograph synthesis for small aridland watershed, *Agric. Engng.*, **40**, 608–11, 615.

HIGGINS, G. M., R. SHABBIR BAIG and R. BRINKMAN, 1974, The sands of Thal: wind regimes and sand ridge formations, *Zeit. für Geomorph.*, **18**, 272–90.

HINZE, J. O., 1959, *Turbulence – An Introduction to its Mechanism and Theory* (McGraw-Hill, New York).

HOLEMAN, J. N., 1968, The sediment yield of the major rivers of the world, *Wat. Resour. Res.*, **4**, 737–47.

HOLLAND, J. L., and J. C. STEVENSON, 1963, Foundation problems in arid-climate silts, *Geol. Soc. Am. Spec. Paper*, **76**, 276–7.

HOLWAY, J. V., G. E. CORDY and T. L. PÉWÉ, 1978, Geology for land use in the Paradise Valley Quadrangle of Central Arizona: a useful tool for city planners, *Geol. Soc. Am. Cordilleran Section Program Abstracts*, **109**.

HONDA, H., 1974, Fundamental study on the planting and space effects in public nuisance prevention in the city: iii Dust catching ability of plant foliage, *Tech. Bull. Fac. Hort. Chiba Univ.*, **22**, 81–8.

HOOKE, R. LeB., 1968, Steady state relationships on arid region alluvial fans in closed basins, *Am. J. Sci.*, **266**, 609–29.

HÖRNER, N. G., 1936, Geomorphic processes in continental basins of central Asia, *Int. Geol. Congr. Rep. 16th Session* (Washington, DC), **2**, 721–35.

HORTON, R. E., 1945, Erosional development of streams and their drainage basins: hydrophysical approach to quantitative morphology, *Bull. Geol. Soc. Am.*, **56**, 275–370.

HOUSEMAN, J., 1961, Dust haze at Bahrain, *Meteorology Mag.*, **91**, 50–2.

HOWARD, A. D., J. B. MORTON, M. GAL-EL-HAK and D. B. PIERCE, 1977, Simulation model of erosion and deposition on a barchan dune, *NASA* **CR-2838** (*University of Virginia, Charlottesville*).

HOWE, G. M., *et al.*, 1968, Classification of world desert areas, *US Army Natick Lab. Tech. Rep.*, **69-38-ES**.

HUDSON, N. W., 1971, *Soil Conservation* (Batsford, London).

HUDSON, N. W., and D. C. JACKSON, 1959, *Erosion Research*, Henderson Research Station, Report of progress 1958–59 (Fed. of Rhodesia and Nyasaland., Min. of Agric).

HUNT, C. B., and A. L. WASHBURN, 1966, Patterned ground, in *US Geol. Surv. Prof. Paper*, **494-B**, B104–B133.

HUNTING SURVEYS LTD, 1977, *Technical report – sand dune movement study south of Umm Said, 1963–1976, for the Government of Qatar, Ministry of Public Works* (Hunting Surveys Ltd, Borehamwood, UK).

HUSCHKE, R. E. (ed.), 1959, *Glossary of Meteorology* (American Meteorological Society, Boston. Mass.).

ICONA, 1969, Projecto de restauraçion hidrologica-forestal del embalse de Beninar, *Inst. Nacional para el Conservacion de la Naturaleza* (Granada, Spain).

IDSO, S. B., 1976, Dust storms, *Sci. Am.*, **235**, 108–11, 113–14.

INBAR, M., 1972, A geomorphic analysis of a catastrophic flood in a Mediterranean basaltic watershed, paper presented to *22nd Int. Geogr. Congr.*, Montreal, Canada, 1972.

INGLE, J. C., 1966, The movement of beach sand – an analysis using fluorescent grains, *Developments in Sedimentology*, **5** (Elsevier Publishing Co., Amsterdam).

INTERNATIONAL ASSOCIATION OF HYDROLOGICAL SCIENCES, 1974, The effects of man on the interface of the hydrological cycle with the physical environment, *Proc. Paris Symposium*, Publ. **113**.

INTERNATIONAL ASSOCIATION OF HYDROLOGICAL SCIENCES, 1979, The hydrology of areas of low precipitation, *Proc. Canberra Symp.*

INTERNATIONAL GEOGRAPHICAL UNION SUBCOMMISSION, 1968, Project of the unified key to the detailed geomorphological map of the world, *Folia Geographica, Series Geographica – Physica*, **II** (Polska Akademia Nauk, Krakaw).

IRAN, DEPARTMENT OF THE ENVIRONMENT, 1977, *Case study on desertification; Iran: Turan, An Associated Case Study*, International Cooperation to Combat Desertification, General Assembly Resolution 3337 (XXIX), *UN Conference on Desertification*, 29 Aug–9 Sept, 1977, Nairobi, Kenya.

ISACHENKO, A. G., 1973, *Principles of Landscape Science and Physical Geographic Regionalisation*, J. S. Massey and N. J. Rosengren (eds.), trans. R. J. Zatorski (Melbourne University Press, Melbourne).

JACKSON, E. A., 1957, Soil features in arid regions with particular reference to Australia, *Aust. Inst. Agric. Sci. J.*, **25**, 196–208.

JENNINGS, J. N., 1955, Le complex des sabkhas: un commentaire provenant des antipodes, *Revue Géomorph. Dyn.*, **6**, 69–72.

JENSEN, M., 1954, *Shelter Effect – Investigations into the Aerodynamics of Shelter and its Effects on Climate and Crops* (Danish Technical Press, Copenhagen).

JOFFE, J. S., 1949, *Pedology* (Pedology Publications, New Brunswick), 2nd ed.

JOHNSON, W. C., 1965, Wind in the Southwestern Great Plains, *Agricultural Research Service, Conservation Research Rep.,* **6** (US Dept of Agriculture).

JONES, D. K. C., 1980, British applied geomorphology: an appraisal, *Zeit. für Geomorph.,* **Suppl. 36,** 48–73.

JONES, E., 1964, Aspects of urbanization in Venezuela, *Ekistics,* **18,** 420–5.

JUTSON, J. T., 1918, The influence of salts in rock weathering in sub-arid Western Australia, *Proc. R. Soc. Vict.,* **30** (n.s.), 165–72.

KADIB, A.-L., 1963, Sand transport by wind, studies with sand (0.145 mm diameter), *Wave Research Project* **HEL-2-5,** Hydraulic Engineering Lab., University of California.

KADIB, A.-L., 1964, Calculation procedure for sand transport by wind in natural beaches, *US Army Coastal Engineering Reseach Center, Miscellaneous Paper,* **2-64** (Washington, DC).

KANTEY, B. A., 1971, Terrain evaluation – a problem in whole engineering, *The Civil Engineer in South Africa* (Nov.), 407–11.

KANTEY, B. A., and M. J. MOUNTAIN, 1968, Terrain evaluation for civil engineering projects in South Africa, *Aust. Road Res. Board.,* **4,** 1613–23.

KANTEY, B. A., and A. A. B. WILLIAMS, 1962, The use of soil engineering maps for road projects, *The Civil Engineer in South Africa,* **4,** 8.

KAO, D. T. Y. (ed.), 1974, *National Symposium on Urban Rainfall and Runoff and Sediment Control* (University of Kentucky, Lexington).

KAPPUS, U., J. M. BLECK, and S. H. BLAIR, 1978, Rainfall frequencies for the Persian Gulf coast of Iran, *Hydrol. Sci. Bull.,* **23** (1), 119–29.

KATES, R. W., D. L. JOHNSON, and K. JOHNSON HARING, 1976, *Population, Society and Desertification* (Clark University, Program in International Development and Social Change).

KAUL, R. N., 1970, *Afforestation in Arid Zones* (Dr W. Junk, N.K. Publishers, The Hague).

KAZI, A., and Z. R. AL-MANSOUR, 1980, Empirical relationship between Los Angeles abrasion and Schmidt hammer strength tests with application to aggregates around Jeddah, *Q. J. Engng. Geol.,* **13,** 45–52.

KELLER, C., 1946, *El Departamento de Arica* (Min. Econ. y Com., Santiago de Chile).

KERR, R. C., and J. O. NIGRA, 1952, Eolian sand control, *Bull. Am. Ass. Petrol. Geol.,* **36,** 1541–73.

KESSELI, J. E., and C. B. BEATY, 1959, Desert flood conditions in the White Mountains of California and Nevada, *US Army Quartermaster Research and Engineering Center, Tech. Rep.,* **EP-108.**

KING, D., 1960, The sand-ridge deserts of South Australia and related aeolian landforms of Quaternary arid cycles, *Trans. R. Soc. S. Aust.,* **83,** 99–108.

KING, L. C., 1953, Canons of landscape evolution, *Bull. Geol. Soc. Am.,* **64,** 721–52.

KING, L. C., 1962, *The Morphology of the Earth* (Oliver and Boyd, Edinburgh).

KING, L. C., 1963, *South African Scenery* (Oliver and Boyd, Edinburgh), 3rd edn.

KING, R. B., 1970, A parametric approach to land system classification, *Geoderma,* **4,** 37–46.

KIRKBY, M. J., 1969, Erosion by water on hillslopes, in R. J. Chorley (ed.), *Water, Earth and Man* (Methuen, London), 229–38.

KIRKBY, M. J., 1976, Hydrograph modelling strategies, in R. F. Peel, M. C. Chisholm, and P. Haggett, *Processes in Physical and Human Geography* (Heinemann, London), 69–90.

KIRKBY, M. J., 1978, Implications for sediment transport, in M. J. Kirkby (ed.), *Hillslope Hydrology* (Wiley, Chichester).

KLIMASZEWSKI, M., 1956, The principles of the geomorphological survey of Poland, *Przegl. Geogr.,* **28,** (Suppl.), 32–40.

KLIMASZEWSKI, M., 1961, The problems of the geomorphological and hydrographic map of the Upper Silesian industrial disrict, *Problems of Applied Geography, Geographical Studies* (Polish Acad. Sciences, Institute of Geography, Warsaw), **25**, 73–81.

KRAMMES, J. S., and J. OSBORN, no date, Water-repellant soils and wetting agents as factors influencing erosion, *US Pacific Southwest Forest and Range Experiment Station, Notes,* 177–87.

KRINSLEY, D. B., 1970, A geomorphological and palaeoclimatological study of the playas of Iran, *US Geol. Surv. Final Scientific Rep., Contract PRO CP70-800,* 2 vols.

KUWAIT INSTITUTE OF SCIENTIFIC RESEARCH, 1978, *Proposal for a Geological and Geophysical Exploration for Gravel and Sand in Northern Kuwait* (KISR, Kuwait).

KWAAD, F. J. P. M., 1970, Experiments on the granular disintegration of granite by salt action, *Fys. Geogr. en Bodemkundig Lab., From Field to Laboratory, Publ.,* **16**, 67–80.

LAKE EYRE COMMITTEE, 1950, *Lake Eyre, South Australia – the great flooding of 1949– 50* (R. Geogr. Soc. Aust.).

LANE, K. S., and D. E. WASHBURN, 1946, Capillary tests by capillarimeter and by soil filled tubes, *Proc. Highway Res. Board,* **26**, 460–73.

LANGBEIN, W. B., and S. A. SCHUMM, 1958, Yield of sediment in relation to mean annual precipitation, *Trans. Am. Geophys. Un.,* **30**, 1076–84.

LANGER, A. M., and P. F. KERR, 1966, Mojave playa crust: physical properties and mineral content, *J. Sedim. Petrol.,* **36**, 377–96.

LANGFORD-SMITH, T., 1962, Riverine plains chronology, *Aust. J. Sci.,* **25**, 96–7.

LANGFORD-SMITH, T., 1978, Arid land settlement today: some problems and prospects, *Aust. Geogr. Studies,* **16**, 3–14.

LAPIDUS, I. M. (ed.), 1969, *Middle Eastern Cities: A Symposium on Ancient, Islamic and Contemporary Middle East Urbanism* (University of California Press, Berkeley).

LAWRANCE, C. J., 1972, *Terrain Evaluation in West Malaysia. Part 1 Terrain classification and survey methods* (Transport and Road Research Laboratory, Crowthorne, UK), Report, **LR 506.**

LAWSON, T. V., 1968, Methods of producing velocity profiles in wind tunnels, *Atmospheric Environments,* **2**, 73–6.

LEGGET, R. F., 1973, *Cities and Geology* (McGraw Hill, New York).

LEOPOLD, L. B., 1951, Rainfall frequency: an aspect of climatic variation, *Trans. Am. Geophys. Un.,* **32**, 347–57.

LEOPOLD, L. B., 1962, The VIGIL network, *Int. Ass. Sci. Hydrol. Bull.,* **7**, 5–9.

LEOPOLD, L. B., 1968, Hydrology for urban land planning – a guidebook on the hydrologic effects of urban land use, *US Geol. Surv. Circ.,* **554.**

LEOPOLD, L. B., W. W. EMMETT, and R. M. MYRICK, 1966, Channel and hillslope processes in a semiarid area, New Mexico, *US Geol. Surv. Prof. Paper,* **352-G**, 193– 253.

LEOPOLD, L. B., and T. MADDOCK, 1953, The hydraulic geometry of stream channels and some physiographic implications, *US Geol. Surv. Prof. Paper,* **252.**

LEOPOLD, L. B., and J. P. MILLER, 1956, Ephemeral streams – hydraulic factors and their relation to the drainage net, *US Geol. Surv. Prof. Paper,* **282-A.**

LETTAU, K., and H. LETTAU, 1969, Bulk transport of sand by the barchans of the Pampa La Joya in southern Peru, *Zeit. für Geomorph.,* **13**, 182–95.

LEVESON, D., 1980, *Geology and the Urban Environment* (OUP, New York).

LIBYAN ARAB REPUBLIC, NATIONAL HOUSING CORPORATION, 1975, *Housing Design Manual* (Colin Buchanan and Partners, London), vols. 1 (Planning) and 4 (Specifications).

LINSLEY, R. K., M. A. KOHLER, and J. L. B. PAULHUS, 1949, *Applied Hydrology* (McGraw-Hill, New York).

LLOYD, A., 1966, The generation of shear flow in a wind tunnel, *J. Fluid Mech.*, **25**, 79–96.

LOFGREN, B. E., 1969, Land subsidence due to the application of water, in D. J. Varnes and G. Kiersch (eds.), *Reviews in Engineering Geology* (Geol. Soc. Am., Colorado), **2**, 271–303.

LONGWELL, C. R., R. F. FLINT and J. E. SANDERS, 1969, *Physical Geology* (Wiley, New York).

LOS ANGELES COUNTY BOARD OF SUPERVISORS, 1915, *Report of the Board of Engineers – Flood Control.*

LOS ANGELES COUNTY BOARD OF SUPERVISORS, 1971, *Environmental Development Guide* (Regional Planning Commission, Los Angeles).

LUEDER, D. R., 1959, *Aerial Photographic Interpretation: principles and application* (McGraw-Hill, New York).

LUSTIG, L. K., 1965, Sediment yield of the Castaic watershed, western Los Angeles County, California – a quantitative geomorphic approach, *US Geol. Surv. Prof. Paper*, **422-F**.

LUSTIG, L. K., 1969, Quantitative analysis of desert topography, in W. G. McGinnies and B. J. Goldman (eds.) *Arid Lands in Perspective* (University of Arizona Press, Tucson), 47–58.

LYLES, L., 1976, Wind patterns and soil erosion on the Great Plains, *Proc. Symp: Shelter belts on the Great Plains, Denver, Colorado, 20–22 April, Great Plains Agric. Council Publ.*, **78**, 22–9.

LYLES, L., D. V. ARMBRUST, J. D. DICKERSON and N. P. WOODRUFF, 1969, Spray-on adhesives for temporary wind erosion control, *J. Soil Wat., Conserva.*, **24**, 190–3.

LYLES, L., and N. P. WOODRUFF, 1960, Abrasive action of windblown soil on plant seedlings, *Agron. J.*, **52**, 533–6.

MABBUTT, J. A., 1966, Mantle-controlled planation of pediments, *Am. J. Sci.*, **264**, 78–91.

MABBUTT, J. A., 1977, *Desert Landforms* (MIT Press, Cambridge, Mass.).

MABBUTT, J. A., and G. A. STEWART, 1963, The application of geomorphology in resource surveys in Australia and New Guinea, *Revue Géomorph. Dyn.*, **14**, 97–109.

MACGREGOR, M. T. G. de, and V. CARMEN VALVERDE, 1975, Evaluation of the urban population in the arid zones of Mexico, 1900–1970, *Geogr. Rev.*, **65**, 214–28.

MACVEHIL, G. E., G. R. LUDWIG, and T. R. SUNDARAM, 1967, On the feasibility of modelling small-scale atmospheric motions, *Cornell Aeronautical Lab. Report*, **2B2328-P-1** (Buffalo, N.Y.).

MADER, G. G., and D. F. CROWDER, 1969, An experiment in using geology for city planning – the experience of the small community of Portola Valley, California, in *Environmental Geology and Planning* (US Dept. Housing and Urban Development), 176–89.

MAINGUET, M., 1976, Trade winds topography and sandflow in the Sahara and Australia, in A. T. Grove (ed.), *Cambridge Meeting on Desertification*, 22–26 Sept. 1975, Mimeo Rep. (Dept. Geogr., University of Cambridge), 40–3.

MANNER, S. B., 1958, Factors affecting sediment delivery rates in the Red Hills physiographic area, *Trans. Am. Geophys. Un.*, **39**, 669–75.

MANOHAR, M., and P. BRUUN, 1970, Mechanics of dune growth by sand fences, *Dock Harb. Auth.*, **51**, 243–52.

MARÇAIS, J., 1977, The geological world atlas, *Nature and Resources*, **13**, 15–17.

MARSHALL, J. K., and H. M. CHURCHWOOD, 1974, Dust abatement at Agnew, Western Australia, *CSIRO, Division of Land Resources Management, Management Report*, **1**.

MCGILL, J. T., 1964, The growing importance of urban geology, *US Geol. Surv. Circ.*, **487**.

McHarg, I., 1968, A comprehensive route selection method, *Highways Res. Rec.,* **246**, 1–15.

McKee, E. D., and C. Breed, 1973, An investigation of major sand seas in desert areas throughout the world, *Third ERTS Symposium Abstracts* (NASA, Goddard Space Flight Center, Washington).

McKee, E. D., and J. R. Douglass, 1971, Growth and movement of dunes at White Sands National Monument, New Mexico, *US Geol. Surv. Prof. Paper,* **750-D**, D108–D114.

McLaughlin, W. T., and R. L. Brown, 1942, Controlling coastal sand dunes in the Pacific North West, *US Dept. Agric. Circ.,* **660**.

Meginnis, H. G., 1935, Effect of cover on surface run-off and erosion in the loessial uplands of Mississippi, *US Dept. of Agric. Circ.,* **347**.

Mehra, S. R., L. R. Chadda amd R. N. Kapur, 1955, Role of detrimental salts in soil stabilization with and without cement – 1. Effect of sodium sulphate, *Indian Concr. J.,* **29**, 336–7.

Meigs, P., 1953, World distribution of arid and semi-arid homoclimates, *Reviews of Research in Arid Zone Hydrology* (UNESCO, Paris), 203–9.

Mein, R. G., and C. L. Larsen, 1973, Modelling infiltration during a steady rain, *Wat. Resour. Res.,* **9**, 384–94.

Melton, M. A., 1965, The geomorphic and palaeoclimatic significance of alluvial deposits in southern Arizona, *J. Geol.,* **73**, 1–38.

Meyer, L. D., and J. E. Monke, 1965, Mechanics of soil erosion by rainfall and overland flow, *Trans. Am. Soc. Agr. Engng.,* **8**. 572–7.

Miller, J. P., 1958, High mountain streams: effect of geology on channel character- istics and bed material, *New Mexico State Bur. Mines Min. Resources Memo,* **4**.

Miller, V. C., 1961, *Photogeology* (McGraw-Hill, New York).

Miller, V. C., 1968, Aerial photographs and land forms (photogeomorphology), *Aerial Survey Integrated Studies – Proc. Toulouse Conf. 1964 UNESCO Nat. Resources Res. Publ.,* **6**, 41–69.

Minty, E. J., 1965, Preliminary report on an investigation into the influence of several factors on the sodium sulphate test for aggregate, *Aust. Road Res.,* **2**, 49–52.

Minty, E. J., and K. Monk, 1966, Predicting for durability of rock, *Proc. 3rd Conf. Aust. Road Res. Board,* **3** (2), 1316–33.

Mitchell, C. W., 1973, *Terrain Evaluation* (Longman, London).

Mitchell, C. W., and J. A. Howard, 1978. Final Summary report on the application of LANDSAT imagery to the soil degradation mapping at 1:5,000,000, UNFAO *AGLT Bulletin,* 1/78.

Mitchell, C. W., and R. M. S. Perrin, 1967, The subdivision of hot deserts of the world into physiographic units, *Actes du IIème Symposium Internat. de Photo- Interpretation* (Paris, 1966), **IV.1**, 90–106.

Mitchell, C. W., and R. Webster, P. H. T. Beckett and B. Clifford, 1979, An analysis of terrain classification for long-range prediction of conditions in deserts, *Geogr. J.,* **145**, 72–85.

Monod, Th., 1958, Majabât Al-Koubrâ, *Mém. Inst. Fr. Afr. Noire,* **52**.

Moore, D. D., 1968, Estimating mean runoff in ungaged semiarid areas, *Bull. Int. Ass. Sci. Hydrol.,* **13**, 29–39.

Morales, C., 1977, Saharan Dust: mobilisation, transport, deposition, *Review and Recommendations from a Workshop held in Gothenburg, Sweden, 25–28 April, ERC (SIES)/NFR-SCOPE/MARC Workshop.*

Morales, C. (ed.), 1979, *Saharan Dust* (SCOPE 14) (Wiley, Chichester).

Motts, W. S., 1965, Hydrologic types of playas and closed valleys and some relations of hydrology to playa geology in J. T. Neal (ed.), *Geology, Mineralogy and Hydrology of US Playas Air Force Cambridge, Environmental Research Paper,* **96**, 73–104.

MOUNTJOY, A. B., 1976, Urbanization, the equation and development in the Third World, *Tijdschr. Econ. Soc. Geogr.*, **67**, 130–7.

MUELLER, G., 1960, The theory of formation of north Chilean nitrate deposits through 'capillary concentration', *Rep. Int. Geol. Congr. 19th, Nordern Part I*, 76–86.

MUELLER, G., 1968, Genetic histories of nitrate deposits from Antarctica and Chile, *Nature*, **219**, 113–16.

MULHEARN, P. J., and E. F. BRADLEY, 1977, Secondary flows in the lee of porous shelterbelts, *Boundary Layer Meteorology*, **12**, 75–92.

MULVANEY, T. J., 1851, On the use of self-registering rain and flood gauges in making observations of the relation of rainfall and flood discharges in a given catchment, *Proc. Inst. Civ. Engrs, Ireland*, **4**, 1850–1.

MUSGRAVE, G. W., 1947, Quantitative evaluation of factors in water erosion – a first approximation, *J. Soil Wat. Conserva.*, **2**, 133–8.

MYERS, V. A., 1967, The estimation of extreme precipitation as the basis for design floods. Resumé of practice in the United States, *IASH – UNESCO Leningrad Symposium on floods and their computation*, 84–101.

NATIONAL INSTITUTE FOR ROAD RESEARCH, 1971 (revised draft dated 1976), The production of soil engineering maps for roads and the storage of materials data (Nat. Inst. for Road Research, Pretoria, South Africa), *Tech. Recomm. for Highways.*, **2**.

NATURE AND RESOURCES, 1974, Survey on water balance of lakes and reservoirs of the world, *Nature and Resources*, **10**, 18–20.

NEAL, J. T. (ed.), 1965, Geology, mineralogy and hydrology of US playas, *US Air Force Cambridge Res. Lab. Environ. Res. Paper*, **96**.

NEAL, J. T., 1969, Playa variation, in W. G. McGinnies and B. J. Goldman (eds.), *Arid Lands in Perspective* (University of Arizona Press, Tucson), 13–44.

NEAL, J. T. (ed.), 1975, *Playas and Dried Lakes* (Dowden, Hutchinson & Ross, Stroudsburg, Penn.).

NEAL, J. T., A. M. LANGER, and P. F. KERR, 1968, Giant desiccation polygons of Great Basin playas, *Bull. Geol. Soc. Am.*, **79**, 69–90.

NETTERBERG, F., 1967, Some roadmaking properties of South African calcretes, *Proc. 4th Reg. Conf. for Africa on Soil Mech. and Foundation Eng.* (Cape Town), **1**, 77–81.

NETTERBERG, F., 1970, Occurrence and testing for deleterious salts in road construction materials with particular reference to calcretes, *Symposium on soils and earth structures in arid climates, Adelaide, Paper No. 2856*, 87–92.

NETTERBERG, F., 1971, Calcrete in road construction, *Nat. Inst. Road Res. Bull.*, **10** CSIR Res. Rept. 286 (Pretoria, South Africa).

NETTERBERG, F., G. E. BLIGHT, P. F. THERON and G. P. MARRAIS, 1974, Salt damage to roads with bases of crusher-run Witwatersrand Quartzite, *Proc. 2nd Conf. Asphalt Pavements, S. Africa, Session 7*, 34–53.

NEWMAN, A. J., 1971, Problems of concreting in arid climates, *Overseas Building Notes No. 139* (HMSO, London).

NILES, J. S., 1961, A universal equation for measuring wind erosion, *Agric. Res. Serv. Special Rep. 22–69.*

OLLIER, C. D., 1977, Terrain classification: principles and applications, in J. R. Hails (ed.), *Applied Geomorphology* (Elsevier, Amsterdam), 277–316.

OLLIER, C. D., R. WEBSTER, C. J. LAWRANCE and P. H. T. BECKETT, 1967, The preparation of a land classification map at 1/1 000 000 of Uganda, *Actes du IIème Symp. Internat. de Photo-Interpretation* (Paris), Sect. IV.1, 115–22.

OPEN UNIVERSITY, 1974, *Constructional and Other Bulk Material* (Open University Press, Milton Keynes).

OSBORN, H. B., and R. B. HICKOK, 1968, Variability of rainfall affecting runoff from a semi-arid rangeland watershed, *Water Resour. Res.*, **4**, 199–203.

OSBORN, H. B., and L. LANE, 1969, Precipitation – runoff relations for very small semi-arid rangeland watersheds, *Water Resour. Res.*, **5**, 419–25.

PACK, F. J., 1923, Torrential potential of desert waters, *Pan-Am. Geologist*, **40**, 349–56.

PALMER, L., 1976, Application of land-use constraints in Oregon, in D. R. Coates (ed.), *Urban Geomorphology* (Geological Society of America, Colorado), *Special Paper*, **174**, 61–84.

PANIZZA, M., 1978, Analysis and mapping of geomorphological processes in environmental management, *Geoforum*, **9**, 1–15.

PARCHER, J. V., 1971, Foundation problems, *Ann. Oklahoma Acad. Sci.*, **2**, 15–22.

PARKER, G. G., 1963, Piping, a geomorphic agent in landform development of the drylands, *Int. Ass. Sci. Hydrol. Publ.*, **65**, 103–13.

PARKER, G. G., and E. A. JENNE, 1967, Structural failure of Western US highways caused by piping, *US Geol. Soc. Water Research Div.*, Washington, D.C., Jan. 18th, 1967, for 46th Ann. Meeting Highways Res. Board.

PARKER, G. G., L. M. SHOWN and K. W. RATZLAFF, 1964, Officer's cave, a pseudokarst feature in altered tuff and volcanic ash of the John Day Formation in eastern Oregon, *Bull. Geol. Soc. Am.*, **75** (5), 393–402.

PAYLORE, P., and E. A. HANEY (eds.), 1976, *Desertification: process, problems, perspectives* (University of Arizona, Tucson).

PEÇSI, M., 1970, *Geomorphological Regions of Hungary* (Akademici Kiado, Budapest).

PEEL, R. F., 1974, Insolation weathering: some measurements of diurnal temperature changes in exposed rocks in the Tibesti region, central Sahara, *Zeit für Geomorph., Supp. Band*, **21**, 19–28.

PERRIN, R. M. S., and C. W. MITCHELL, 1969, 1971, An Appraisal of physiographic units for predicting site conditions in arid areas, *MEXE Report IIII*, 2 vols. (vol. 1, 1969; vol. 2, 1971).

PERRY, R. A., et al., 1962, *General Report on Lands of the Alice Springs Area, Northern Territory, 1956–57* (CSIRO, Australia), *Land Research Series*, **6**.

PETROV, M. P., 1976, *Deserts of the World* (John Wiley, New York).

PÉWÉ, T. L., 1978, Geologic hazards in the Phoenix area, Arizona, *Ariz. Bur. Geol. Min. Tech., Field Notes* **8**, 18–20.

PHILLIP, J. R., 1957, The theory of infiltration, *Soil Science*, **83**, 345–57, 435–48; **84**, 163–77, 257–64, 329–39; **85**, 278–86, 333–6.

PHILLIPS, D. G., 1958, Rural-to-urban migration in Iraq, *Economic Development and Cultural Change*, 7, 405–21.

PLATE, E. J., and J. E. CERMAK, 1963, Micro-meteorological wind tunnel facility, *Report CER 63EJP-JEC9 Fluid Dynamics and Diffusion Lab. (Colorado State University, Fort Collins, Colo.)*.

PORATH, A., and A. P. SCHICK, 1974, The use of remote sensing systems in monitoring desert floods, *Int. Ass. Sci. Hydrol. Sympo. Flash Floods* (Paris, 1974) *Publication*, **112**, 133–8.

PROKOPOVICH, N. P., 1969, *Some Geologic Problems in Reclamation of Arid Lands* (US Dept. Inter. Bur. Reclam., Sacramento).

QASHU, H. K., and D. D. EVANS, 1969, Practical implications of water repellency of soils, *Prog. Agric. Ariz.*, **21**, 3–5.

RAHN, P., 1967, Sheetfloods, streamfloods, and the formation of pediments, *Ann. Ass. Am. Geogr.*, **57**, 593–604.

RAINE, J. K., 1974, Wind protection by model fences in a simulated atmospheric boundary layer, 5th Australasian Conf. on Hydraulics and Fluid Mechanics (Canterbury, Christchurch, NZ), 200–10.

RANTZ, S. E., 1970, Urban sprawl and flooding in southern California, *US Geol. Surv. Prof. Paper, 601–B.*

RAPP, A., 1974, A review of desertification in Africa – water, vegetation and man, *Secretariat for International Ecology, Stockholm, SIES Report No. 1.*

READ, R. A., 1961, Bibliography of Great Plains forestry, *Station Paper Rocky Mtn. Forestry Experimental Station No. 58.*

RENARD, K. G., 1970, Hydrology of semi-arid rangeland watersheds, *US Dept. Agric. Research Service Publ. ARS 41–162.*

RENARD, K. G., 1972, Sediment problems in the arid and semi-arid Southwest, *Proc. 27th Ann. Meeting of Soil Conservation Soc. of Am.,* Portland, Oregon, 225–32.

RENARD, K. G., and R. V. KEPPEL, 1966, Hydrographs of ephemeral streams in the southwest USA, *Proc. Am. Soc. Civ. Engng., 92 (HYZ)*, 33–52.

RIM, M., 1958, Simulation, by dynamic model, of sand track morphologies occurring in Israel, *Bull. Res. Counc. Israel, 7G,* 123–36.

ROBERTS, D. V., and G. E. MELICKIAN, 1970, Geologic and other natural hazards in desert areas, *Dames and Moore Engineering Bulletin,* 37, 1–12.

ROBINOVE, C. J., 1979, Integrated terrain mapping with digital LANDSAT images in Queensland, Australia, *US Geol. Surv. Prof. Paper, 1102.*

ROBINSON, E., 1968, Effects of air pollution on visibility, in A. C. Stern (ed.), *Air Pollution* (Academic Press, New York), Ch. 11.

ROBINSON, G. M., and D. E. PETERSON, 1962, Notes on earth fissures in southern Arizona, *US Geol. Surv. Circ., 466.*

RODDA, J. C., 1969, The flood hydrograph, in R. J. Chorley (ed.), *Water, Earth and Man* (Methuen, London), 405–18.

ROMANOVA, M. A., 1964, Regional stability of sand deposits of central Karakums by spectral luminance, *Dokl. Akad. Nauk SSSR,* 156, 1095.

ROMANOVA, M. A., 1968, Spectral luminance of sand deposits as a tool in land evaluation, in G. A. Stewart (ed.), *Land Evaluation* (Macmillan, Australia), 342–8.

ROSTLER, F. S., and W. M. KUNKEL, 1964, Soil stabilisation, *Ind. Engng. Chem.,* 56, 27–33.

ROSTLER, F. S., and B. A. VALLERGA, 1960, *A Manual on Control of Dust and Wind Erosion with Coherex (Golden Bear Oil Co.,* Bakersfield, California).

RUBIN, J., 1966, Theory of rainfall uptakes by soils initially drier than their field capacities, and its applications, *Water Resour. Res.,* 2, 739–49.

RUBIN, J., and STEINHARDT, R., 1964, Soil-water relations during rain infiltration. 3. Water uptakes at incipient ponding, *Proc. Soil Sci. Soc. Am.,* 28, 614–19.

RUBY, E. C., 1973, *Sediment Trend Study – Los Angeles River Watershed* (USDA Forest Service, Angeles National Forest, California Region).

RUHE, R. V., 1967, Geomorphic surfaces and surficial deposits in southern New Mexico, *State Bur. Mines Miner. Res., New Mexico Inst. Min. Tech. Mem., 18.*

RUSSELL, R. B. C., 1974, Chemical and physical properties of sabkha-type materials, *Transp. Road Res. Lab. Gt Br., Rep. 79UC.*

SALEM, A. M. S., and M. A. SOWELIM, 1967, Dust deposits in the city of Cairo, *Atmospheric Environment,* 1, 211–20.

SALISHCHEV, K. A., 1972, National atlases of natural resources, *Nature and Resources,* 8, 20–7.

SAVAGE, R. P., 1963, Experimental study of dune building with sand fences, *Proc. 8th Conf. Coastal Engineering, Mexico City,* 380–96.

SAVAGE, R. P., and W. W. WOODHOUSE, 1968, Creation and stabilisation of coastal barrier dunes, *Proc. 11th Conf. on Coastal Engineering, Am. Soc. Civ. Engng., London,* vol. 1, 671–700.

SAVIGEAR, R. A. G., 1956, Technique and terminology in the investigation of slope forms, *Premier Rapport de la Commission pour l'Etude des Versants* (IGU Rio de Janeiro), 1, 66–75.

SAVIGEAR, R. A. G., 1961, Slopes and hills in West Africa, *Zeit für Geomorph.*, **Suppl.**, 1, 56–71.

SAVIGEAR, R. A. G., 1965, A technique of morphological mapping, *Ann. Ass. Am. Geogr.*, **55** (*3*), 514–38.

SCHAEFFER, R. J., 1932, *The Weathering of Natural Building Stones* (HMSO, London).

SCHICK, A. P., 1970a, East Sinai desert floods research area, *IASH-UNESCO Symposium on Representative and Experimental Basins, Wellington, New Zealand.*

SCHICK, A. P., 1970b, Desert floods: interim results of observations in the Nahal Yael watershed, southern Israel, 1965–1970, *IASH-UNESCO Symposium on Representative and Experimental Basins, Wellington, New Zealand*, 478–93.

SCHICK, A. P., 1971, A desert flood: physical characteristics; effects on man, geomorphic significance, human adaptation – A case study of the Southern Arava watershed, *Jerusalem Studies in Geography*, 2, 91–155.

SCHICK, A. P., 1974a, Alluvial fans and desert roads – a problem in applied geomorphology, *Abhandlungen der Akademie Wissenschalten in Göttingen, Mathematisch-Physikatische, Klasse III, Folge No.* **29**, 418–25.

SCHICK, A. P., 1974b, Formation and obliteration of desert stream terraces – a conceptual analysis, *Zeit. für Geomorph.*, **Suppl. 21**, 88–105.

SCHICK, A. P., 1977, A tentative sediment budget for an extremely arid watershed in the southern Negev, *Proc. 8th Ann. Geomorph. Symp., State University of New York, Sept. 23–24.*

SCHICK, A. P., 1978, Field experiments in arid fluvial environments: considerations for research design, *Zeit. für Geomorph.*, **Suppl. 29**, 22–8.

SCHICK, A. P., 1979, Fluvial processes and settlements in arid environments, *Geol. J.*, 3, 351–60.

SCHICK, A. P., and D. SHARON, 1974, Geomorphology and climatology of arid watersheds, *Mimeogr. Rep., Dept Geogr., Hebrew Univ., Jerusalem.*

SCHULTZ, C. B., and J. C. FRYE (eds.), 1965, *Loess and Related Eolian Deposits of the World, VII INQUA Congr.* (University Nebraska Press, Lincoln, Nebraska), vol. 12.

SCHUMM, S. A., 1960, The shape of alluvial channels in relation to sediment types, *US Geol. Surv. Prof. Paper, 352–B*, 17–30.

SCHUMM, S. A., 1961a, Effect of sediment characteristics on erosion and deposition in ephemeral-stream channels, *US Geol. Surv. Prof. Paper, 352–C*, 31–70.

SCHUMM, S. A., 1961b, Dimensions of some stable alluvial channels, *US Geol. Surv. Prof. Papers, 424–B*, 26–7.

SCHUMM, S. A., 1963, A tentative classification of alluvial river channels, *US Geol. Surv. Circ., 477.*

SCHUMM, S. A., 1964, Seasonal variations of erosion rates and processes on hillslopes in western Colorado, *Zeit. für Geomorph.*, **Suppl. Bd.**, 5, 215–38.

SCHUMM, S. A., 1968, River adjustment to altered hydrologic regimen – Murrumbidgee River and palaeochannels, Australia, *US Geol. Surv. Prof. Paper, 598.*

SCHUMM, S. A., 1969, River metamorphosis, *J. Hydrol. Div., Proc. Am. Soc. Civ. Enging.*, **95**, 255–73.

SCHUMM, S. A., and R. J. CHORLEY, 1964, The fall of threatening rock, *Am. J. Sci.*, **262**, 1041–54.

SCHUMM, S. A., and R. J. CHORLEY, 1966, Talus weathering and scarp recession in the Colorado plateaux, *Zeit. für Geomorph.*, **10**, 11–36.

SCHUMM, S. A., and R. R. HADLEY, 1961, Progress in the application of landform analysis in studies of semi-arid erosion, *US Geol. Surv. Circ., 437*.

SCOGING, H., 1978, The relevance of time and space in modelling potential sheet erosion from semi-arid fields, *Proc. Workshop Soil Erosion in Europe and the USA* (Ghent, Belgium).

SCOGING, H., and J. B. THORNES, 1979, Infiltration characteristics in a semiarid environment, *IAHS-AISH Pub.*, **128**, 159–68.

SCOTT, R. F., and J. J. SCHOUSTRA, 1968, *Soil Mechanics and Engineering* (McGraw-Hill, New York).

SEEVERS, P. M., and J. V. DREW, 1973, Evaluation of ERTS-1 imagery in mapping and managing soil and range resources in the Sand Hills Region of Nebraska, *Symposium on Significant Results obtained from the Earth Resources Technology Satellite – 1, vol. 1, Sec. A, NASA, Goddard Space Flight Center, New Carrollten, Maryland*, 87–9.

SEEVERS, P. M., and J. V. DREW, 1974, Evaluation of ERTS-1 imagery in mapping and managing soil and range resources in the Sand Hills Region of Nebraska, *ERTS Experiment Type II Report, NASA, Goddard Space Flight Center, Greenbelt, Maryland*.

SETHURAMAN, S., and R. M. BROWN, 1976, A comparison of turbulence measurements made by a hot-film probe, a bivane, and a directional vane in the atmospheric surface layer, *J. Appl. Met.*, **15**, 138–44.

SHARP, R. P., 1942, Mudflow levees, *J. Geomorph.*, **5**, 222–7.

SHARP, R. P., 1964, Wind-driven sand in the Coachella Valley, California, *Bull. Geol. Soc. Am.*, **75**, 785–804.

SHARP, R. P., 1966, Kelso Dunes, Mohave Desert, California, *Bull. Geol. Soc. Am.*, **77**, 1045–74.

SHARP, R. P., and L. H. NOBLES, 1953, Mud-flow of 1941 at Wrightwood, southern California, *Bull. Geol. Soc. Am.*, **64**, 547–60.

SHARPE, C. F. S., 1938, *Landslides and Related Phenomena* (Columbia University Press, New York).

SHERMAN, L. K., 1932, Stream flow from rainfall by the unit-graph method, *Engng. News Rec.*, **108**, 501–5.

SHERWOOD, P. T., 1962, The effect of sulphates on cement and lime stabilized soils, *Roads and Road Constr.*, **40**, *No. 470*, 34–40.

SIMONETT, D. S., 1976, Roles for space sensing in studying desertification, *in Desertification: Process, Problems, Perspectives* (University of Arizona Press, Tucson), 53–8.

SIMONETT, D. S., G. R. COCHRANE, S. A. MORAIN and D. E. EGBERT, 1969, Environmental mapping with spacecraft photography: A central Australian example, *US Geol. Surv. Interagency Rep., NASA-172*, 1–47.

SINGH, S. (ed.), 1977, Geomorphological investigation of the Rajasthan Desert, *Central Arid Zone Research Institute Jodhpur Monograph*, **7**.

SKIDMORE, E. L., 1965, Assessing wind erosion forces: directions and relative magnitudes, *Proc. Soil Sci. Soc. Am.*, **29**, 587–90.

SKIDMORE, E. L., P. S. FISHER, and N. P. WOODRUFF, 1970a, Computer equation aids wind erosion control, *Crops and Soils*, **22**, 19–20.

SKIDMORE, E. L., P. S. FISHER, and N. P. WOODRUFF, 1970b, Wind erosion equation: computer solution and application, *Proc. Soil Sci. Soc. Am.*, **34**, 931–5.

SKIDMORE, E. L., and N. P. WOODRUFF, 1968, Wind erosion forces in the United States and their use in predicting soil loss, *Agriculture Handbook No. 346, Agric. Res. Serv., US Dept. of Agric.*

SKUTSCH, M. M., and R. T. N. FLOWERDEW, 1976, Measurement techniques in environmental impact assessment, *Environmental Conservation*, **3**, 209–17.

SLATYER, R. O., and J. A. MABBUTT, 1964, Hydrology of arid and semiarid regions, in V. T. Chow (ed.), *Handbook of Applied Hydrology* (Mcgraw-Hill, New York) Ch. 24, 24–46.

SLOSSON, J. E., 1969, The role of engineering geology in urban planning, in *The Governor's Conference on Environmental Geology* (Colorado Geological Survey, Denver), 8–15.

SMITH, H. T. U., 1963, Eolian geomorphology, wind direction and climatic change in North Africa, *US Air Force, Cambridge Research Labs., Contract No. AF 19 (628)– 298.*

SMITH, H. T. U., 1969, *Photo-interpretation Studies of Desert Basins in Northern Africa.* Mimeographed rept. prepared for Air Force Cambridge Res. Labs., Office of Aerospace Res., USAF (Bedford, Mass.).

SMITH, K. W. G., 1962, Some problems of salt in semi-arid soils for stabilisation with cement, *Proc. 1st Aust. Road Res. Board Conf., Melbourne, Sept. 1962*, **1**, (2), 1078– 84.

SMITH, R. E., and D. L. CHERRY, 1973, Rainfall excess models for soil water flow theory, *J. Hydrol. Div., Am. Soc. Civ. Engrs.*, **99**, HY9, 1337–51.

SMITH, R. E., and H. A. SCHREIBER, 1973, Point processes of seasonal thunderstorm rainfall: 1 Distribution of rainfall events, *Water Resources Res.*, **9**, 87–84.

SMITH, R. K., and L. M. LESLIE, 1976, Thermally driven vortices: a numerical study with application to dust-devil dynamics, *Quart. J. R. Met. Soc.*, **102**, 791–804.

SMITH, R. M., and P. C. TWISS, 1965, Extensive gauging of dust deposition rates, *Trans. Kansas Acad. Sci.*, **68**, 311–21.

SOLENTSEV, N. A., 1962, Basic problems in Soviet landscape science, *Soviet Geography*, **3**, 3–15.

SOLIMAN, M. M., 1974, Urbanization and the processes of erosion and sedimentation in the River Nile, *Int. Ass. Sci. Hydrol. Publ.*, **133**, 123–9.

SOUTH AFRICAN BUREAU OF STANDARDS, SABS 830, *Standard Specification for Chloride Content in Building Materials.*

SOUTH AFRICAN BUREAU OF STANDARDS, SABS 841, *Standard Specification for Determining the Aggregate Crushing Value.*

SOUTH AFRICAN BUREAU OF STANDARDS, 10% FACT value, *SABS 842.*

SOUTH AFRICAN BUREAU OF STANDARDS, 1976, Aggregates from natural sources, *SABS 1083.*

SPANGLE, W. and Associates, 1974, *Application of Earth Science Information in Urban Land-Use Planning, State of the Art Review and Analysis* (US Geol. Surv., Washington, D.C.).

SPANGLE, W. and Associates, F. B. LEIGHTON and Associates, and BAXTER, MCDONALD and Co., 1976, Earth-science information in land-use planning – guidelines for earth scientists and planners, *US Geol. Surv. Circ., 721.*

SPEIGHT, J. G., 1968, Parametric description of land form, in G. A. Stewart (ed.), *Land Evaluation* (Macmillan, Melbourne), 239–60.

SPEIGHT, J. G., 1974, A parametric approach to landform regions, *Inst. Brit. Geogr. Sp. Publ.*, **7**, 213–30.

SPEIGHT, J. G., 1976, Numerical classification of landform elements from air photo data, *Zeit für Geomorph.* **Suppl.**, **25**, 154–68.

SPEIGHT, J. G., 1977, Towards explicit procedures for mapping natural landscapes, *CSIRO, Division of Land Use Research* (Canberra) *Tech. Memo., 77/3.*

STAPLEDON, D. H., 1970, Changes and structural defects developed in some South Australian clays, and their engineering consequences, *Inst. Engrs* (Australia), *Soils Symp.*, **1970**, 39–48.

STERN, A. C. (ed.), 1968, *Chemical Analysis of Inorganic Particulate Pollutants* (Academic Press, New York).

312   URBAN GEOMORPHOLOGY IN DRYLANDS

STEWART, G. A. (ed.), 1968, *Land Evaluation* (Macmillan, Melbourne).

STEWART, G. A. *et al.*, 1970, Lands of the Ord-Victoria areas, Western Australia and Northern Territories, *CSIRO Land Res. Ser., 28.*

STOCKING, M. A., 1977, Rainfall energy in erosion: some problems and applications, *University of Edinburgh, Res. Disc. Paper 13.*

STOECKELER, J. H., 1962, Shelterbelt influence on Great Plains field environment and crops, *Production Research Report No. 62, US Dept of Agriculture.*

STONE, K. H., 1961, World air photo coverage, *Photogrammetric Engng.*, **27**, 605–10.

SVASEK, J. N., and J. H. J. TERWINDT, 1974, Measurements of sand transport by wind on a natural beach, *Sedimentology*, **21**, 311–22.

SWARTZENDRUBER, D., and O. HILLEL, 1975, Infiltration and runoff for small field plots under constant intensity rainfall, *Water Resour. Res.*, **11**, 445–51.

TASK COMMITTEE ON PREPARATION OF SEDIMENTATION MANUAL, COMMITTEE ON SEDIMENTATION OF THE HYDRAULICS DIVISION, 1970, Sediment measurement techniques: E. Airborne sediment, *J. Hydrol. Div., Proc. Am. Soc. Civ. Engrs.*, **96**, 29–41.

TATOR, B. A., 1958, The aerial photograph and applied geomorphology, *Photogramm. Engng.*, **24** (*4*), 549–61.

TERZAGHI, K., and R. B. PECK, 1948, *Soil Mechanics in Engineering Practice* (Wiley, New York).

TEYCHENNE, D. C., 1968, Lightweight aggregates; their properties and use in the UK, *1st Int. Congr. Lightweight Aggregates, Lond.*, vol. 1, 23–37.

THOMAS, M. F., 1969, Geomorphology and land classification in tropical Africa, in M. F. Thomas and G. Whittington (eds.), *Environment and Land Use in Africa* (Methuen, London), 103–45.

THOMAS, M. F., 1976, Purpose, scale and method in land resource surveys, *Geographia Polonica*, **34**, 207–23.

THORNES, J. B., 1976, Semi-arid erosional systems: case studies from Spain, *London School of Economics Geog. Dept. Paper, 7.*

THORNES, J. B., 1977, Channel changes in ephemeral streams – observations, problems and models, in K. J. Gregory (ed.) *River Channel Changes* (Wiley, Chichester), 317–35.

THORNTHWAITE, C. W., 1948, An approach toward a rational classification of climate, *Geogr. Rev.*, **38**, 55–94.

TINUS, R. W., (ed.), 1976, Shelterbelts on the Great Plains, *Proc. Symp. on Shelterbelts on the Great Plains, Denver, Colorado, April 20–22, Great Plains Agric. Council Publ. No. 78.*

TOMLINSON, M. J., 1978, Middle East – Highway and airfield pavements, *Q. J. Engng Geol.*, **11**, 65–73.

TRICART, J., 1959, Présentation d'une feuille de la carte géomorphologique du delta du Sénégal au 1:50 000, *Revue Géomorph. Dyn.*, **11**, 106–16.

TRICART, J., 1961, Notice explicative de la carte géomorphologique du delta du Sénégal, *Bureau Recherches Géologiques et Minières*, **8**, 1–137.

TRICART, J., 1965, *Principes et Méthodes de la Géomorphologie* (Masson, Paris).

TRICART, J., 1966, Geomorphologie at amenagement rural (example du Venezuela), *Cooperation Technique*, **44–45**, 69–81.

TSOAR, H., 1974, Desert dunes morphology and dynamics, El Arish (Northern Sinai), *Zeit. für Geomorph., Suppl.*, **20**, 41–61.

TWIDALE, C. R., 1964, A contribution to the general theory of domed inselbergs; conclusions derived from observations in South Australia, *Trans. Inst. Brit. Geogr.*, **34**, 91–113.

ULLMAN, E., 1954, Amenities as a factor in regional growth, *Geogr. Rev.*, **44**, 119–32.

UNESCO, 1968, *Geology of Saline Deposits* (UNESCO, Paris).

UNESCO, 1970, *International Legend for Hydrogeological Maps* (UNESCO, Paris).

UNESCO, 1973, *International Classification and Mapping of Vegetation* (UNESCO, Paris).

UNESCO, 1976, *Engineering Geological Maps: a guide to their preparation* (UNESCO, Paris).

UNESCO, 1979, Map of the world distribution of arid regions, *MAB Technical Notes*, 7.

UNITED NATIONS, STATISTICAL OFFICE, 1977, *Demographic Yearbook 1976* (United Nations, New York).

UPPAL, I.S., and B. P. KAPUR, 1957, Role of detrimental salts in soil stabilisation with cement – 3. Effect of magnesium sulphate, *Indian Concr. J.*, **31**, 228–331.

US CORPS OF ENGINEERS, Los Angeles District, 1969, *Report on Floods of January and February, 1969 in southern California*.

US DEPARTMENT OF AGRICULTURE, 1975, *Present and Prospective Technology for Predicting Sediment Yields and Sources* (USDA Agric. Res. Service, Washington).

US DEPARTMENT OF HOUSING AND URBAN DEVELOPMENT, *et al.*, 1969, *Environmental Planning and Geology* (US Govt. Printing Office, Washington).

US DEPARTMENT OF HOUSING AND URBAN DEVELOPMENT, 1970, *Proceedings of the National Conference on Sediment Control, Washington D. C. Sept. 14–16, 1969* (HUD, Washington, D.C.).

VALLEJO, G. DE, 1977, Engineering geology for urban planning and development with an example from Tenerife (Canary Islands), *Bull. Int. Ass. Engng. Geol.*, **15**, 37–43.

VAN EIMERN, J., K. KARSCHON, L. A. RAZUMORA, and G. W. ROBERTSON (eds.) 1964, Windbreaks and shelterbelts, *WMO Tech. Note, No. 59*.

VAN LOHUIZEN DE LEEUW, J. E., 1973, Mohenjo Daro – a cause of common concern, in J. E. van Lohuizen de Leeuw and J. M. M. Ubaghs (eds.), *South Asian Archaeology*, 1–11.

VANNEY, J. R., 1960, Pluie et crue dans le Sahara Nord-occidental, *Mem. Reg. Inst. Rech. Sah.*, **4**.

VERSTAPPEN, H. TH., 1959, The role of aerial survey in applied geomorphology, *Revue Géomorph. Dyn.*, **10**, 156–62.

VERSTAPPEN, H. TH., 1970, Introduction to the I.T.C. system of geomorphological survey, *Koninklijk Nederlands Aardrijkskundig Genootschap Geografisch Tijdschrift Nieuwe Recks*, **4.1**, 85–91.

VERSTAPPEN, H. TH., 1977, *Remote Sensing in Geomorphology* (Elsevier, Amsterdam).

VERSTAPPEN, H. TH., and R. A. VAN ZUIDAM, 1968, ITC system of geomorphological survey, *ITC Textbook of Photo-Interpretation (Delft)*, Ch. VII.2.

VINOGRADOV, B. V., 1976, Establishment of critical zones of desertification from aero- and space-survey data, in Problems in the Development and Conservation of Desert and Semidesert Lands, *Working Group on Desertification in and around Arid Lands, 23rd Int. Geographical Congress, Pre-Congress Symposium K26 Ashkhabad USSR*, July 20–26, 75–8.

WALLÉN, C. C., 1956, Fluctuations and variability in Mexican rainfall, in G. F. White (ed.), *The 'Future of Arid Lands* (Am. Ass. Advancement of Science, Washington), **43**, 141–55.

WALLINGTON, C. E., 1968, A method of reducing observing and procedure bias in wind-direction frequencies, *Meteorology Mag.*, **97**, 293–302.

WARD, R. C., 1967, *Principles of Hydrology* (McGraw-Hill, London).

WATERS, R. S., 1958, Morphological mapping, *Geography*, **43**, 10–17.

WAY, D. S., 1973, *Terrain Analysis* (Dowden, Hutchinson & Ross, Stroudsberg, Penn.).

WEBSTER, R., 1963, The use of basic physiographic units in air photo interpretation, *Arch. Int. des Photogramme.*, **14**, 143–8.

WEBSTER, R., and P. H. T. BECKETT, 1970, Terrain classification and evaluation using air photography: A review of recent work at Oxford, *Photogrammetria*, **26**, 51–7.

WEINERT, H. H., 1976, Road building materials, in C. B. Coetzee (ed.) *Mineral Resources of the Republic of South Africa, Handbook 7* (5th edn) (Dept. of Mines, Geol. Survey, Pretoria), 407–13.

WEINERT, H. H., and K. A. CLAUSS, 1967, Soluble salts in road foundations, *4th Reg. Conf. Africa on Soil Mechanics and Foundation Eng., Cape Town, South Africa*, 213–18.

WELLMAN, H. W., and A. T. WILSON, 1965, Salt weathering, a neglected erosive agent in coastal and arid environments, *Nature*, **205**, 1097–8.

WHITNEY, M. I., 1978, The role of vorticity in developing lineation by wind erosion, *Bull. Geol. Soc. Am.*, **89**, 1–18.

WILLIS, R. P., 1967, Geology of the Arabian Peninsula – Bahrain, *US Geol. Surv. Prof. Paper, 560–E.*

WILSON, A. W., 1960, Urbanization of the arid lands, *Professional Geographer*, **12**, 4–7.

WILSON, A. W., 1973, The larger urban centers of the coastal deserts, in D. H. K. Amiran and A. W. Wilson (eds), *Coastal Deserts: their Natural and Human Environments* (University of Arizona Press, Tucson), 33–6.

WILSON, A. W., 1974, Urban growth and manufacturing employment in the arid zone, in *Urbanization in the Arid Lands: a Symposium*, Lubbock, Texas, *Icasals Publication 75–1*, 95–105.

WILSON, A. W., 1977, Technology, regional interdependence and population growth: Tucson, Arizona, *Econ. Geogr.*, **53**, 388–92.

WILSON, E. M., 1969, *Engineering Hydrology* (Macmillan Press, London).

WILSON, I. G., 1971, Desert sandflow basins and a model for the development of ergs, *Geogr. J.*, **137**, 180–97.

WINCH, K. L., 1976, *International Maps and Atlases in Print* (Bowker, London), 2nd edn.

WINKLER, E. M., and E. J. WILHELM, 1970, Salt burst by hydration pressures in architectural stone in urban atmosphere, *Geol. Soc. Am. Bull.*, **81**, 567–72.

WISCHMEIER, W. H., and O. D. SMITH, 1958, Rainfall energy and its relationship to soil loss, *Trans. Am. Geophys. Un.*, **39**, 285–91.

WISCHMEIER, W. H., and O. D. SMITH, 1965, Predicting rainfall erosion losses from cropland east of the Rocky Mountains, *US Dept. of Agric., Handbook 282.*

WITKIND, I. J., 1972, Map showing construction materials in the Henrys Lake quadrangle, Idaho and Montana, *US Geol. Surv. Map I–781–F.*

WOLMAN, M. G., and R. GERSON, 1978, Relative scales of time and effectiveness of climate in watershed geomorphology, *Earth Surface Processes*, **3**, 189–208.

WOLMAN, M. G., and J. P. MILLER, 1960, Magnitude and frequency of forces in geomorphic processes, *J. Geol.*, **68**, 54–74.

WOODRUFF, N. P., R. A. READ, and W. S. CHEPIL, 1959, Influence of a field windbreak on summer wind movement and air temperature, *Kansas Agricultural Experiment Station, Tech. Bull. 100.*

WOODRUFF, N. P., and F. H. SIDDOWAY, 1965, A wind erosion equation, *Proc. Soil Sci. Soc. Am.*, **29**, 602–8.

WOOLLEY, R. R., 1964, Cloudburst floods in Utah, 1850–1938, *US Geol. Surv. Water Supply Paper No. 994.*

WORLD BANK, 1972, *Urbanization Sector Working Paper* (World Bank, Washington, D.C.).

WORLD METEOROLOGICAL ORGANISATION, 1973, Manual for estimation of probable maximum precipitation, *Operational Hydrology Report No. 1, Geneva, Switzerland.*

WRIGHT, A. C. S., and H. URZÚA, 1963, Meteorizaçion en la region costera del desierto del Norte de Chile, *Com. y Res. y Trabajos, Conf. Latino-americana para el estudo de las regiones aridas*, 26–8.

YAALON, D. H., 1963, On the origin and accumulation of salts in groundwater and soils in Israel, *Bull. Res. Counc. of Israel*, **11G**, 105–31.

YAALON, D. H., and E. GANOR, 1966, Chemical composition of dew and dry fallout in Jerusalem, Israel, *Nature*, **217**, 1139–40.

YAALON, D. H., and O. KALMAR, 1972, Vertical movement in an undisturbed soil: continuous measurement of swelling and shrinkage with a sensitive apparatus, *Geoderma*, **8**, 231–40.

YAALON, D. H., and J. LOMAS, 1970, Factors controlling the supply and the composition of aerosols in a near-shore coastal environment, *Agric. Met.*, **7**, 443–54.

YAIR, A., 1972, Observation sur les effets d'un ruissellement dirigé selon la pente des interfluves dans une region semi-aride d'Israel, *Revue Géogr. Phys. Géol. Dyn.*, **14** (*5*), 537–48.

YAIR, A., 1974, Sources of runoff and sediment supplied by the slopes of a first order drainage basin in an arid environment (Northern Negev, Israel), *Abhandlungen der Akademie der Wissenschaften in Göttingen, Mathematisch – Physikalische, Klasse III, Folge No. 29*, 403–16.

YAIR, A., and M. KLEIN, 1973, The influence of surface properties on flow and erosion processes on debris covered slopes in an arid area, *Catena*, **1**, 1–18.

YAIR, A., and H. LAVEE, 1974, Areal contribution to runoff on scree slopes in an extreme arid environment – a simulated rainstorm experiment, *Zeit. für Geomorph.*, **Suppl.**, **21**, 106–21.

YAIR, A., D. SHARON, and H. LAVÉE., 1978, An instrumented watershed for the study of partial area contribution of runoff in the arid zone, *Zeit. für Geomorph.*, **Suppl.**, **29**, 71–82.

YOUNG, A., 1973, Soil survey procedures in land development planning, *Geogr. J.*, **139** (*1*), 53–64.

ZINGG, A. W., 1949, A study of the movement of surface wind, *Agric. Engng.*, **30**, 11–13, 19.

ZINGG, A. W., 1950, The intensity-frequency of Kansas winds, *Soil Conservation Service, SCS-TP-88, US Dept. of Agric.*

INDEX